Statistics for Imaging, Optics, and Photonics

Statistics for Imaging, Optics, and Photonics

PETER BAJORSKI

A JOHN WILEY & SONS, INC., PUBLICATION

Published by John Wiley & Sons, Inc., Hoboken, New Jersey
Published simultaneously in Canada

For general information on our other products and services or for technical support, please contact our Customer Care Department within the United States at (800) 762-2974, outside the United States at (317) 572-3993 or fax (317) 572-4002.

Wiley also publishes its books in a variety of electronic formats. Some content that appears in print may not be available in electronic formats. For more information about Wiley products, visit our web site at www.wiley.com.

Library of Congress Cataloging-in-Publication Data:

Bajorski, Peter, 1958-
  Statistics for imaging, optics, and photonics / Peter Bajorski.
      p. cm. – (Wiley series in probability and statistics ; 808)
    Includes bibliographical references and index.
    ISBN 978-0-470-50945-6 (hardback)
  1. Optics–Statistical methods. 2. Image processing–Statistical methods.
3. Photonics–Statistical methods. I. Title.
    QC369.B35 2012
    621.3601'5195–dc23

                                                      2011015224

Printed in the United States of America

oBook ISBN: 9781118121955
ePDF ISBN: 9781118121924
ePub ISBN: 9781118121948
eMobi ISBN: 9781118121931

10 9 8 7 6 5 4 3 2 1

To my Parents

&

To Grażyna, Alicja, and Krzysztof

To my Parents

&

To Christine, Alleah and Kayumal

Contents

Preface

This book grew out of my lecture notes for a graduate course on multivariate statistics for imaging science students. There is a growing need for statistical analysis of data in imaging, optics, and photonics applications. Although there is a vast literature explaining statistical methods needed for such applications, there are two major difficulties for practitioners using these statistical resources. The first difficulty is that most statistical books are written in a formal statistical and mathematical language, which an occasional user of statistics may find difficult to understand. The second difficulty is that the needed material is scattered among many statistical books.

The purpose of this book is to bridge the gap between imaging, optics, and photonics, and statistics and data analysis. The statistical techniques are explained in the context of real examples from remote sensing, color science, printing, astronomy, and other related disciplines. While it is important to have some variety of examples, I also want to limit the amount of time needed by a reader to understand the examples' background information. Hence, I repeatedly use the same or very similar examples, or data sets, for a discussion of various methods.

I emphasize intuitive and geometric understanding of concepts and provide many graphs for their illustration. The scope of the material is very broad. It starts with rudimentary data analysis and ends with sophisticated multivariate statistical methods. Necessarily, the presentation is brief and does not cover all aspects of the discussed methods. I concentrate on teaching the skills of statistical thinking, and providing the tools needed the most in imaging, optics, and photonics.

Some of the covered material is unique to this book. Due to applications of kurtosis in image analysis, I included Section 2.8, where a new perspective and additional new results are shown. In order to enhance interpretation of principal components, I introduced impact plots in Section 7.2.3. The traditional stopping rules in principal component analysis do not work well in imaging application, so I discuss a new set of stopping rules in Section 7.3. There are many other details that you will not find in most statistical textbooks. They enhance the reader's understanding and answer the usual questions asked by students of the subject.

Specific suggestions about the audience for this book, its organization, and other practical information are given in Chapter 1.

Many people have contributed to this book, and I would like to thank them all. Special thanks go to John Schott, who introduced me to remote sensing and continued to support me over the years in my work in statistical applications to imaging science. I am also indebted to my long-time collaborator Emmett Ientilucci, who explained to me many intricacies of remote sensing and provided many examples used in my research, teaching, and also in this book.

I have also enjoyed tremendous help from my wife, Grażyna Alina Bajorska, who shared her statistical expertise and provided feedback, corrections, and suggestions to many parts of this book.

Many other people provided data sets used in this book, and generously devoted their time to reading sections of the manuscript and providing valuable feedback. I thank them all and list them here in alphabetical order: Clifton Anderson, Jason Babcock, Alicja Bajorska, John Grim, Jared Herweg, Joel Kastner, Thomas Kinsman, Trine Kirkhus, Matthew McQuillan, Rachel Obajtek, Jeff Pelz, Jonathan Phillips, Paul Romanczyk, Joseph Voelkel, Chris Wang, Jens Petter Wold, and Jiayi Zhou.

PETER BAJORSKI

Fairport, New York
March 2011

CHAPTER 1

Introduction

Things vary. If they were all the same, we would not need to collect data and analyze them. Some of that variability is not desirable, but we have tools to recognize that and constructively deal with it. A typical example is an imaging system, starting with your everyday camera or a printer. Manufacturers put a lot of effort into minimizing noise and maximizing consistency of those devices. How is that done? The best way is to start with understanding the system and then measuring its variability. Once you have your measurements, or data, you will need statistical methods to understand and analyze them, so that proper conclusions can be drawn. This is where this book becomes handy. We will show you how to deal with data, how to distinguish between different types of variability, and how to separate the real information from noise.

Statistics is the science of the collection, modeling, and interpretation of data. In this book, we are going to demonstrate how to use statistics in the fields of imaging, optics, and photonics. These are very broad fields—not easy to define. They deal with various aspects of the generation, transmission, processing, detection, and interpretation of electromagnetic radiation. Common applications include the visible, infrared, and ultraviolet ranges of the electromagnetic spectrum, although other wavelengths are also used. This plethora of different measurements makes it difficult to extract useful information from data. The strength of statistics is in describing large amounts of data in a concise way and then drawing general conclusions, while minimizing the impact of data noise on our decisions.

Here are some examples of real, practical problems we are going to deal with in this book.

Example 1.1 (Eye Tracker Data). Eye tracking devices are used to examine people's eye movements as they perform certain tasks (see Pelz et al. (2000)). This information is used in research on the human visual system, in psychology, in product design, and in many other applications. In eye tracking experiments, a lot of data are collected. In a study of 30 shoppers, lasting 20 min per shopper, over one million video frames are generated. In order to reduce the amount of data, fixation periods are identified when a shopper fixes her gaze at one spot. This reduces the number of

Statistics for Imaging, Optics, and Photonics, Peter Bajorski.

Figure 1.1   Shampoo bottles on a store shelf. The cross shows the spot the shopper is looking at.

frames to under 100,000, but those images still need to be labeled in order to describe what the shoppers are looking at. Many of those images are fixations on the same product, but possibly from a different angle. The image frame might also be slightly shifted. Our goal is to find the groups of images of the same product. One approach could be to compare the images pixel by pixel, but that would not work well when the image is shifted. One could also try to segment the image into identifiable objects and then compare the objects from different images, but that would require a lot of computations. Another approach is to ignore the spatial structure of the image and describe the image by how the three primary colors mix in the image.

Figure 1.1 shows a sample fixation image used in a paper by Kinsman et al. (2010). The cross in the image shows the spot the shopper is looking at. This 128 by 128 pixel image was recorded with a camcorder in the RGB (red, green, and blue) channels. This means that each pixel is represented by a mixture of the three colors. Mathematically, we can describe the pixel with three numbers, each representing the intensity of one of the colors. For educational purposes, we select here a small subset of all pixels and use only the red and green values. Figure 1.2a shows a scatter plot of this small subset. We can see some clusters, or concentrations, of points. Each cluster corresponds to a group of pixels with a given mix of color. The group in the top right corner of the graph is a mix of a large amount of red with a large amount of green.

Our goal is to find those clusters automatically and describe them in a concise way. This is called unsupervised learning because we learn about the clusters without prior information (supervision) about the groups. One possible solution is shown in Figure 1.2b, where five clusters are identified and described by the elliptical shapes. This provides a general structure for the data. In a real implementation, this needs to be done on all 16,384 pixels in a three-dimensional space of the red, green, and blue intensity values. Methods for efficient execution of such tasks will be shown in this book.

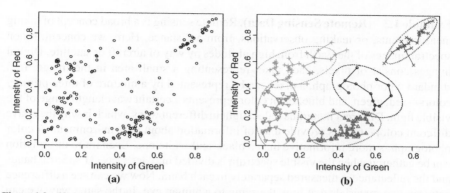

Figure 1.2 Understanding structure of data. Original data are shown in panel (a) and clusters of the same data with elliptical descriptors are shown in panel (b).

Example 1.2   (Printing Data). Printer manufacturers want to ensure high consistency of printing by their devices. There are various types of calibrations and tests that can be done on a printer. One of them is to print a page of random color patches such as those shown in Figure 1.3. The patches are in four basic colors of the CMYK color model used in printing: cyan, magenta, yellow, and black. In a given color, there are several gradations, from the maximum amount of ink to less ink, where the patch has a lighter color if printed on a white background. For a given gradation of color, there are several patches across the page printed in the same color. Our goal is to measure the consistency of the color in all those patches. We also want to monitor printing quality over time, including possible changes in quality after the printer's idle time. An experiment was performed to study these issues, and the resulting data set is used throughout this book. Methods for exploratory analysis of such data and then for statistical inference will be discussed.

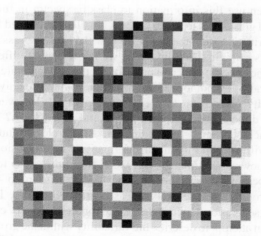

Figure 1.3 Random color patches for printing-quality testing.

Example 1.3 (Remote Sensing Data). Remote sensing is a broad concept of taking measurements, or making observations, from a distance. Here, we concentrate on spectral images of the Earth from high altitudes by way of aircraft or satellite. Digital images consist of pixels, each pixel representing a small area in the image. In a standard color photograph, a pixel can be represented by a mixture of three primary colors—red, green, and blue. Each color represents a certain wavelength range of the visible light. Different materials reflect light in different ways, which is why they have different colors. Colors provide a lot of information about our environment. A color photograph is more informative than a black-and-white one. Even more information can be gathered when the visible spectrum is divided into, let's say, 31 spectral bands and the reflectance is measured separately in each band. Now we can see a difference between two materials that look the same to a human eye. In the same way, we can measure reflectance of electromagnetic waves in other (invisible) wavelengths, including infrared, ultraviolet, and so on. The amount of information increases considerably, but this also creates many challenges when analyzing such data. Each pixel is now represented by a spectral curve describing reflectance as a function of wavelength. The spectral curves are often very spiky with not much smoothness in them. It is then convenient to represent them in their original digitized format, that is, as p-dimensional vectors, where p is the number of spectral wavelengths. The number p is often very large, sometimes several hundred or even over a thousand. This creates major difficulties with visualization and analysis of such data. In Figure 1.2, we saw a scatter plot of two-dimensional data, but what do we do with 200-dimensional data? This book will show you how to work in very high dimensional spaces and still be able to extract the most important information.

Remote sensing images are used in a wide range of applications. In agriculture, one can detect crop diseases from aerial images covering large areas. One example that we are going to use in this book is an image of grass area, where a part of the image was identified as representing diseased grass. Our goal is to learn how to recognize diseased grass based on a 42-dimensional spectral vector representing a pixel in the image. We can then use this information to classify spectra in future images into healthy or diseased grass. This learning process is called supervised learning because we have prior information from the image on how the healthy grass and the diseased grass look in terms of their spectrum. Once we know how to differentiate the two groups based on the spectra, we can apply the method to large areas of grass.

The diseased grass does not look much different from the healthy grass, if you are assessing it visually or looking at a color photograph. However, there is more information in 42 dimensions, but how can we find it and see it? In this book, we will show you methodologies for finding the most relevant information in 42 dimensions. We will also find the most informative low-dimensional views of the data. Figure 1.4 shows an optimal way of using two dimensions for distinguishing between three types of grass pixels—the healthy grass (Group 1), less severely diseased grass (Group 2), and severely diseased grass (Group 3). The straight lines show the optimum separation between the groups for the purpose of classification, and the ellipses show an attempt to describe the variability within the groups. In this book, we will show you how to construct such separations, how to evaluate their efficiency,

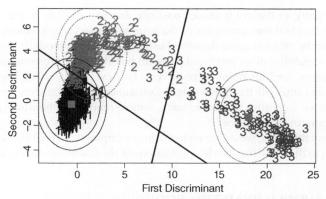

Figure 1.4 A two-dimensional representation of a 42-dimensional set of image pixels representing healthy grass (Group 1), less severely diseased grass (Group 2), and severely diseased grass (Group 3).

how to describe the variability within groups, and then check how reliable such descriptions are.

Example 1.4 (Statistical Thinking). Even before any data are collected, we need to utilize statistical thinking so that our study is scientifically valid and the conclusions are representative of the intended scope of the study. Whenever we use data and try to analyze them, we need to take the following three steps:

1. Formulate the practical problem at hand as a statistical problem.
2. Solve the problem using statistics. This usually involves the collection and analysis of data.
3. Translate the problem solution back to the real-world application.

The purpose of this book is to show you how to solve practical problems by using this statistical approach. Let's say you are a quality engineer at Acme Labs producing plastic injection molding parts. You are part of a team assigned to provide a sensor for automatically detecting whether the produced parts have an acceptable shade of a chosen color. Many steps are needed to accomplish the task, but here we give an example of two steps where statistics would be useful:

1. Define what it means that a color shade is acceptable or not.
2. Find and test an instrument that would measure the color with sufficient precision at a reasonable cost.

The color shade acceptability is somewhat subjective and will depend on the observer and viewing conditions when the material is compared visually. See Berns (2000) for a more detailed discussion of color and color measurement. In this book, we will focus on instrumental color measurement. The produced parts of nominally the same color will vary slightly in the color shade, possibly due to variation in the production process. Instrumental measurements of the color will also vary. All those sources of variability can be measured and described using statistical methods. It would be best to know the variability of all produced parts, all possible measurements made with a given instrument, and all possible observers. However, it is either impossible or impractical to gather all that knowledge. Consequently, in statistics we deal with samples, and we determine to what extent a sample represents the whole population that it is attempting to describe.

Throughout this book, we are going to use the examples described above, as well as many others, to illustrate real-world applications of the discussed statistical methods.

1.1   WHO SHOULD READ THIS BOOK

This book is primarily intended for students and professionals working in the fields of imaging, optics, and photonics. Hence, all examples are from these fields. Those are vast areas of research and practical applications, which is why the examples are written in a simplified format, so that nonexperts can relate to the problem at hand. Nevertheless, this book is about statistics, and the presented tools can be potentially useful in any type of data analysis. So, practitioners in other fields will also find this book useful.

The reader is expected to have some prior experience with quantitative analysis of data. We provide a gentle and brief introduction to data analysis and concentrate on explaining the associated concepts. If a reader needs more practice with those tools, it is recommended that other books, with a more thorough coverage of fundamentals, are studied first.

Some experience with vector and matrix algebra is also expected. Familiarity with linear algebra and some intuition about multidimensional spaces are very helpful. Some of that intuition can be developed by working slowly through Chapter 5.

This book is not written for statisticians, although they may find it interesting to see how statistical methods are applied in this book.

1.2   HOW THIS BOOK IS ORGANIZED

This chapter is followed by two chapters that review the fundamentals needed in subsequent chapters. Chapter 2 covers the tools needed for exploratory data analysis as well as the probability theory needed for statistical inference. In Chapter 3, we briefly introduce the fundamental concepts of statistical inference. The regression models covered in Chapter 4 are very useful in statistical analysis, but that material is not necessary for understanding the remaining chapters. On the other hand, two supplements to that chapter provide the fundamental information about vector and matrix algebra as well as random vectors, all needed in the following chapters.

Starting with Chapter 5, this book is about multivariate statistics dealing with various structures of data on multiple variables. We lay the foundation for the multidimensional considerations in Chapter 5. This is where a reader comfortable with univariate statistics could start reading the book. Chapter 6 covers basic multivariate statistical inference that is needed in specific scenarios but is not necessary for understanding the remaining parts of the book. Principal component analysis (PCA) discussed in Chapter 7 is a very popular tool in the fields of imaging, optics, and photonics. Most professionals in those fields are familiar with PCA. Nevertheless, we recommend reading that chapter, even for those who believe they are familiar with this methodology. We are aware of many popular misconceptions, and we clarified them in that chapter. Each of the remaining chapters moves somewhat separately in three different directions, and they can be read independently. Chapter 8 covering canonical correlation analysis is difficult technically. In Chapter 9, we describe classification, also called supervised learning, which is used to classify objects into populations. Clustering, or unsupervised learning, is discussed in Chapter 10, which can be read independently of the majority of the book material.

1.3   HOW TO READ THIS BOOK AND LEARN FROM IT

Statistics is a branch of mathematics, and it requires some of the same approaches to learning it as does mathematics. First, it is important to know definitions of the terms used and to follow the proper terminology. Knowing the proper terminology will not only make it easier to use other books on statistics, but also enable easier communication with statisticians when their help is needed. Second, one should learn statistics in a sequential fashion. For instance, the reader should have a good grasp of the material in Chapters 2 and 3 before reading most of the other parts of this book. Finally, when reading mathematical formulas, it is important to understand all notation. You should be able to identify which objects are numbers, or vectors, or matrices—which are known, which are unknown, which are random or fixed (nonrandom), and so on. The meaning of the notation used is usually described, but many details can also be guessed from the context, similar to everyday language. When writing your own formulas, you need to make sure that a reader will be able to identify all of the features in your formulas.

As with all areas of mathematics and related fields, it is critical to understand the basics before the more advanced material can be fully mastered. The particular difficulty for many nonstatisticians is the full appreciation of the interplay between the population, the model, and the sample. Once this is fully understood, everything else starts to fall into place.

Each chapter has a brief list of problems to practice the material. The more difficult problems are marked by a star. We recommend that the readers' main exercise be the recreation of the results shown in the book examples. Once the readers can match their results with ours, they would most likely master the mechanics of the covered methodologies, which is a prerequisite for their deeper understanding. Most concepts introduced in this book have very specific geometric interpretations that help in their understanding. We use many figures to illustrate the concepts and elicit the geometric interpretation. However, readers are encouraged to sketch their own graphs when reading this book, especially some representations of vectors and other geometric figures.

Real-world applications are usually complex and require a considerable amount of time to even understand the problem. For educational purposes, we show simplified versions of those real problems with smaller data sets and straightforward descriptions, so that nonexperts can easily relate to them.

We often provide references where the proofs of theorems can be found. This is not meant as a recommendation to read those proofs, but simply as potential further reading for more theoretically inclined readers. We provide derivations and brief proofs only in cases when they are simple and provide helpful insight and illustration of the introduced concepts.

In this book, we try to keep the mathematical rigor at an intermediate level. For example, the main statistical theme of distinguishing between the population and the sample quantities is emphasized, but only in places where it is necessary. In other places, readers will need to keep track of those subtleties on their own, using their statistical thinking skills, hopefully developed by that time.

We use mathematical notation and formulas generously, so readers are encouraged to overcome their fear of formulas. We treat mathematical language as an indispensable tool to describe things precisely. As with any other language learning, it becomes easier with practice. And once you know it, you find it useful, and you cannot resist using it.

We abstain from a mathematical tradition of reminding the reader that the introduced objects must exist before one can use them. We usually skip the assumptions that sets are nonempty and the numbers we use are finite. For example, if we write a definite integral, we implicitly assume that it exists and is a finite number.

1.4   NOTE FOR INSTRUCTORS

The author has used the multivariate material of this book in a 10-week graduate course on multivariate statistics for imaging science students. With the additional material developed for this book and the review of the univariate statistics, the book is also suitable for a similar 15-week course. The author's experience is that some review of the material in Chapters 2, 3 and 4 is very helpful for students for a better understanding of the multivariate material. The computational results and graphs in this book were created with the powerful statistical programming language R (see R Development Core Team (2010)). However, students would usually use their preferred software, such as ENVI/IDL or MATLAB. It is our belief that students benefit from implementing statistical techniques in their own computational environment rather than using a statistical package that is chosen for the purpose of the course and possibly never again used by the students. This is especially true for students dealing with complex data such as those used in imaging, optics, and photonics.

1.5 BOOK WEB SITE

The web site for this book is located at

http://people.rit.edu/~pxbeqa/ImagingStat

It contains data sets used in this book, color versions of some of the book figures (if the color is relevant), and many other resources.

CHAPTER 2

Fundamentals of Statistics

This chapter is a brief review for readers with some prior experience with quantitative analysis of data. Readers without such experience, or those who prefer more thorough coverage of the material, may refer to the textbooks by Devore (2004) or Mendenhall et al. (2006).

2.1 STATISTICAL THINKING

Statistics is a branch of mathematics, but it is not an axiomatic science as are many other of its branches (where facts are concluded from predetermined axioms). In statistics, the translation of reality to a statistical problem is a mix of art and science, and there are often many possible solutions, each with a variety of possible interpretations.

The science of statistics can be divided into two major branches—descriptive statistics and inferential statistics. Descriptive statistics describes samples or populations by using numerical summaries or graphs. No probabilistic models are needed for descriptive statistics. On the other hand, in inferential statistics, we draw conclusions about a population based on a sample. Here we build a probabilistic model describing the population of interest, and then draw information about the model from the sample. When analyzing data, we often start with descriptive statistics, but most practical applications will require the use of inferential statistics. This book is primarily about inferential statistics.

In Chapter 1, we emphasized that variability is everywhere, and we need to utilize statistical thinking to deal with it. In order to assess the variability, we first need to define precisely what we are trying to measure, or observe. We can then collect the data and analyze them. Let us describe that process, and on the way, introduce definitions of some important concepts in statistics.

Statistics for Imaging, Optics, and Photonics, Peter Bajorski.
© 2012 John Wiley & Sons, Inc. Published 2012 by John Wiley & Sons, Inc.

Definition 2.1. A *measurement* is a value that is observed or measured.

Definition 2.2. An *experimental unit* is an object on which a measurement is obtained.

Definition 2.3. A *population* is often defined as a set of experimental units of interest to the investigator. Sometimes, we take repeated measurements of one characteristic of a single experimental unit. In that case, a *population* would be a set of all such possible measurements of that experimental unit, both the actual measurements taken and those that can be taken hypothetically in the future.

Definition 2.4. A *sample* is a subset selected from the population of interest.

When designing a study, one should specify the population that addresses the question of interest. For example, when investigating the color of nominally red plastic part #ACME-454, we could define a population of experimental units as all parts #ACME-454 produced in the past and those that will be produced in the future at a given plant of ACME Labs.

We can say that this population is hypothetical because it includes objects not existing at the time. It is often convenient to think that the population is infinite. This approach is especially useful when dealing with repeated measurements of the same object. Infinite populations are also used as approximations of populations consisting of a large number of experimental units. As you can see, defining a population is not always exact science.

Once we know the population of interest, we can identify a suitable sampling method, which describes how the sample will be selected from the population. Our goal is to make the sample to be representative of the population, that is, it should look like the population, except for being smaller. The closer we get to this ideal, the more precise are our conclusions from the sample to the population. There are whole books describing how to select samples (see Thompson (2002), Lohr (2009), Scheaffer et al. (2011), and Levy and Lemeshow (2009)).

If a data set was given to you, you need to find out how the data were collected, so that you can identify the population it represents. The less we know about the sampling procedure used, the less useful the sample is. In extreme cases, it might be prudent to use the old adage "garbage in–garbage out," and try to collect new data instead of using unreliable data.

Let's say, you were given data on color measurements of 10 parts #ACME-454 that were taken from the current production process. However, there is no information about the process of selecting the 10 parts. They all might have been taken from one batch produced within 1 h or each part might have been produced on a different day. They could also be rejects from the process. In this case, it would be more productive to design a new study of those parts in order to collect new data.

The purpose of this section is to give the reader a general overview of the principles of statistical thinking and a sense of the nuances associated with statistics. If reading it led you to having even more questions than you started with, then continue to the following sections and chapters, where you will find many answers.

2.2 DATA FORMAT

Data are often organized in a way that is convenient for data collection. In order to implement statistical thinking and better understand the data, we usually find it convenient to organize the data into the format of a traditional statistical database. The format consists of a spreadsheet, where *observations* are placed in rows and *variables* are placed in columns. Example 2.1 illustrates this traditional formatting technique.

Example 2.1 Optical fibers permit transmission of signals over longer distances and at higher bandwidths than other forms of communication. An experiment was performed in order to find out how much power is lost when sending signals through optical fiber. Five pieces of 100 m length of optical fiber were tested. A laser light signal was sent from one end through each piece of optical fiber, and the output power was measured at the other end. The power of the laser source was 80 mW. The results are shown in Table 2.1, where each row represents a set of results for a single piece of optical fiber. Each unique optical fiber is identified by a number recorded in the first column of the table. The remaining columns contain the variables from the experiment. The Input Power (P_{in}) is the nominal value of 80 mW, which is the same for all observations. The Output Power (P_{out}) given in the next column is a quantity that was measured in the experiment. The Power Loss (L_{power}) in the last column was calculated in decibels (dB) according to the following formula:

$$\text{Power Loss (dB)} = 10\log_{10}\frac{\text{Output Power}}{\text{Input Power}}. \tag{2.1}$$

Organized in this way, the data are easily analyzed. For a small data set like this one, we can often draw some conclusions directly from the table, but for larger data sets, we will need some summary statistics and graphs to understand the data.

Since the Power Loss is calculated from the Output Power (with constant Input Power), the two variables convey the same information (within this data set). So, if we

Table 2.1 Experimental Results on Five Pieces of Optical Fiber

Optical Fiber Number	Input Power (mW)	Output Power (mW)	Power Loss (dB)
1	80	72.8	−0.4096
2	80	70.0	−0.5799
3	80	72.0	−0.4576
4	80	68.8	−0.6550
5	80	73.6	−0.3621

Negative dB means that there is loss of power.

are trying to characterize a typical fiber based on the five pieces, which of the two variables should we use? This question will be addressed in the next section on descriptive statistics. □

The data are not always as neatly organized as those in Table 2.1. At the same time, it is not always necessary to have an actual statistical database in the Table 2.1 format. However, in the process of statistical thinking, we want to identify what the observations and variables are in a given context, since this will be crucial in our statistical analysis.

2.3 DESCRIPTIVE STATISTICS

When dealing with data, especially with large amounts of data, we find it useful to summarize them with some appropriately chosen summary (or descriptive) statistics. We will now concentrate on the values of one variable and will denote the n observations of that variable by $x_1, x_2, \ldots, x_n$. Note that the subscript index does not imply any particular order in those values. The first step in understanding the data is to describe the magnitude of the observations. When we think of data as numbers on the number axis, the magnitude will tell us a general location of the data on the axis. In the following subsection, we discuss various statistics for describing the data location.

2.3.1 Measures of Location

The most popular descriptive statistic is the *sample mean* defined by

$$\overline{x} = \frac{1}{n}\sum_{i=1}^{n} x_i, \tag{2.2}$$

which describes the general (on average) location of the data. One appealing property of the sample mean is a physical property that it is the balance point for a system of equal weights placed at the points x_i, $i = 1, \ldots, n$, on the number axis. Figure 2.1 shows an example of five data points with equal weights, which are balanced at the $\overline{x}$ point.

Example 2.1 (cont.) For the data in Table 2.1, we can calculate the sample means of all three variables. For the Input Power variable, we get its sample mean $\overline{P}_{in} = 80\,\text{mW}$, of course. For Output Power, we obtain $\overline{P}_{out} = 71.44\,\text{mW}$, and for the Power Loss, $\overline{L}_{power} = -0.4928$. The means are supposed to represent a typical or an average optical fiber. Let us assume that an optical fiber regarded as average has the

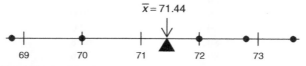

Figure 2.1 Five Output Power values balanced at the sample mean point (see Example 2.1).

Output Power value of $P_{out} = 71.44$ mW, that is, the same as the previously calculated mean. According to formula (2.1), its power loss would be described as -0.4915 dB, which is different from the previously calculated average Power Loss of $\overline{L}_{power} = -0.4928$. The question is which of the two values should be regarded as a typical power loss value. There is an easy mathematical explanation for why the two numbers differ. Let us say that a variable y is calculated as a function of another variable x, that is, $y = f(x)$. In this case, Power Loss is calculated as a function of Output Power. This means that for observations x_i, $i = 1, \ldots, n$, we have $y_i = f(x_i)$, $i = 1, \ldots, n$. What we have just observed in our calculations simply means that $\overline{y} \neq f(\overline{x})$. In other words, a transformation of the mean is not necessarily the same as the mean of the transformed values. A special case is when the function f is linear, and we do get an equality $\overline{y} = f(\overline{x})$, that is, for $y_i = ax_i + b$, we have $\overline{y} = a\overline{x} + b$.

Despite the above explanation, we still do not know which of the two power loss values we should regard as typical for the type of optical fiber used in the experiment. The answer will depend on how such a number would be used. Here we give two possible interpretations. If the five measurements were performed on the same piece of optical fiber, then the sample mean $\overline{P}_{out}$ would estimate the "true" output power of the fiber. The true power loss for that fiber should then be calculated as $10 \log_{10}(\overline{P}_{out}/80) = -0.4915$ dB. An alternative scenario would be when the five different pieces tested in the experiment represent an optical fiber used in an existing communication network, and we are trying to characterize a typical network power loss (over 100 m). In this case, it would be more appropriate to use the value of $\overline{L}_{power} = -0.4928$. To understand this point, imagine the five pieces being connected into one 500 m optical fiber. Its power loss would then be calculated as the sum of the five power loss values in Table 2.1, resulting in the total power loss of -2.4642 dB. The same value (up to the round-off error) can be obtained by multiplying the typical value of $\overline{L}_{power} = -0.4928$ by 5.

We now need to introduce the concept of ordered statistics. Let's say we have n observations x_i, $i = 1, \ldots, n$, of a given variable. We order those numbers from the smallest to the largest, and call the smallest one the value of the first-order statistic denoted by $x_{(1)}$. The second smallest value becomes the second-order statistic denoted by $x_{(2)}$, and so on until the largest value becomes the nth-order statistic denoted by $x_{(n)}$. We can now introduce the *sample median*, which is the middle value in the data set defined as

$$\tilde{x} = \begin{cases} x_{(k)} & \text{for odd } n = 2k-1, \\ (x_{(k)} + x_{(k+1)})/2 & \text{for even } n = 2k. \end{cases} \quad (2.3)$$

In Example 2.1, $n = 5$ is odd, hence $k = 3$, and for the Output Power variable, we have $\tilde{x} = x_{(3)} = 72$. The sample median is called a robust statistic because it is not impacted by unusual observations called outliers. It is also useful for skewed data, where the mean is pulled away from the bulk of data because of being influenced by a few large values. Figure 2.2 shows an example where the bulk of the data is in the range between 0 and 2, but the sample mean is above 2 because of two outliers.

Figure 2.2   A data set skewed to the right due to two outliers. The sample mean does not represent the bulk of data as well as the sample median does.

The sample median can be regarded as too robust in the sense that it depends only on the ordered statistics in the middle of the data. As a compromise between the mean and the median, we can define a trimmed mean, where a certain percent of the lowest and highest values are removed, and the mean is calculated from the remaining values. Note that the median is an extreme case of the trimmed mean, where the same number of the lowest and highest values are removed until only one or two observations are left.

The sample median divides the data set into two halves. For a more detailed description of the data distribution, we can divide data into one hundred parts and describe the position (or location) of each part. To this end, we can define a *sample* $(100p)th$ *percentile*, where p is a fraction $(0 \leq p \leq 1)$, as a number x such that approximately $(100p)\%$ of data is below x and the remaining $(100(1-p))\%$ of data is above x. A $(100p)$th percentile is also called a pth quantile. Percentiles are often used in reporting results of standardized tests, because they tell us how a person performed in relation to all other test takers. Of course, it is not always possible to divide the data into an arbitrary fraction, so we need a more formal definition. We first assign the kth-order statistic $x_{(k)}$ as the $(k-1)/(n-1)$ quantile. When a different-level quantile is needed, it is interpolated from the two nearest quantiles previously calculated as the ordered statistics. The sample percentiles are best calculated for large samples, but here we give an educational example for the five observations of the Output Power variable in Example 2.1. For $n = 5$, the five ordered statistics are assigned as 0th, 25th, 50th, 75th, and 100th percentiles. A 90th percentile is calculated by a linear interpolation as the weighted average of the two ordered statistics, that is,

$$\frac{100-90}{100-75} x_{(4)} + \frac{90-75}{100-75} x_{(5)}, \tag{2.4}$$

which gives 73.28 for the Output Power variable (given as Problem 2.1). There are many other ways of calculating percentiles, and the best way may depend on the context of data. For large n, all methods give similar results.

It is easy to see that the sample median is the 50th percentile. We also define the *first* and *third quartiles* as the 25th and 75th percentiles, respectively. The two quartiles together with the median, which is also the second quartile, divide the data set into four parts with approximately even counts of points.

2.3.2   Measures of Variability

In the previous subsection, we discussed the location aspect of data. Another important feature of data is their variability. The simplest measure of variability is the *range*, which is defined as the difference between the maximum and minimum values, that is, $x_{(n)} - x_{(1)}$ for a sample of size n. A significant disadvantage of the range is its dependence on the two most extreme observations, which makes it sensitive to outliers.

A different way to describe variability is to use deviations from a central point, such as the mean. The deviations from the mean, defined as $d_i = x_i - \overline{x}$, have the property that they sum up to zero (see Problem 2.2). Hence, the measures of variability typically consider magnitudes of deviations and ignore their signs. The most popular measures of variability are the *sample variance* defined as

$$s^2 = \frac{1}{n-1} \sum_{i=1}^{n} d_i^2 = \frac{1}{n-1} \sum_{i=1}^{n} (x_i - \overline{x})^2 \qquad (2.5)$$

and the associated *sample standard deviation* defined as $s = \sqrt{s^2}$. They both convey the equivalent information, but the advantage of the standard deviation is that it is expressed in the units of the original observations, while the variance is in squared units, which are difficult to interpret.

Let us now consider a linear transformation of x_i defined as $y_i = ax_i + b$ for $i = 1, \ldots, n$. Using some algebra, one can check that the sample variance of the transformed data is equal to $s_y^2 = a^2 s_x^2$ and the sample standard deviation is $s_y = |a| s_x$ (see Problem 2.3). This means that both statistics are not impacted by a shift in data, and scaling of data by a positive constant results in the same scaling of the sample standard deviation.

Another measure of variability is the *interquartile range* (IQR), defined as the difference between the third and first quartiles, which is the range covering the middle 50% of the data.

2.4 DATA VISUALIZATION

We all know that a picture is worth a thousand words. In the statistical context, it means that valuable information can be extracted from graphs representing data—information that might be difficult to notice and convey when reporting only numbers. For an efficient graphical presentation, it is important that the maximum amount of information is conveyed with the minimum amount of ink. This allows representations of large data sets and at the same time keeps the graphs clear and easy to interpret. This concept has been popularized by Tufte (2001), who used the information-to-ink ratio as a measure of graph efficiency. In those terms, bar charts and pie charts are very inefficient, and indeed they are of very little value in data analysis.

2.4.1 Dot Plots

One of the simplest graphs is a dot plot, where one dot represents one observation, and one axis (such as the horizontal axis as in Figure 2.3) is devoted to showing the range

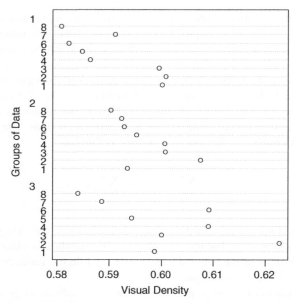

Figure 2.3   Dot plot for Visual Density of eight patches of cyan printed on three different pages (groups).

of values. The second axis may not be used at all (with all dots lined up along a horizontal line), or it can be used to show additional information such as grouping of observations, or their order. One advantage of a dot plot is that it can be created in any software program capable of plotting dots in a system of coordinates.

Example 2.2   As part of a printing experiment described in Appendix B, three pages were printed with an identical pattern of color patches, such as the one shown in Figure 1.3 in the context of Example 1.2. On each page, there were eight patches of cyan (at maximum gradation, or amount, of the cyan ink). For each patch, Visual Density was measured as a quality control metric. Figure 2.3 shows a dot plot of Visual Density for the three pages as three groups. The horizontal lines within each group represent eight patches. The three groups of data (as pages) seem to be somewhat different, but it is unclear if the differences could have happened by chance or if they manifest a real difference. No real difference would be good news because it would mean consistent printing from page to page. This question would need to be addressed by statistical inference discussed in Chapters 3 and 4.

In Figure 2.3, we may have an impression of a slanted shape of points within each group, where the patches with a higher identification number tend to give lower densities. This suggests a possible pattern from patch to patch. In order to test this hypothesis, we can group data into eight groups (for eight patches) of three observations each and create a dot plot with patches as groups. In that case, the number of groups is fairly large, and it makes sense to use a different version of a dot plot, where each group is plotted along one horizontal line as in Figure 2.4. We can

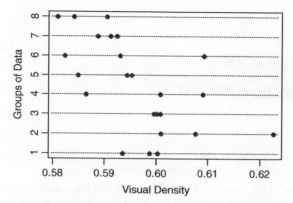

Figure 2.4 Dot plot for Visual Density of eight patches of cyan (as groups) printed on three different pages.

now see that Patches 5, 7, and 8 tend to have lower Visual Density values than some other patches, especially Patch 2. Since we have only three observations per patch, it is unclear if this effect is incidental, or if there is a real systematic difference among patches. Again, this question needs to be answered with some formal statistical methods that will be discussed in Chapter 3. □

2.4.2 Histograms

Dot plots are convenient for small to medium-sized data sets. For large data sets, we start getting significant overlap of dots, which can be dealt with by stacking the points, but this requires extra programming or a specialized function. Also, it becomes difficult to assess the shape of the distribution with too many points. In those cases, we can use a *histogram*, which resembles a bar chart, except that the bars represent adjacent bins or subintervals of equal length defined within the range of given data. For example, the histogram in Figure 2.5 uses bins of width 0.05. The tallest bar represents the bin from 0 to 0.05, the next bin to the right is from 0.05 to 0.1, and so on. The height of the bar shows the number of points (frequency) in the bins. In this

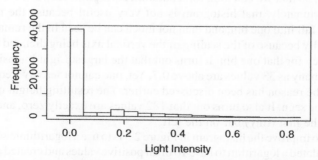

Figure 2.5 A histogram of the Light Intensity values from an image of a fish as used in Example 2.3.

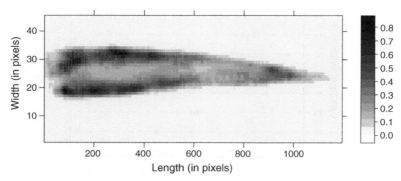

Figure 2.6   Light Intensity values from an image of a fish as used in Example 2.3.

example, there are almost 40,000 observations in the bin from 0 to 0.05. The bins in a histogram are adjacent with no gaps between them. Consequently, there are usually no gaps between the bars. If there is a gap in the bars, it means that the respective bin had zero frequency and was not plotted (or had zero height). In very large data sets, the height of a bar might be larger than zero but still be so small (in relation to the vertical scale of frequencies) that the bar is not visible.

Example 2.3   Consider Fish Image data set representing an image of a fish on a conveyer belt as explained in Appendix B. The average transflected Light Intensity over 15 image channels was calculated for each image pixel and plotted in Figure 2.6. We use a convention that higher values are shown in darker colors. This produces better displays in most cases than the traditional approach in imaging to use white for the highest values. Using white for largest values may seem logical from the point of view of color management, but it usually produces poor quality displays.

There are 45 pixels along the width of the conveyer belt and 1194 pixels along its length, for a total of 53,730 pixels. In a paper by Wold et al. (2006), a threshold on the Light Intensity was used to distinguish between the fish and non-fish pixels, but no details were provided as to the process of selecting the threshold. In order to determine the threshold, it is helpful to perform exploratory analysis of the data. To this end, we can create a histogram of all 53,730 Light Intensity values as shown in Figure 2.5, so that we can look for a natural cutoff point between the two sets of pixels. Unfortunately, that histogram is not very useful because the majority of observations fall into one bin, and then not much can be seen in the remaining bins. This is partially because of the scaling of the vertical axis being dictated by the very high frequency for that one bin. It turns out that the largest Light Intensity is above 0.82, and as many as 33 values are above 0.7. Yet, one cannot see any frequency bars above 0.7. The reason has been discussed earlier. The resulting height of the bar is too small to be seen. It also turns out that 182 values are exactly zero, and they were included in the first (tiny) bar on the left.

One way to improve the histogram in Figure 2.5 is to use a logarithmic scale. To this end, we calculated a logarithm to base 10 of all positive values and created a histogram shown in Figure 2.7. A larger number of bins were used, so that finer details of the

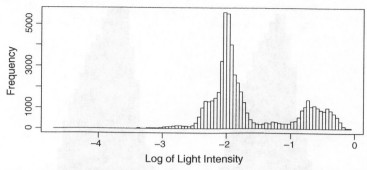

Figure 2.7 A histogram of a base 10 logarithm of the Light Intensity values from an image of a fish as used in Example 2.3.

distribution could be seen. The computer software for creating histograms usually has a built-in algorithm for a default number of bins, but users often have an option to specify their own preference. Some experimentation may be needed to find a suitable number of bins.

Based on the data in Figure 2.6, we know that there are more pixels representing the conveyer belt than those representing the fish. We also know that the higher values represent the fish. This information, together with Figure 2.7, suggests the threshold value identifying the fish pixels to be somewhere between -1.5 and -1 for $\log_{10}$(Light Intensity), which corresponds to $0.0316 < $ Light Intensity < 0.1. However, it is unclear which exact value would be best. In order to find a good threshold value, we can look at spatial patterns of pixels identified as fish. Since each image pixel represents an area within the viewing scene, it is often represented as a rectangle, like those in Figure 2.8. We could require that the set of selected pixels forms a connected set because the image represents a fish in one piece. In the context of a pixilated image, we define a set A of pixels as a *connected set*, if for any pair of pixels from A, one can find a path connecting the pixels. The path can directly connect two pixels only when they are neighbors touching at the sides (but not if they only touch at

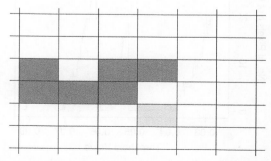

Figure 2.8 The darker shaded area is a connected set, but when the lighter shaded pixel is added, the set of pixels is not connected.

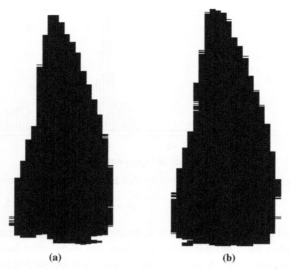

(a)   (b)

Figure 2.9   Dark areas show connected sets of pixels with Light Intensity above 0.08104 (a) and above 0.03809 (b), based on Fish data from Example 2.3.

corners). The darker shaded area in Figure 2.8 is a connected set, but when the lighter shaded pixel is added, the set of pixels is not connected.

When selecting all pixels with Light Intensity above 0.08104, one obtains a connected set of pixels shown as the black area in Figure 2.9a. Reducing the threshold below 0.08104 adds additional pixels that are not connected with the main connected set. An algorithm was used, where the threshold value was lowered, and the number of

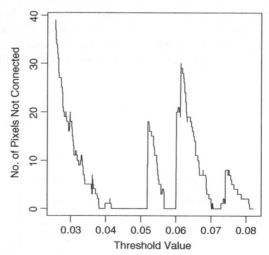

Figure 2.10   The number of pixels not connected to the main connected set shown as a function of the threshold value (for Fish data from Example 2.3).

pixels not connected to the main connected set was recorded and shown in Figure 2.10 as a function of the threshold value. We can see that for thresholds slightly above 0.07, the selected pixels again form a connected set (because the number of pixels not connected equals zero). This happens again at several ranges of the smaller threshold value until the smallest such value at 0.03809 (the place most to the left in Figure 2.10 where the function value is still zero). Below that value, the number of pixels not connected goes to very high values (beyond the range shown in Figure 2.10). Clearly, a good choice for the threshold value would be the one for which the number of pixels not connected is zero. However, Figure 2.10 still leaves us with a number of possible choices. Further investigation could be performed by looking at the type of graphs shown in Figure 2.9 and assessing the smoothness of the boundary lines.   □

2.4.3 Box Plots

Another useful graph for showing the distribution of data is a *box plot* (sometimes called a box-and-whisker plot). An example of a box plot is shown in Figure 2.11, where a vertical axis is used for showing the numerical values. The box is plotted so that its top edge is at the level of the third quartile, and the bottom edge is at the level of the first quartile. A horizontal line inside the box is drawn at the level of the median. In the simplest version of a box plot, vertical lines (called whiskers) extend from the box to the minimum and maximum values. Some box plots may show outliers with special symbols (stars, here), and the whiskers extending only to the highest and lowest values that are not outliers (called upper and lower adjacent values). Clearly, this requires an automated decision as to which observations are outliers. Computer software often uses some simplified rules based on the interquartile range. For example, an observation might be considered an outlier when it is above the third quartile or below the first quartile by more than $1.5 \cdot IQR$. However such rules are potentially misleading because any serious treatment of outliers should also take into account the sample size. We discuss outliers and their detection in Section 3.6.

Example 2.4 In Example 2.2, we discussed the Visual Density of cyan patches on three pages printed immediately after the printer calibration. In the experiment described in Appendix B, the printer was then idle for 14 h, and a set of 30 pages

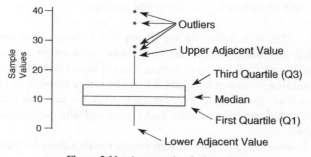

Figure 2.11 An example of a box plot.

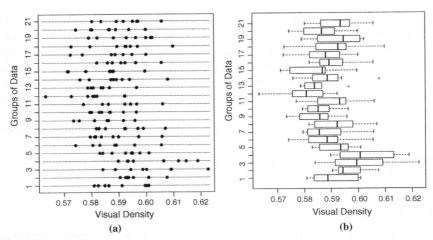

Figure 2.12   Visual Density of cyan printed on 21 pages shown as groups in the dot plots (a) and the box plots (b).

was printed, of which 18 pages were measured by a scanning spectrophotometer. This gives us a total of 21 pages with eight measurements of cyan patches in each page. Figure 2.12 shows the data in 21 groups using the dot plots (panel (a)) and the box plots (panel (b)). The box plots are somewhat easier to interpret, and this advantage increases with the increased number of groups and observations per group.

In Figure 2.12, we cannot see any specific patterns in Visual Density changes from page to page, which means that the idle time and subsequent printing of 30 pages had no significant impact on the quality of print as measured by the Visual Density of cyan patches.

2.4.4 Scatter Plots

When two characteristics, or variables, are recorded for each observation, or row, in the statistical database, we can create a two-dimensional *scatter plot* (as shown in Figure 2.13), where each observation is represented as a point with the two coordinates equal to the values of the two variables. A specific application of a scatter plot is best illustrated by the following example.

Example 2.5   This is a follow-up on Example 1.1, where you can find some background information about eye tracking. Here we want to consider an RGB image obtained in an Eye Tracking experiment as explained in Appendix B. This is a 128 by 128 pixel image (shown in Figure 2.14). The image consists of 16,384 pixels, which are treated as observations here. For each pixel, we have the intensity values (ranging from 0 to 1) for the three colors: Red, Green, and Blue, which can be regarded as three variables.

Figure 2.13 shows a scatter plot of Red versus Green values for that image. The pixels (observations) are represented as very small dots, so that thousands of them can

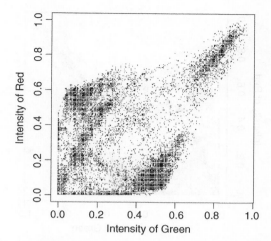

Figure 2.13 A scatter plot of intensities from the Eye Tracking image discussed in Example 2.5 and shown in Figure 2.14.

be seen as separate dots in the graph. A scatter plot is intended for continuous variables, and a primary color intensity is a continuous variable in principle. However, the three colors in the RGB image were recorded using 8 bits, which means that there are only 256 gradations of each color. This causes some discreteness of values, which can be seen as a pattern of dots lining up horizontally and vertically in Figure 2.13. It also turns out that there are many pixels in this image with exactly the same combination of gradations for the two colors. That is, some dots in the scatter plot represent more than one pixel. In order to deal with this issue, a technique of random jitter can be used, which amounts to adding a small random number to each point coordinate, before the points are plotted. This way, the dots do not print on the top of

Figure 2.14 An RGB image from the Eye Tracking data set.

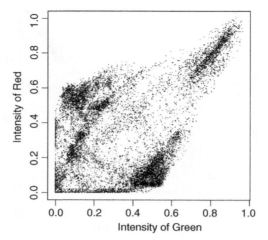

Figure 2.15   A scatter plot of color intensities from the Eye Tracking image shown in Figure 2.14. A small amount of random jitter was added to each dot.

each other. In Figure 2.15, a jitter in the amount equal to $(U-0.5)/256$ was used, where U is a random variable with the uniform distribution on the interval $(0, 1)$. The jitter improved the image, which no longer exhibits granulation, and we can better see where the larger concentrations of dots are. The use of jitter becomes even more important for highly discrete data.

The scatter plot shown in Figure 2.15 tells us that many pixels have high values both in Red and in Green. There is also a large group of pixels with approximately 50% of red and a small amount of green and then another group of pixels with approximately 50% of green and a small amount of red. There are no pixels with a very large value in one color and a low value in the other color, which is why the top left corner and the bottom right corner are both empty.   □

2.5   PROBABILITY AND PROBABILITY DISTRIBUTIONS

2.5.1   Probability and Its Properties

In statistics, we typically assume that there is some randomness in the process we are trying to describe. For example, when tossing a coin, the outcome is considered random, and one would expect to obtain heads or tails with the same probability of 0.5. On the other hand, a physicist may say that there is nothing random about tossing a coin. Assuming full knowledge about the force applied to the coin, one should be able to calculate the coin trajectory as well as its spin, and ultimately predict heads or tails. However, it is usually not practical to collect that type of detailed information about the coin toss, and the assumption of 50–50 chances for heads or tails is regarded as sufficient, given lack of additional information. In general, one can say that randomness is a way of dealing with insufficient information. This would explain why, for a given process, one can build many models depending on the available information. Also, the more information we have, the more likely we are to reduce the randomness in our model.

In order to calculate a probability of an event, we need to assume a certain probabilistic model, which involves a description of basic random events we are dealing with and a specification of their probabilities. For example, when assuming 50–50 chances for heads or tails, we are saying that each of the two events, heads and tails, has the same probability of 0.5. We can call this simple model a fair-coin model. Assuming this model, one can then calculate the probability of getting 45 tails and 55 heads in 100 tosses of the coin.

In statistics, we use this information in order to deal with an inverse problem. That is, let's assume we observe 45 tails and 55 heads in 100 tosses of a coin, but we do not know if the coin is fair with the same chances of heads or tails. Statistics would tell us, with certain confidence, what the probabilities are for heads or tails in one toss. It would also tell us if it is reasonable to assume the same probability of 0.5 for both events. If you think we can safely conclude, based on these 100 tosses, that the coin is fair, you are correct. What would be your answer if you observed 450 tails in 1000 tosses? If you are not sure, you can continue reading about the tools that will allow you to do the calculations needed to answer this question.

Before we introduce a formal definition of probability, we need to define a sample space as follows.

Definition 2.5. A *sample space* is the set of all possible outcomes of interest in a given situation under consideration.

The outcomes in a sample space are mutually exclusive, that is, only one outcome can occur in a given situation under consideration. For example, when a coin is tossed three times, the outcome is a three-element sequence of heads and tails. When we take 10 measurements, the outcome is a sequence of 10 numbers.

Definition 2.6. An *event* is a subset of a sample space.

When a coin is tossed three times, observing heads in the first toss is an event consisting of four outcomes: (H, H, H), (H, H, T), (H, T, H), and (H, T, T), where H stands for heads and T stands for tails. In a different example, when we take 10 measurements on a continuous scale, we can define an event that all of those measurements are between 20 and 25 units.

Definition 2.7. *Probability* is a function assigning a number between 0 and 1 to all events in a sample space such that these two conditions are fulfilled:

1. The probability of the whole sample space is always 1, which acknowledges the fact that one of the outcomes always has to happen.
2. For a set of mutually exclusive events A_i, we have $P\left(\bigcup_{i=1}^{k} A_i\right) = \sum_{i=1}^{k} P(A_i)$, where k is the number of events, which may also be infinity.

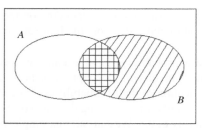

Figure 2.16   A Venn diagram showing two intersecting events. The probability $P(A|B)$ equals $P(A \cap B)$ as a fraction of $P(B)$.

We can say that probability behaves like the area of a geometric object on a plane. The sample space can be thought of as a rectangle with an area equal to 1, and all events as subsets of that square. Many properties of probability can be better understood through such geometric representation. Figure 2.16 discussed below shows an example of such representation called a Venn diagram.

When the sample space is finite, we often try to construct it so that all outcomes are equally likely. In this way, the calculation of probability is reduced to the task of counting the number of cases, such as permutations, combinations, and other combinatorial calculations. More on these rudimentary topics in probability can be found in most books on the fundamentals of statistics such as Devore (2004) or Mendenhall et al. (2006).

Definition 2.8. For any two events A and B, where $P(B) > 0$, the *conditional probability of A given that B has occurred* is defined by

$$P(A|B) = \frac{P(A \cap B)}{P(B)}. \tag{2.6}$$

Without any information about B, we would use the unconditional probability $P(A)$ as a description of the probability of A. However, once we find out that B has happened, we should use the conditional probability $P(A|B)$ to describe the probability of A. One can think of the conditional probability as probability defined on the subset B as the whole sample space, and consequently, we consider only that part of A that also belongs to B as shown in Figure 2.16.

If A and B are disjoint events, then $P(A|B) = 0$, which means that A cannot happen if B has already occurred. A different concept is that of independence of events, which can be defined as follows.

Definition 2.9. Two events A and B are *independent* if and only if $P(A \cap B) = P(A) \cdot P(B)$.

When $P(B) > 0$, the events A and B are independent if and only if $P(A|B) = P(A)$, which means that the probability of A does not change once we find out that B has occurred. Some people confuse independent events with disjoint events, but the two concepts are very different. If the events A and B are both independent and disjoint, then $0 = P(A|B) = P(A)$, which means that this can happen only for an uninteresting case when one of the sets has probability zero.

The event that B has not occurred is denoted as a complement set $B^c = \mathbb{S} \setminus B$, where $\mathbb{S}$ is the whole sample space. When $P(B^c) > 0$, the events A and B are independent if and only if $P(A|B^c) = P(A)$, which means that knowing that B has not occurred also does not change the probability of A happening. We can say that knowing whether B has occurred or not is not helpful in predicting A. The following theorem is often useful for calculating conditional probabilities.

Theorem 2.1 (Bayes' Theorem). Let $A_1, \dots, A_k$ be a set of mutually exclusive events such that $P(A_i) > 0$ for $i = 1, \dots, k$ and $\bigcup_{i=1}^{k} A_i$ is equal to the whole sample space. For any event B such that $P(B) > 0$, we have

$$P(A_i|B) = \frac{P(A_i \cap B)}{P(B)} = \frac{P(B|A_i) \cdot P(A_i)}{\sum_{j=1}^{k} P(B|A_j) \cdot P(A_j)} \quad \text{for } i = 1, \dots, k. \quad (2.7)$$

This theorem is often used to calculate the probabilities $P(A_i|B)$, when we know the conditional probabilities $P(B|A_i)$. The following example illustrates such an application.

Example 2.6 Medical imaging is often used to diagnose a disease. Consider a diagnostic method based on magnetic resonance imaging (MRI), which was tested on a large sample of patients having a particular disease. This method confirmed the disease in 99% of cases of the disease. Consider a randomly chosen person from the general population, and define A as the event that the person has the disease and B as the event that the person tested positive. Based on the above testing, we say that the probability $P(B|A)$ can be estimated as 0.99. This probability is called the sensitivity of the diagnostic method. The high sensitivity may seem like a proof of the test's good performance. However, we also need to know how the test would perform on people without the disease. So, the MRI diagnostic method was also tested on a large sample of people not having the disease. Based on the results, the probability $P(B^c|A^c)$ of testing negative for a healthy person was estimated as 0.9. This probability is called the specificity of the diagnostic method. Again, this may seem like a well performing method.

In practice, when using MRI on a patient, we do not know if the patient has the disease, so we are interested in calculating the probability $P(A|B)$ that a person testing positive has the disease. In order to apply Bayes' theorem, we also need to know $P(A)$, that is, the prevalence of the disease in the general population. In our example, it turns out that approximately 0.1% of the population has the disease, that is, $P(A) = 0.001$. Under these assumptions, the probability $P(A|B)$ can be calculated as 0.0098, which is surprisingly low (see Problem 2.4). The key to understanding why this happens is to consider all people not having the disease. They constitute 99.9% of the general population, and about 10% of them may test positive. On the other hand, only 0.1% of

Table 2.2   Examples of Probabilities of Disease if Tested Positive as a Function of Sensitivity, Specificity, and Disease Prevalence

Disease Prevalence $P(A)$	Sensitivity $P(B\|A)$	Specificity $P(B^c\|A^c)$	Probability of Disease if Tested Positive $P(A\|B)$
0.5	0.9	0.9	0.9
0.01	0.99	0.9	0.0909
0.001	0.99	0.9	0.0098
0.001	0.99	0.99	0.0902
0.001	0.99	0.999	0.4977
0.001	0.99	0.9999	0.9083
0.001	0.99	0.99999	0.9900

all people have the disease, which is a very small fraction (approximately 1%) of all people testing positive. This explains why most people testing positive do not have the disease. Table 2.2 shows some other interesting scenarios on how the probability $P(A|B)$ that a person with positive test result has the disease depends on sensitivity, specificity, and disease prevalence. $\qquad\Box$

2.5.2   Probability Distributions

We can now precisely define a random variable as a function assigning a number to each outcome in the sample space $\mathcal{S}$, that is, $X : \mathcal{S} \to \mathbb{R}$, where $\mathbb{R}$ is the set of real numbers. A value of the random variable is called a realization of X. For example, when a coin is tossed three times, define X as the number of times we observe heads. For each possible outcome, that is, a three-element sequence of heads and tails, we can count the number of heads. This will be the value of the random variable X.

Each random variable defines a probability measure on the set of real numbers $\mathbb{R}$. For each subset $A \subset \mathbb{R}$, we define $P(A) = P_{\mathcal{S}}(X^{-1}(A))$, where $P_{\mathcal{S}}(\cdot)$ is the probability defined on the sample space $\mathcal{S}$ and $X^{-1}(A)$ is the set of those outcomes in $\mathcal{S}$ that are assigned a value belonging to the set A (note that $X^{-1}(\cdot)$ is the inverse function). This probability measure is called the probability distribution of X. Continuing our example with X being the number of heads in three tosses, and taking A consisting of one number, say $A = \{2\}$, we obtain $P(A) = P_{\mathcal{S}}((H, H, T), (H, T, H), (T, H, H))$, which is equal to 3/8. This probability is more conveniently denoted by $P(X = 2)$.

In scientific applications, it is often impractical to list all possible events leading to a given value of X. For example, let X be the reflectance of a ceramic tile as measured in the spectral wavelength band between 400 and 410 µm. The random variable X will be subject to variability due to many factors such as the condition of the instrument, the process followed by the instrument operator, and so on. It would be difficult to describe all possible events that can happen during such measurements. For all practical purposes, it is sufficient to deal with the probability distribution of X without explicitly defining sample space and probability on it.

We now need to introduce some mathematical tools in order to describe probability distributions. It is convenient to distinguish two types of distributions: discrete distributions for discrete random variables, and continuous distributions for continuous random variables.

Definition 2.10. A random variable is *discrete* when all of its possible values can be counted using whole numbers.

Definition 2.11. A random variable is *continuous* when all of its possible values consist of an interval or a union of intervals on the real line $\mathbb{R}$.

A discrete probability distribution is described by a *probability mass function* $p(x) = P(X = x)$ defined for each possible value x of the random variable X. For example, if X is the number of heads in three tosses,

$$p(x) = \begin{cases} 1/8 & \text{for } x = 0 \text{ or } 3, \\ 3/8 & \text{for } x = 1 \text{ or } 2. \end{cases} \tag{2.8}$$

Property 2.1 A function defined on a discrete set D is a probability mass function of a certain distribution if and only if $p(x) \geq 0$ and $\sum_{x \in D} p(x) = 1$.

The set D in the above definition is the set of all possible values of X. Examples of some useful discrete distributions are shown in Appendix A.

A continuous probability distribution is described by *a probability density function* $f(x)$ such that for any two numbers a and b with $a \leq b$

$$P(a \leq X \leq b) = \int\limits_{a}^{b} f(x)dx. \tag{2.9}$$

An example of a probability density function is plotted in Figure 2.17 as a bold bell-shaped curve. This is a density function of a *normal* distribution that approximates the distribution of data from two different samples. Each sample was generated randomly from the normal distribution. For the sample size of $n = 40$ in the left panel, the sampling variability is fairly large, and the histogram is not very well approximated by the density function. For the large sample size of $n = 400$, the approximation is much better, and it gets even better with larger samples. One can think of a density function as an idealized histogram for a very large or infinite sample size.

Property 2.2 A function $f : \mathbb{R} \to \mathbb{R}$ is a probability density function of a certain distribution if and only if $f(x) \geq 0$ and $\int_{-\infty}^{\infty} f(x)dx = 1$.

Examples of some useful continuous distributions and their density functions are shown in Appendix A. For continuous random variables, $P(X = x)$ is always equal

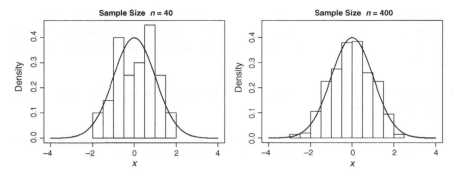

Figure 2.17 Histograms of two samples approximated by the normal density function describing the model from which the data were generated.

to zero, so a probability mass function would not be useful for describing such distributions.

Another way to describe any distribution (including a discrete or continuous one) is to use a cumulative distribution function (CDF) defined as

$$F(x) = P(X \le x). \tag{2.10}$$

For any continuous distribution, the derivative of the CDF is equal to the density function, that is, $F'(x) = f(x)$ for any point x such that the derivative $F'(x)$ exists.

We can calculate probabilities of events associated with a given random variable X with the help of the CDF. Often, we also need to solve a reverse problem, that is, to find x such that $F(x)$ is equal to a given probability.

Definition 2.12. Let p be a number between 0 and 1. The $(100p)$th *percentile* of the distribution defined by $F(x)$ is a number η_p such that $p = F(\eta_p)$.

Often, it is convenient to define the upper percentile as follows.

Definition 2.13. Let p be a number between 0 and 1. The $(100p)$th *upper percentile* of the distribution defined by $F(x)$ is a number τ_p such that $p = 1 - F(\tau_p)$.

It is easy to see that the $(100p)$th percentile η_p is equal to the $(100(1-p))$th upper percentile τ_{1-p} of the same distribution. For continuous distributions, the percentile η_p exists for any value $p \in (0, 1)$. Tables of percentiles for some important statistical distributions can usually be found in statistical textbooks. These days, one can often obtain percentiles from computer software, but we still provide some percentile values in Appendix A for added convenience. Appendix A shows the notation used throughout this book for percentiles of a wide range of distributions.

Even though a distribution is precisely defined by its cumulative distribution function or by a density or mass function (for continuous or discrete distributions, respectively), it is often beneficial to characterize distributions by using single numbers or parameters. Some important characteristics are the first and the third quartile (the 25th and 75th percentiles, respectively) and the median (the 50th percentile). Other characteristics of distributions are defined in the next section.

2.5.3 Expected Value and Moments

The expected or mean value of a random variable X is defined as

$$E(X) = \begin{cases} \int_{-\infty}^{\infty} x \cdot f(x)dx & \text{if } X \text{ is continuous,} \\ \sum_{x \in D} x \cdot p(x) & \text{if } X \text{ is discrete.} \end{cases} \quad (2.11)$$

The expected value describes an average outcome based on a theoretical distribution. It is different from the sample mean $\bar{x}$ calculated from data. If data are generated from the distribution of X, the sample mean $\bar{x}$ should be close to $E(X)$ and it will get closer, on average, as the sample size increases. The expected value $E(X)$ is often denoted by μ, but a subtlety here is that μ should be considered as a parameter, while $E(X)$ is an operation on the distribution of X that produces a number.

Based on the linear property of integrals and summations, one can show (see Problem 2.5) that for any constants a and b

$$E(aX + bY) = aE(X) + bE(Y). \quad (2.12)$$

For any natural number k, the kth moment of X is defined as the expectation of X^k

$$E(X^k) = \begin{cases} \int_{-\infty}^{\infty} x^k \cdot f(x)dx & \text{if } X \text{ is continuous,} \\ \sum_{x \in D} x^k \cdot p(x) & \text{if } X \text{ is discrete,} \end{cases} \quad (2.13)$$

and the central moments are defined as moments centered around the mean, that is, $E\left[(X-E(X))^k\right]$. The mean value is interpreted as a position parameter or a "central" point, because it is an average of possible values of X weighted by their probabilities or by density. The second central moment, called variance, is denoted by

$$\text{Var}(X) = E\left[(X-E(X))^2\right] = E(X^2) - [E(X)]^2. \quad (2.14)$$

The variance measures variability around the mean value, while the noncentral moment $E(X^2)$ measures variability of X around zero. By using property (2.12), one can show that for any constants a and b

$$\text{Var}(aX + b) = a^2 \text{Var}(X), \quad (2.15)$$

which means that the variance is not affected by a shift (adding a constant). This makes sense because a simple shift does not impact variability. Since the variance is expressed in the squared units of X, it is convenient to introduce the concept of standard deviation defined as the square root of variance and denoted by

$$\text{StDev}(X) = \sqrt{\text{Var}(X)}. \tag{2.16}$$

The standard deviation is a measure of variability expressed in the units of X, and its interpretation is further explained in Section 2.5. From equation (2.15), we obtain

$$\text{StDev}(aX + b) = |a| \cdot \text{StDev}(X), \tag{2.17}$$

which means that multiplying X by a positive constant results in the same multiplication of the standard deviation. The standard deviation is often denoted by σ, but again we have a subtlety here, where σ should be thought of as a parameter, while $\text{StDev}(X)$ is an operation on the distribution of X that produces a number.

The standard deviation as a parameter is often considered a scale parameter. We can use the standard deviation σ and the mean (expected value) $\mu = E(X)$ to standardize X, that is, we define the standardized variable $Z = (X-\mu)/\sigma$. It is easy to see that $E(Z) = 0$ and $\text{Var}(Z) = 1$. Since $(X-\mu)$ and σ are in the same units, the variable Z has no units.

2.5.4 Joint Distributions and Independence

Consider two random variables X and Y. We can study their relationship by considering a random vector (X, Y). This random vector can also be treated as a random point (X, Y) on the plane $\mathbb{R}^2$. Assume that we observe a large number of values, or realizations, of (X, Y). Each realization or data point can be plotted in the system of x and y coordinates as a point. Figure 2.18a shows a scatter plot of such points as an example. The relationship between X and Y is fully described by the joint

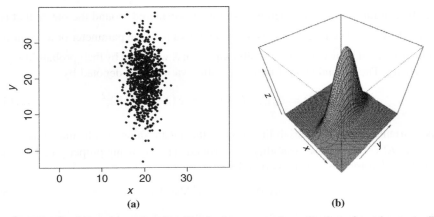

Figure 2.18   Panel (a) shows a scatter plot of (x, y) values generated as realizations of a random vector (X, Y) with the joint density function shown in panel (b).

distribution of these variables on the plane $\mathbb{R}^2$. The joint distribution, in turn, can be fully described by the cumulative bivariate distribution function defined as

$$F(x, y) = P(X \le x \text{ and } Y \le y). \tag{2.18}$$

For a continuous bivariate distribution, there exists a bivariate density function $f(\cdot, \cdot)$ such that

$$P((X, Y) \in A) = \iint_A f(s, t) ds \, dt \quad \text{for any } A \subset \mathbb{R}^2. \tag{2.19}$$

In particular, we have this property of the cumulative distribution function

$$F(x, y) = \int_{-\infty}^{y} \int_{-\infty}^{x} f(s, t) ds \, dt. \tag{2.20}$$

Figure 2.18b shows a density function of the form $f(s, t) = f_1(s)f_2(t)$, where f_1 is the density function of the normal distribution $N(20, 2)$ with the mean of 20 and standard deviation of 2, and f_2 is the density function of $N(20, 6)$ (see Appendix A for the specific formula). The points in Figure 2.18a are values or realizations generated from the distribution defined by $f(s, t)$. The higher concentration of points around the center $(20, 20)$ corresponds to the higher value of the joint density function shown in Figure 2.18b. The range of x coordinates is smaller than the one for the y coordinates because of the smaller standard deviation in the x direction as seen in the elongated shape of the density in panel (b).

In the context of the bivariate distribution of (X, Y), the distributions of X and Y are called marginal distributions. For a continuous bivariate distribution of (X, Y), the marginal density function of one of the variables, let's say X, can be calculated by "summing up" the probabilities associated with the other variable, say Y, that is,

$$f_X(x) = \int_{-\infty}^{\infty} f(x, y) dy. \tag{2.21}$$

As another example, define a bivariate density function

$$f_0(x, y) = \begin{cases} 0.5 & \text{if } -1 \le y + x \le 1 \text{ and } -1 \le y - x \le 1, \\ 0 & \text{otherwise,} \end{cases} \tag{2.22}$$

which is positive inside of a rotated square shown in Figure 2.19b.

The marginal distributions are obtained by "projecting" the bivariate density on the x or y axes, respectively. This is best understood by projecting the points in Figure 2.19a on one of the axes. Figure 2.20a shows a histogram of projections of

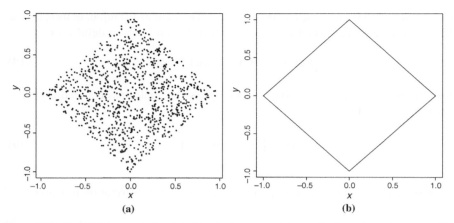

Figure 2.19   Panel (a) shows a scatter plot of (x, y) values generated as realizations of a random vector (X, Y) with the joint density function equal to 0.5 inside of the rotated square shown in panel (b) and zero outside of the square.

those points onto the x-axis. Figure 2.20b shows a theoretical distribution of X derived from (2.21) and (2.22) and given by the formula

$$f_X(x) = \begin{cases} 1 - |x| & \text{if } |x| \le 1, \\ 0 & \text{otherwise.} \end{cases} \qquad (2.23)$$

When dealing with a bivariate distribution of (X, Y), we might be interested in knowing the distribution of Y given an observed value of $X = x$, which represents a new piece of information. That distribution is called a *conditional distribution* of Y

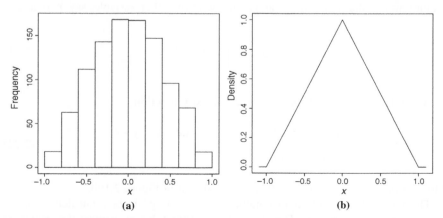

Figure 2.20   Panel (a) shows a histogram of projections of points from Figure 2.19a onto the x-axis. Panel (b) shows a theoretical distribution of the projection on the x-axis, that is, the marginal distribution of X.

given $X = x$. While values of the random vector (X, Y) lie on the plane $\mathbb{R}^2$, the conditional distribution of Y given $X = x$ is concentrated on the subset of the plane, namely, vertical line crossing x-axis at x.

For a continuous bivariate distribution of (X, Y), the conditional probability density function of Y given $X = x$ is defined as

$$f_{Y|X}(y|x) = \frac{f(x,y)}{f_X(x)} \quad \text{for } -\infty < y < \infty, \tag{2.24}$$

where f_X is the marginal density function defined in (2.21) and x is any value such that $f_X(x) > 0$.

Intuitively, the conditional distribution is obtained by taking a cross section of the distribution of (X, Y), such as the one depicted in Figure 2.19b, along the line $X = x$. Dividing by $f_X(x)$ in formula (2.24) reflects the fact that the conditional distribution is a probability measure defined on a smaller space determined by $X = x$. This also makes the resulting function a density function, but it does not change the shape of the function.

Consider the bivariate density function $f_0(x,y)$ defined by (2.22). From Figure 2.19b, we can see that that for any $|x| \le 1$, $f_0(x,y)$ as a function of y is positive and constant on the interval $[-(1-|x|), \ 1-|x|]$ and zero outside this interval. Therefore, for any $|x| \le 1$, the conditional distribution of Y given $X = x$ is the uniform distribution concentrated on the interval $[-(1-|x|), \ 1-|x|]$. Since the length of this interval is $2(1-|x|)$, the conditional density function is given by the formula

$$f_{Y|X}(y|x) = \begin{cases} 1/(2(1-|x|)) & \text{if } -(1-|x|) \le y \le 1-|x|, \\ 0 & \text{otherwise.} \end{cases} \tag{2.25}$$

This formula can also be derived directly from definition (2.24) and formulas (2.22) and (2.23). Notice that in this example, the conditional distribution of Y depends on the observed value $X = x$. This information changes the range of possible values of Y from the general range $[-1, \ 1]$, without any knowledge of X, to the narrower range $[-(1-|x|), \ 1-|x|]$ when the value of $X = x$ is already known. This means that Y is dependent on X.

We now extend the definition of independence from random events to random variables.

Definition 2.14. The random variables X and Y are called *independent* when the events associated with those variables are independent, that is, for any sets $A, B \subset \mathbb{R}$

$$P(X \in A \text{ and } Y \in B) = P(X \in A) \cdot P(Y \in B). \tag{2.26}$$

For a continuous bivariate distribution of (X, Y), X and Y are independent if and only if

$$f(x,y) = f_X(x)f_Y(y) \quad \text{for all pairs of } x \text{ and } y \text{ values.} \tag{2.27}$$

As expected, independence of random variables is closely related to the conditional distributions. For a continuous bivariate distribution of (X, Y), X and Y are independent if and only if

$$f_{Y|X}(y|x) = f_Y(y) \quad \text{for all pairs of } x \text{ and } y \text{ values such that } f_X(x) > 0. \quad (2.28)$$

We can say that X and Y are independent if and only if information contained in X is not helpful in predicting Y. For example, the random variables X and Y with the joint distribution shown in Figure 2.19b are not independent because the conditional distribution shown in formula (2.25) depends on x, as we discussed previously. In Figure 2.18b, we depicted the joint distribution of two independent variables X and Y. We can imagine that although the cross sections of the surface taken at various values of x are different, smaller for x farther from the mean value of 20, they have the same bell shape and after normalizing by the marginal density $f_1(x)$, they all are identical to $f_2(y)$, the marginal density function of Y.

2.5.5 Covariance and Correlation

In order to capture an important property of the joint distribution, it is useful to define *covariance* of the random variables X and Y as

$$\text{Cov}(X, Y) = E[(X - E(X))(Y - E(Y))]. \quad (2.29)$$

With some algebra, one can show that

$$\text{Cov}(X, Y) = E(XY) - E(X)E(Y). \quad (2.30)$$

Using equation (2.12), one can show that for any constants a and b

$$\text{Cov}(aX + bY, Z) = a\,\text{Cov}(X, Z) + b\,\text{Cov}(Y, Z) \quad (2.31)$$

for any random variable Z, which means that the covariance is linear with respect to its first argument. From symmetry, the same property holds for the second argument of the covariance. Since $\text{Var}(X) = \text{Cov}(X, X)$, we obtain

$$\text{Var}(aX + bY) = a^2\,\text{Var}(X) + b^2\,\text{Var}(Y) + 2ab\,\text{Cov}(X, Y). \quad (2.32)$$

Let us now take $Y \equiv 1$, that is, a random variable equal to a constant 1. Then $Y = E(Y)$ and $\text{Var}(Y) = 0$. We can also see from (2.29) that the covariance of a constant variable Y with an arbitrary random variable Z is zero, that is, $\text{Cov}(Y, Z) = 0$. We can now write formula (2.31) as

$$\text{Cov}(aX + b, Z) = a\,\text{Cov}(X, Z), \quad (2.33)$$

which means that the covariance is not affected by a shift of X (adding a constant), but it is affected by the scale, that is, when X is multiplied by a constant. In the same way, we could obtain property (2.15) as a special case of (2.32).

The covariance $\text{Cov}(X, Y)$ measures a degree of linear association between X and Y. Unfortunately, that measure is distorted by the impact of a scale change. To make the measure scale independent, we introduce the *correlation coefficient* defined by the covariance scaled by the standard deviations of the variables as follows:

$$\text{Corr}(X, Y) = \frac{\text{Cov}(X, Y)}{\text{StDev}(X)\text{StDev}(Y)}, \qquad (2.34)$$

where $\text{StDev}(X) > 0$ and $\text{StDev}(Y) > 0$. It can be proven that $|\text{Corr}(X, Y)| \leq 1$, and the equality holds if and only if there exist constants $a \neq 0$ and b such that $Y = aX + b$ with probability 1, which means that X and Y are perfectly collinear. The correlation coefficient $\text{Corr}(X, Y)$ is often denoted by $\rho_{X,Y}$ or simply ρ.

Definition 2.15. The random variables X and Y are called *uncorrelated* when $\text{Corr}(X, Y) = 0$.

From (2.32), we conclude that the random variables X and Y are uncorrelated, if and only if

$$\text{Var}(X + Y) = \text{Var}(X) + \text{Var}(Y). \qquad (2.35)$$

From (2.30), we conclude that the random variables X and Y are uncorrelated, if and only if

$$E(XY) = E(X)E(Y). \qquad (2.36)$$

When two random variables are independent, they are also uncorrelated. However, in general, the reverse implication is not true. We have already discussed the random variables X and Y with the joint distribution shown in Figure 2.19 as being dependent. One can show that they are also uncorrelated. Another example of uncorrelated dependent variables is shown in Figure 2.21, where Y clearly depends on X, but not in a

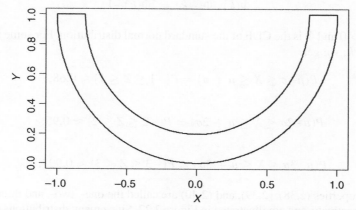

Figure 2.21 An example of uncorrelated and dependent random variables X and Y. The joint density function of the random variables X and Y is equal to a positive constant within the U-shaped area shown here and is equal to zero outside of that area.

linear fashion. In both cases, the lack of correlation can be concluded from the following property (see Problem 2.10 for an outline of the proof).

Property 2.3 If the joint distribution of X and Y is symmetric with respect to a vertical or horizontal straight line, and the correlation $\mathrm{Corr}(X, Y)$ exists, then $\mathrm{Corr}(X, Y) = 0$.

2.6 RULES OF TWO AND THREE SIGMA

In Section 2.2, we introduced the sample standard deviation as a measure of sample variability. In Section 2.4, we discussed the population standard deviation σ as a measure of variability in a random variable, say X, where $\sigma = \mathrm{StDev}(X) = \sqrt{\mathrm{Var}(X)}$. We now want to provide interpretation of the standard deviation σ by describing what the knowledge of σ can tell us about the variability in X. We will start by assuming that X follows the normal (Gaussian) distribution $N(\mu, \sigma)$ defined in Appendix A. The normal distribution is the most important distribution in probability and statistics, because data distributions and some theoretical distributions are often well approximated by the normal distribution. The reasons for that will be discussed in Section 2.7.

Property 2.4 If X follows the normal (Gaussian) distribution, then for any constants $a \neq 0$ and b, the variable $aX + b$ also follows the normal distribution. (See Problem 2.11 for a hint on the proof.)

We standardize X by defining $Z = (X-\mu)/\sigma$. It is easy to see (from (2.12) and (2.15)) that $E(Z) = 0$ and $\mathrm{Var}(Z) = 1$. From Property 2.4, the standardized variable Z has the normal distribution $N(0, 1)$, which is called the *standard normal distribution*.

$$P(|X-\mu| \leq k\sigma) = P(\mu-k\sigma \leq X \leq \mu + k\sigma) = P(-k \leq Z \leq k)$$

$$= \Phi(k)-\Phi(-k) = 2\Phi(k)-1, \tag{2.37}$$

where $k > 0$ and Φ is the CDF of the standard normal distribution. For some specific values of k, we get

$$P(\mu-\sigma \leq X \leq \mu + \sigma) = P(-1 \leq Z \leq 1) \approx 0.68, \tag{2.38}$$

$$P(\mu-2\sigma \leq X \leq \mu + 2\sigma) = P(-2 \leq Z \leq 2) \approx 0.95, \tag{2.39}$$

$$P(\mu-3\sigma \leq X \leq \mu + 3\sigma) = P(-3 \leq Z \leq 3) \approx 0.997. \tag{2.40}$$

The properties (2.38), (2.39), and (2.40) are called the one-, two-, and three-sigma rules, respectively, and are illustrated in Figure 2.22. Since many distributions are well approximated by the normal distribution, these rules are widely used, especially for a

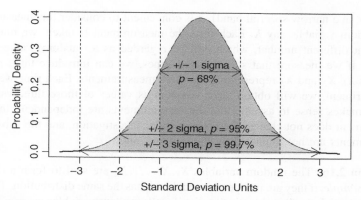

Figure 2.22 One-, two-, and three-sigma rules shown as areas under the normal density curve.

quick and intuitive understanding of the amount of variability associated with a given value of the standard deviation σ. For example, the two-sigma rule tells us that approximately 95% of the distribution lies within two standard deviations from the mean.

Even though the approximation by the normal distribution works quite well in many contexts, it would be good to know the significance of σ in other types of distributions. The following theorem addresses this issue in a general context.

Theorem 2.2 (Chebyshev's Inequality). For a random variable X with a finite mean μ and standard deviation σ, we have

$$P(\mu - k\sigma < X < \mu + k\sigma) \geq 1 - \frac{1}{k^2}, \qquad (2.41)$$

where $k > 0$ is an arbitrary constant.

The proof can be found in Ross (2002). When applying (2.41) with $k = 2$, we can see that at least 75% of the distribution lies within two standard deviations from the mean, compared to the 95% based on normality. With $k = 3$, we obtain 8/9 or at least 88.8% of the distribution being within three standard deviations from the mean, compared to the 99.7% based on normality.

2.7 SAMPLING DISTRIBUTIONS AND THE LAWS OF LARGE NUMBERS

In Section 2.2, we discussed a sample $x_1, x_2, \ldots, x_n$ of n measurements or observations as a set of specific numbers. However, before the observations are collected, there is uncertainty about their values. Also, if another set of observations were collected from the same unchanged process, the values would be somewhat different due to natural variability. This is why we often treat observations as random variables, so that we can study their properties in repeated sampling. For example, if we want to measure reflectance of a given surface as a single number (let's say, in a narrow spectral band), it is convenient to consider this measurement as a random variable, say X. Each time the measurement is taken, we may get a somewhat different number, which will be regarded as a (random) value of that variable. If we measure that surface three times, we can introduce three random variables $X_1, X_2,$ and X_3 representing the three measurements. Each time we repeat the experiment, we will obtain three numbers as values of those three variables. It often makes sense to assume that the measurements are independent, that is, a measurement does not change the process under investigation, and the subsequent measurements are not impacted by the previous ones.

Definition 2.16. The random variables $X_1, X_2, \ldots, X_n$ are said to form a *(simple) random sample*, if they are independent, and each has the same distribution. They are called i.i.d. (independent, identically distributed) random variables.

For each sample, we can calculate a statistic, such as the sample mean, which can be treated as a random variable $\overline{X}$, since its values will vary in repeated samples. The distribution of $\overline{X}$ is called its sampling distribution in order to emphasize the fact that it describes the behavior of $\overline{X}$ over repeated samples.

Consider a random sample $X_1, X_2, \ldots, X_n$ from an arbitrary distribution G with a finite mean μ. The law of large numbers tells us that $\overline{X}$ approaches μ as n tends to infinity. Technical details about this convergence can be found in Ross (2002) and Bickel and Doksum (2001). The convergence means that we can draw conclusions about the population (represented by the distribution G) based on the sample $X_1, X_2, \ldots, X_n$, and there is a benefit from having larger samples. For very large n, the mean $\overline{X}$ will be very close to μ. Another far-reaching consequence can be concluded from the following construction. Let A be an arbitrary probabilistic event with a certain probability $P(A)$. The event A could be "obtaining heads in a single toss of a coin." Consider repeated independent trials (coin tosses), where the event A can happen with probability $P(A)$. For the ith trial, define Y_i as equal to 1 when A happens and 0 otherwise. Note that $P(Y_i = 1) = P(A) = E(Y_i) = \mu$. The sample mean $\overline{Y}$ is the relative frequency of the event A in n trials (fraction of *heads* in n tosses). The law of large numbers tells us that the fraction of trials when A happens (fraction of heads) in n trials approaches the probability $P(A)$ of the event (heads) as n tends to infinity. This may seem intuitively obvious, but it is good to have a confirmation of this fact as a basis for this interpretation of probability.

The law of large numbers tells us that $\overline{X}$ approaches μ as n tends to infinity, but it does not tell us how fast it is approaching. This information would be very useful from a practical point of view, so that we know the consequences of using a specific sample size n. From properties (2.15) and (2.35), one can show that

$$\text{StDev}(\overline{X}) = \frac{\sigma}{\sqrt{n}}, \tag{2.42}$$

where $\sigma = \text{StDev}(X_i)$, $i = 1, \ldots, n$. This means that we can standardize $\overline{X}$ by defining $Z_n = (\overline{X} - \mu)/(\sigma/\sqrt{n}) = \sqrt{n}(\overline{X} - \mu)/\sigma$, such that $\text{Var}(Z_n) = 1$. We know that $(\overline{X} - \mu)$ converges to 0. When it is multiplied by $\sqrt{n}$, it no longer converges to 0, nor does it go to infinity (since $\text{Var}(Z_n) = 1$). We could say that $\sqrt{n}$ is just the right multiplier to make Z_n "stable." For example, if we used the multiplier n^a, then $n^a(\overline{X}-\mu)/\sigma$ would approach infinity for $a > 0.5$, and it would approach 0 for $a < 0.5$.

If the random sample $X_1, X_2, \ldots, X_n$ comes from the normal distribution $N(\mu, \sigma)$, the distribution of Z is standard normal $N(0, 1)$ (see Property 2.4). This allows us to tell how close $\overline{X}$ is to μ with the probability given by

$$P\left(|\overline{X}-\mu| < k\frac{\sigma}{\sqrt{n}}\right) = P(|Z| < k) = 2\Phi(k)-1. \tag{2.43}$$

When the distribution G of the sample is not normal, the distribution of Z often is not easy to calculate, and it also depends on n. Fortunately, the following theorem allows an approximation of the distribution of Z for large n.

Theorem 2.3 (The Central Limit Theorem, CLT). Let $X_1, X_2, \ldots$ be a sequence of independent, identically distributed random variables, each having a finite mean μ and standard deviation σ. Then the distribution of Z approaches the standard normal distribution as n tends to infinity, that is,

$$\lim_{n \to \infty} P\left(\frac{\overline{X}-\mu}{\sigma/\sqrt{n}} < k\right) = \Phi(k). \tag{2.44}$$

The proof can be found in Ross (2002).

The CLT allows us to use equation (2.43) as an approximation in cases of samples from non-normal distributions. Various sources give some rules of thumb (e.g., $n \geq 30$) as to how large n is needed for the normal approximation. This could be potentially misleading. The precision of the normal approximation depends on the shape of the X_i's distribution. For example, the convergence is generally slower for nonsymmetric distributions. Figure 2.23 shows an example of the density functions of Z, when the distribution of X_i's is chi-squared with one degree of freedom and n is equal to 3, 10, and 30, respectively. The density of the standard normal distribution is also shown for comparison. The CLT approximation using (2.44) can be better assessed based on Figure 2.24, where the CDFs of the same distributions are shown. Precision of the normal approximation is further discussed in Chapter 3.

The CLT explains why real data often follow the normal distribution (approximately). Many characteristics are sums of a large number of small independent factors. For example, height in a large population depends on influences of particular genes, elements in the diet, and other factors. Hence, the height distribution is typically well approximated by the normal distribution. Another example is when we take multiple measurements of the same object. The measurement error usually depends on many independent small factors (environmental conditions, gauge conditions, operators' impact, etc.) that add up to the final result. Again, the measurement error is typically well approximated by the normal distribution.

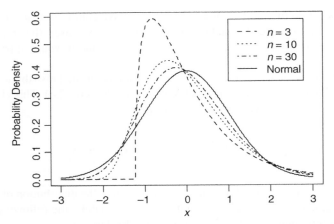

Figure 2.23   The density functions of Z, when the distribution of X_i's is chi-squared with one degree of freedom and n is equal to 3, 10, and 30, respectively. The solid line is the density of the standard normal distribution intended as the approximation.

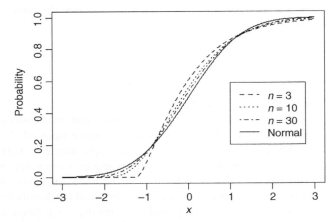

Figure 2.24   The CDF of Z, when the distribution of X_i's is chi-squared with one degree of freedom and n is equal to 3, 10, and 30, respectively. The solid line is the density of the standard normal distribution intended as the approximation.

2.8   SKEWNESS AND KURTOSIS[1]

The first two moments characterize the location and variability in a distribution. In order to characterize the shape of a distribution, it is convenient to consider the standardized variable $Z = (X-\mu)/\sigma$. Since the first two moments of Z are already

---

[1] This section is more technical and is not needed in the remaining part of this book.

determined ($E(Z) = 0$ and $\mathrm{Var}(Z) = 1$), we will use higher order moments in order to elicit the information about the distribution shape.

The lack of symmetry around the mean value in a distribution, that is, skewness, is measured by the *coefficient of skewness* defined as

$$\gamma_1 = E(Z^3) = \frac{E\left[(X - E(X))^3\right]}{[\mathrm{Var}(X)]^{3/2}}. \tag{2.45}$$

For any symmetric distribution, $\gamma_1 = 0$. The fourth moment of Z is defined as *kurtosis* of X, that is,

$$\mathrm{Kurt}(X) = E(Z^4) = \frac{E\left[(X - E(X))^4\right]}{[\mathrm{Var}(X)]^2}. \tag{2.46}$$

For a normal distribution, $\mathrm{Kurt}(X) = 3$, which is why an *excess kurtosis* is often defined as $\gamma_2 = \mathrm{Kurt}(X) - 3$. Since for a normal (Gaussian) distribution, $\gamma_1 = 0$ and $\gamma_2 = 0$, the skewness and kurtosis are sometimes used for checking normality. This approach is utilized in independent component analysis, an advanced multivariate method. A lack of symmetry in a distribution is fairly easy to recognize, but the interpretation of kurtosis is much less obvious. This is why we will detail more information about kurtosis and some related terminology.

A positive value of γ_2 indicates a super-Gaussian distribution (also called leptokurtic), which is often characterized by "fat tails," that is, the density function decreases slowly for large x values. A negative value of γ_2 indicates a sub-Gaussian distribution (also called platykurtic), which is often characterized by "thin tails," that is, the density function decreases rapidly for large x values. The kurtosis is also described as a measure of "peakedness" of a distribution at the center $E(X)$. These interpretations are true only to some extent. We will now discuss a different interpretation that clarifies the matter.

Since Z is a standardized variable, we have $\mathrm{Var}(Z) = E(Z^2) = 1$, and it might be of interest to know how far Z^2 is from 1. This can be measured by the mean square $E(Z^2 - 1)^2$, which is equal to $E(Z^4) - 1 = \mathrm{Kurt}(X) - 1$. We can also write

$$\mathrm{Kurt}(X) = E(Z^2 - 1)^2 + 1. \tag{2.47}$$

If Z^2 is close to 1 (i.e., X is concentrated around $\mu - \sigma$ or $\mu + \sigma$), then $\mathrm{Kurt}(X)$ is small. If $Z^2 \equiv 1$ (which happens for the Bernoulli distribution with only two possible values, each with the same probability $p = 0.5$), then $\mathrm{Kurt}(X) = 1$, which is its smallest possible value (i.e., the excess kurtosis γ_2 is always at least -2). If Z^2 is far from 1, then $\mathrm{Kurt}(X)$ is large. The variable Z^2 can be far from 1 when it is concentrated around 0 (high "peakedness") or concentrated on very large values ("fat tails" in the sense of large probabilities for X much larger than $\mu + \sigma$ in units of σ). Hence, in general, one of these conditions is sufficient to produce large kurtosis, but both of them

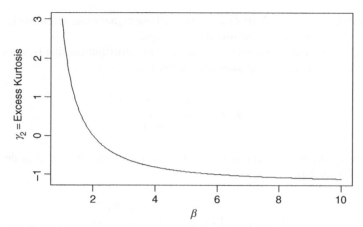

Figure 2.25 Excess kurtosis γ_2 as a function of the shape parameter $1 \leq \alpha \leq 10$ for the exponential power distribution.

give an even larger kurtosis. At the same time, the "peakedness" has small direct impact on the kurtosis because values close to 0 can never be really far from 1. The impact of "peakedness" is indirect. Since we always have $E(Z^2) = 1$, values of Z close to 0 allow some other Z values to be very large (and create "fat tails" in the sense described above).

Example 2.7 Consider the family of exponential power distributions defined in Appendix A. Its excess kurtosis is given by the formula

$$\gamma_2 = \frac{\Gamma(5/\alpha)\Gamma(1/\alpha)}{\Gamma(3/\alpha)^2} - 3, \tag{2.48}$$

where $\alpha > 0$ is the shape parameter. Figure 2.25 shows γ_2 as a function of α on the interval $[1, 10]$. The value $\alpha = 1$ corresponds to the Laplace distribution with $\gamma_2 = 3$, the value $\alpha = 2$ corresponds to the normal distribution with $\gamma_2 = 0$, and with α approaching infinity, the exponential power distribution approaches the uniform distribution having $\gamma_2 = -1.2$. When α approaches 0, the excess kurtosis γ_2 approaches infinity. The exponential power distributions with $\alpha < 2$ are super-Gaussian with "fat tails," while those with $\alpha > 2$ are sub-Gaussian with "thin tails." We illustrate this point in Figure 2.26, where we show densities of the exponential power distributions with $\alpha = 1$ (Laplace), $\alpha = 2$ (normal), and $\alpha = 10$.

Example 2.8 The interpretation of kurtosis as an indication of "fat" versus "thin" tails is not always as clear-cut as shown in Example 2.7. As discussed earlier, a sub-Gaussian distribution is often associated with "thin tails," but here we construct a sub-Gaussian distribution with "fat tails." Consider X following a chi-squared random variable with n degrees of freedom. The excess kurtosis of X is equal to $12/n$, so it is

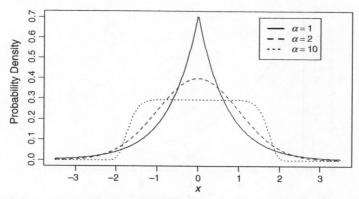

Figure 2.26 The densities of the exponential power distributions with $\alpha = 1$ (Laplace), $\alpha = 2$ (normal), and $\alpha = 10$, chosen so that the mean is 0 and the variance is 1.

considered super-Gaussian with "fat tails." We define a new variable $Y = D(X + a)$, where a is a positive constant and D is a random variable independent of X such that $P(D = 1) = P(D = -1) = 0.5$. The density of the Y variable is symmetric with respect to zero, and it consists of two symmetric shapes of the density of the chi-squared distribution with a gap in between (zero density on the interval $(-a, a)$). We call this distribution *double chi-squared*. Figure 2.27 shows an example of such density for $n = 4$ and $a = 0.41$. If we move the two pieces of the density function farther apart (by increasing a), its general shape does not change. This means that the density of Y has tails that are "fatter" than those of the normal density. However, for $a = 0.41$, one can calculate that $\gamma_2 = 0$. This tells us that the double chi-squared

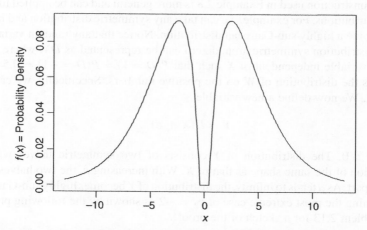

Figure 2.27 Density function of the random variable $Y = D(X + a)$ with zero excess kurtosis ($a \approx 0.41$), where X is a chi-squared random variable with four degrees of freedom and D is a random variable independent of X such that $P(D = 1) = P(D = -1) = 0.5$.

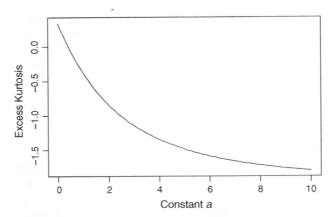

Figure 2.28 The excess kurtosis γ_2 of $Y = D(X + a)$ as a function of a, where X is a chi-squared random variable with four degrees of freedom and D is a random variable independent of X such that $P(D = 1) = P(D = -1) = 0.5$.

distribution shown in Figure 2.27 is an example of a "fat-tailed" distribution with the kurtosis equal to that of the Gaussian (normal) distribution.

Figure 2.28 shows how the excess kurtosis γ_2 of Y depends on a for $n = 4$. Clearly, the distribution becomes highly sub-Gaussian for large a. See Problem 2.12 for directions on how to perform the calculations for this example. □

The excess kurtosis is sometimes used to measure how far a distribution is from the normal distribution. This can be potentially misleading. The double chi-squared distribution introduced in the above example shows an example of a distribution with $\gamma_2 = 0$ (for $a \approx 0.41$), which is far from normal. The density function of that distribution is shown in Figure 2.27.

The construction used in Example 2.8 is more general and can be applied to many other distributions. For example, we can take any symmetric distribution and modify it to become a highly sub-Gaussian distribution. Notice that any random variable W with a distribution symmetric around zero can be represented as DX, where D is a random variable independent of X such that $P(D = 1) = P(D = -1) = 0.5$ and X describes the distribution of W on the positive numbers. Specifically, we can take $X = |W|$. We now define a new variable

$$Y = D(X + a), \tag{2.49}$$

where $a > 0$. The distribution of Y consists of two symmetric halves with the distribution of the same shape as that of X. With increasing a, the two halves move farther apart. As a tends to infinity, the distribution of Y becomes highly sub-Gaussian, approaching the most extreme case of $\gamma_2 = -2$ as shown by the following property (see Problem 2.13 for a sketch of the proof).

Property 2.5 For the excess kurtosis γ_2 of Y defined by (2.49), we have $\lim_{a \to \infty} \gamma_2 = -2$.

PROBLEMS

2.1. For the five observations of the Output Power variable in Example 2.1, find the 90th percentile calculated by a linear extrapolation using formula (2.4).

2.2. Prove that the deviations from the mean, defined as $d_i = x_i - \bar{x}$, have the property that they sum up to zero, that is, $\sum_{i=1}^{n} d_i = 0$.

2.3. Consider n observations x_i, $i = 1, \ldots, n$, and their linear transformations defined as $y_i = ax_i + b$ for $i = 1, \ldots, n$. Prove that $s_y^2 = a^2 s_x^2$ and $s_y = |a|s_x$.

2.4. In the context of Example 2.6, develop a formula for the probability of having the disease if testing positive as a function of sensitivity, specificity, and disease prevalence. Verify the numbers shown in Table 2.2.

2.5. Prove that for any constants a and b and random variables X and Y, we have (formula (2.12))

$$E(aX + bY) = aE(X) + bE(Y).$$

2.6. Prove formula (2.15).

2.7. Prove that for any constants a and b and random variables X, Y, and Z, we have (formula (2.31))

$$\mathrm{Cov}(aX + bY, Z) = a\,\mathrm{Cov}(X, Z) + b\,\mathrm{Cov}(Y, Z).$$

2.8. Prove that for any constants a and b and random variables X and Y, we have (formula (2.32))

$$\mathrm{Var}(aX + bY) = a^2\,\mathrm{Var}(X) + b^2\,\mathrm{Var}(Y) + 2ab\,\mathrm{Cov}(X, Y).$$

2.9. Prove formula (2.33).

2.10. * Prove Property 2.3. *Hint*: Assume that the joint distribution is symmetric with respect to the line $x = \mu$. Define

$$d_+(x, y) = \begin{cases} 1 & \text{for } x > \mu, \\ 0 & \text{otherwise,} \end{cases} \qquad d_-(x, y) = \begin{cases} 1 & \text{for } x < \mu, \\ 0 & \text{otherwise.} \end{cases}$$

From the symmetry assumption, we have

$$E[d_+(X, Y)(X - \mu)(Y - E(Y))] = -E[d_-(X, Y)(X - \mu)(Y - E(Y))].$$

Since

$$\mathrm{Cov}(X, Y) = E[d_+ (X, Y)(X-\mu)(Y-E(Y))] + E[d_-(X, Y)(X-\mu)(Y-E(Y))],$$

we have $\mathrm{Cov}(X, Y) = 0$.

2.11. Let X be a random variable following the normal (Gaussian) distribution $N(\mu, \sigma)$ defined in Appendix A. Show that for any constants $a \neq 0$ and b, the variable $aX + b$ also follows the normal distribution (and the distribution is $N(a\mu + b, |a|\sigma)$). *Hint*: Find the CDF of $aX + b$ from definition and perform integration by substitution.

2.12. * Consider the random variable X following a chi-squared distribution with n degrees of freedom. As in Section 2.8, define $Y = D(X + a)$, where a is a positive constant and D is a random variable independent of X such that $P(D = 1) = P(D = -1) = 0.5$. Find the formula for the kurtosis γ_2 of Y as it depends on a and n. Confirm that $\gamma_2 = 0$ for $n = 4$ and $a \approx 0.41$. Confirm the plots obtained in Figures 2.27 and 2.28. *Hint*: $E(Y^k) = E(D^k)E\left[(X-a)^k\right] = E\left[(X-a)^k\right]$ for k even and $E(Y^k) = E(D^k)E\left[(X-a)^k\right] = 0$ for k odd because $E(D^k) = 1$ for k even and $E(D^k) = 0$ for k odd. The formula for the moments of the chi-squared distribution can be found in Appendix A.

2.13. * Prove Property 2.5. *Hint*: Clearly, $E(Y) = 0$. Show that $E(Y^k) = E\left[(X + a)^k\right]$ for k even. Then show that $E\left[(X + a)^4\right]$ and $\left\{E\left[(X + a)^2\right]\right\}^2$ are four-degree polynomials with respect to a, having the coefficient 1 by the term a^4. This leads to $\lim_{a \to \infty} \mathrm{Kurt}(X) = 1$.

CHAPTER 3

Statistical Inference

This chapter is a brief review for readers with some prior experience with inferential statistics. Readers without such experience, or those who prefer more thorough coverage of the material, may refer to the textbooks by Devore (2004) or Mendenhall et al. (2006). A more engineering oriented approch in the context of signal processing can be found is Schart (1991).

3.1 INTRODUCTION

In Chapter 2, we introduced probability as a way of describing how likely various outcomes are in a given scenario. In Section 2.7, we discussed an example of a sample consisting of three measurements of reflectance of a surface, which were described as values of three random variables X_1, X_2, and X_3. The distribution of these variables describes probabilities of obtaining various measurement values. It is reasonable to assume that such measurements are independent, and each of them follows the normal distribution (as is usually the case for measurement errors) denoted by $N(\mu, \sigma^2)$, where the mean μ represents the "true" reflectance as measured by the spectrometer and the standard deviation σ describes the precision of the measurements. This means that a single measurement is considered as unbiased in the statistical sense. If one suspects that the spectrometer might give readings that are biased with respect to a standard, one can test this hypothesis by comparing μ to the value given by the standard.

A reader might be concerned that our three-element sample is too small to draw any meaningful conclusions. Note that in practice, people often measure the same item only once, and they still obtain useful information. If the measurement error is small, these three measurements of the same object should be sufficient for most practical purposes. A different scenario would be that of a sample describing different objects from a population (one measurement per object). In such cases, we usually need larger samples that can describe the population variability, which is usually larger than the measurement error variability. Ultimately, the proper sample size will depend on the

---

Statistics for Imaging, Optics, and Photonics, Peter Bajorski.
© 2012 John Wiley & Sons, Inc. Published 2012 by John Wiley & Sons, Inc.

population variability and on how much information we expect from the sample. These issues will be discussed throughout this chapter.

If we knew the values of μ and σ, we could tell what to expect from future measurements. The set of all possible measurements, both in the past and in the future, forms a population in this context. The sample consists of the three measurements "selected" from the population. We used "selected" in quotation marks because all elements of the population in this example do not really exist as specific objects. This is an example of an *abstract population* that exists only in the conceptual sense. Since we assumed earlier that the measurements follow the normal distribution, we can say that the population from which the measurements come is described by the normal distribution model $N(\mu, \sigma^2)$.

In inferential statistics, we draw conclusions about a population and its model based on a sample drawn from that population. An example of statistical inference is the estimation of the parameters μ and σ, as well as the assessment of the estimation precision, based on the three measurements of reflectance. In this chapter, we discuss statistical inference in the context of some simple scenarios.

Consider a sample of independent measurements $X_1, X_2, \ldots, X_n$, where all variables X_i follow the same arbitrary distribution described by the cumulative distribution function (CDF) denoted by F. Such variables were referred to as independent and identically distributed in Section 2.7. When we take those measurements, we obtain the values taken on by those variables, and we can denote them with the lowercase letters x_i, $i = 1, \ldots, n$. It makes intuitive sense that the distribution of the points x_i should be somewhat similar to the theoretical distribution of X_i, which describes an infinite population of all possible measurements. To be more precise, we can define an empirical distribution as a discrete distribution on the set of points x_i, $i = 1, \ldots, n$, such that each point is assigned the same probability $1/n$.

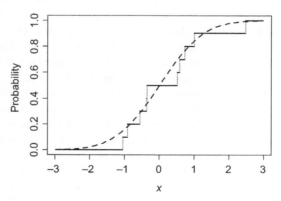

Figure 3.1   The solid line step function shows a realization of the empirical distribution function $F_n(x)$ based on 10 data points simulated from the standard normal distribution. There are 10 steps in this step function, but 2 of them look like one step of double height (around -0.36), because two observations were very close to each other. The dashed line is the theoretical CDF $F(x)$ of the standard normal distribution.

The CDF F_n of this distribution is called the *empirical distribution function*, and it is defined as a step function

$$F_n(x) = \frac{\text{Number of elements in the sample} \leq x}{n}. \tag{3.1}$$

It can be proven that F_n gets closer to F, as n tends to infinity. Figure 3.1 shows a realization of the empirical distribution function $F_n(x)$ based on data simulated from the standard normal distribution. The dashed line shows the theoretical CDF $F(x)$ of the standard normal distribution. The distribution of a random variable is fully described by its CDF, so in further discussion when referring to a distribution, we will use the name of its CDF.

Since the empirical distribution F_n is a good approximation of the theoretical distribution F, the characteristics of F_n should be close to those of F. This often leads to important statistical results. For example, this approach gives reasonable estimators of the parameters of F, as will be discussed in Section 3.2.2.

3.2 POINT ESTIMATION OF PARAMETERS

As discussed in the previous section, the empirical distribution F_n is a good approximation of the theoretical distribution F. Consequently, the mean of the empirical distribution F_n, which is equal to the sample mean $\overline{X}$, should be a good approximation of μ, the mean of the distribution F, that is, the population mean. Hence, $\overline{X}$, treated as a random variable, is called an estimator of μ. In this section, we discuss various methods for finding good estimators and studying their properties.

3.2.1 Definition and Properties of Estimators

Definition 3.1 A random variable that is a function of sample measurements $X_1, X_2, \ldots, X_n$ is called a *statistic*. This function cannot depend on unknown parameters so that its value can be calculated once the measurements' values are known.

Definition 3.2 An *estimator* is a statistic that is used to estimate a parameter of a model describing the population of interest.

An estimator is sometimes called a point estimator to emphasize that its value is a single number or a point on the number line. An estimator is regarded as a random variable, and its value is called an estimate. Estimators are often denoted by capital letters and their values (the estimates) by lowercase letters. However, there are exceptions, and in general, one needs to understand from the context whether the random variables or their values are discussed.

We argued earlier that the sample mean $\overline{X}$, as the mean of the empirical distribution F_n, is an estimator of the population mean μ. However, such construction does not guarantee good properties of the resulting estimators. We will now examine what are some good properties of estimators, and how we may check those properties.

As discussed in Section 2.7, the changing values of $\overline{X}$ in repeated sampling form a distribution of $\overline{X}$, which is called the sampling distribution. The properties of the sampling distribution of an estimator tell us how well the estimator works, such as how often its values are close to the true value of the estimated parameter. The first property that is often checked is whether the estimation is correct on average, that is, we calculate the expected value of the sampling distribution and compare it to the true value of the estimated parameter.

Definition 3.3 An estimator is called *unbiased* if the expected value of its sampling distribution is equal to the true value of the estimated parameter.

Since $E(X_i) = \mu$ for all $i = 1, \ldots, n$, it is easy to see that $E(\overline{X}) = \mu$, that is, $\overline{X}$ is an unbiased estimator of μ. This is a good property to have. However, each observation X_i is also an unbiased estimator of μ. So, is the sample mean a better estimator than a single observation? To answer this question, we need to study the variability of the estimators around μ—the smaller the variability, the better the unbiased estimator. Since all measurements $X_1, X_2, \ldots, X_n$ follow the same distribution, we can denote the variance of a single measurement by $\text{Var}(X_i) = \sigma^2$, $i = 1, \ldots, n$. Recall property (2.42), written again as

$$\text{StDev}(\overline{X}) = \sigma_{\overline{X}} = \frac{\sigma}{\sqrt{n}}. \tag{3.2}$$

Based on the two-sigma rule discussed in Section 2.6, we can tell that $\overline{X}$ is not further than $2\sigma/\sqrt{n}$ from μ with probability of at least 0.75 (Chebyshev's inequality), or perhaps even 0.95, if the normal approximation can be used based on the central limit theorem from Section 2.7. We can also say that increasing the sample size four times increases the precision of $\overline{X}$ two times. Formula (3.2) is clearly useful for assessing the precision of estimation when using $\overline{X}$. However, in order to calculate $\sigma_{\overline{X}}$, we need to estimate the value of σ. To this end, we can use the sample standard deviation introduced in Chapter 2 and calculate the estimated standard deviation of $\overline{X}$ as $s/\sqrt{n}$.

Definition 3.4 The standard deviation of an estimator is called a *standard error*.

In statistical inference, the estimates of standard errors are used in addition to the values of estimators.

The sample mean with its variance inversely proportional to the sample size seems like a reliable estimator. However, it would be good to know if there are estimators with even smaller variances. Within the class of unbiased estimators, we can define the following subclass of optimal estimators.

Definition 3.5 An estimator of parameter θ, a function of the sample $X_1, X_2, \ldots, X_n$, is called the *minimum variance unbiased* (MVU) estimator if it has the minimum variance among all unbiased estimators of θ based on such sample.

The sample mean $\overline{X}$ is an MVU estimator of the population mean for a large class of distributions. Specialized statistical books such as Bickel and Doksum (2001) provide more information on MVU estimators and how they can be constructed in the context of specific distributions and statistical models.

Many useful estimators are not unbiased. For example, the sample standard deviation s is a biased estimator of the population standard deviation σ, even though the sample variance s^2 is an unbiased estimator of σ^2, as will be shown in the next section. The bias of s varies with the distribution of the underlying sample, and hence there is no easy way to eliminate that bias.

There are also situations when a biased estimator might be better than an unbiased one. Consider a parameter θ and its estimator $\widehat{\theta}$, following the traditional statistical notation where an estimator is denoted by adding a "hat" symbol over the parameter symbol. A sampling distribution of an unbiased estimator, say $\widehat{\theta}_1$, is shown in Figure 3.2. Its probability density function shown as a dashed line is rather flat, indicating large variability. Another estimator, say $\widehat{\theta}_2$, is biased (the density shown as a solid line), but it has much lower variability. We may prefer $\widehat{\theta}_2$ because it gives values close to θ more often than the estimator $\widehat{\theta}_1$. This can be formalized by defining the mean squared error as follows.

We want to address the ultimate goal of estimation, which is to have values of $\widehat{\theta}$ close to the true parameter value θ. A useful way to describe the closeness is to calculate the *mean squared error*

$$\mathrm{MSE}(\widehat{\theta}) = E\left[(\widehat{\theta} - \theta)^2\right], \tag{3.3}$$

which can also be expressed as

$$\mathrm{MSE}(\widehat{\theta}) = \mathrm{Var}(\widehat{\theta}) + \left[\mathrm{Bias}(\widehat{\theta})\right]^2, \tag{3.4}$$

where $\mathrm{Bias}(\widehat{\theta}) = E(\widehat{\theta}) - \theta$. See Problem 3.7 for a hint on proving equation (3.4). For unbiased estimators, the second term in (3.4) vanishes, and minimization of the variance is equivalent to the minimization of the MSE. Sometimes we are able to introduce some bias deliberately in a way that reduces the variance significantly, as shown in Figure 3.2, resulting in an overall reduction in the MSE.

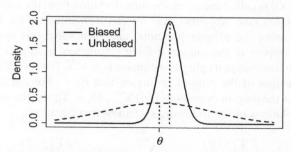

Figure 3.2 A probability density function of a sampling distribution of an unbiased estimator (dashed line) and that of a biased one (solid line) with much lower variability.

3.2.2   The Method of the Moments and Plug-In Principle

We argued earlier that since the empirical distribution F_n is an approximation of distribution F, it is reasonable to use the first moment of F_n, that is, $\overline{X}$, as an approximation of the first moment of F, that is, μ. This reasoning is called the *method of moments*, and it can be used to find estimators of various parameters. For example, to estimate the second central moment of F, the population variance σ^2, we can use the second central moment of the empirical distribution function F_n that is equal to

$$\widehat{\sigma}_n^2 = \frac{1}{n} \sum_{i=1}^{n} (X_i - \overline{X})^2. \tag{3.5}$$

The resulting estimator $\widehat{\sigma}_n^2$ is not unbiased. One can show that $E(\widehat{\sigma}_n^2) = [(n-1)/n]\sigma^2$, which leads to an unbiased estimator

$$s^2 = \frac{n}{n-1} \widehat{\sigma}_n^2 = \frac{1}{n-1} \sum_{i=1}^{n} (X_i - \overline{X})^2, \tag{3.6}$$

which is the sample variance introduced in Chapter 2.

The above-described method of moments can also be used for higher order moments. A more general method, called *a plug-in principle*, allows estimation of any characteristic of F, by using the same characteristic of F_n. For example, the median of F can be estimated by the median of F_n. The advantage of the plug-in principle (including the method of moments) is that it can be used for any distribution F without any additional assumptions.

When we want to assume that F belongs to a certain family of distributions, the plug-in principle can be further generalized as shown in the following example. Let's assume that F belongs to the family of the exponential power distributions defined in Appendix A. That family is parameterized by three parameters μ, α, and β. If X follows the exponential power distributions, then

$$E(X) = \mu, \qquad \text{Var}(X) = \frac{\beta^2 \Gamma(3/\alpha)}{\Gamma(1/\alpha)}, \qquad E|X - E(X)| = \frac{\beta \, \Gamma(2/\alpha)}{\Gamma(1/\alpha)}, \tag{3.7}$$

where $E|X - E(X)|$ is called the mean absolute deviation from the mean and $\Gamma(\cdot)$ is the gamma function (see Appendix A). We can now estimate the left-hand-side expressions by using the plug-in principle, and then solve this system of three equations with respect to the unknown three parameters μ, α, and β. Since only the first equation involves μ, its plug-in estimator is $\widehat{\mu} = \overline{X}$. For $\text{Var}(X)$, we can use a bias-adjusted version of the plug-in estimator, that is, s^2. For the mean absolute deviation, we use the plug-in estimator $(1/n) \sum_{i=1}^{n} |X_i - \overline{X}|$, and the estimators for α and β are the solutions of the following system of two equations:

$$s^2 = \frac{\widehat{\beta}^2 \Gamma(3/\widehat{\alpha})}{\Gamma(1/\widehat{\alpha})}, \qquad \frac{1}{n} \sum_{i=1}^{n} |X_i - \overline{X}| = \frac{\widehat{\beta} \, \Gamma(2/\widehat{\alpha})}{\Gamma(1/\widehat{\alpha})}, \tag{3.8}$$

which would need to be solved numerically. We are presenting this approach as an example only, rather than advocating it as a reliable method of estimation. The method of moments and the plug-in principle sometimes give poor estimators. A method that is often more successful is based on the likelihood principle as discussed in the next section.

3.2.3   The Maximum Likelihood Estimation

So far, the estimation process involved finding a function of data that would get us close to the true value of a parameter. An equivalent, although somewhat reversed, approach would be to try to find the value of a parameter that is most consistent with the data. Here we need to assume that the unknown distribution of our data belongs to a certain family of distributions defined by a set of parameters. Typical examples of such families are the normal (Gaussian), gamma, or the exponential families of distributions (see Appendix A).

For the sake of simplicity, let us assume that our observations come from a normal distribution $N(\mu, 1)$ with an unknown mean μ and the known variance 1. Let us start with the simplest case, when we have only one observation X_1. When deciding what the value of μ should be, we are faced with a choice of one of the density functions (as shown in Figure 3.3). We want to find the value that is the most likely explanation for observing $X_1 = x$, where x is the value that was observed. Since the height of the density function at a point x is proportional to the probability of observing a value close to x, it makes sense to pick the density function, which has the largest value at point x. This results in picking $N(x, 1)$ as our most likely distribution, that is, x is the estimate for μ, and it is called the maximum likelihood estimate (MLE), or estimator if treated as a random variable.

Things get a little bit more complex when we have two observations X_1 and X_2. In this case, we need to consider all possible pairs of values that the vector (X_1, X_2) can take on, and their associated probabilities. This is the same as considering the joint distribution of X_1 and X_2. Since the observations in the sample are assumed to be independent, the joint density is given by

$$f(x_1; \mu) \cdot f(x_2; \mu), \tag{3.9}$$

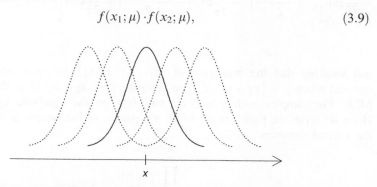

Figure 3.3   Density functions of the normal distributions $N(\mu, 1)$ with varying μ. The solid line shows the density that is most consistent with the observed value x.

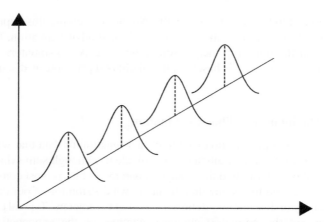

Figure 3.4 Joint density functions (shown symbolically) of the bivariate normal distributions of the form (3.9) with varying μ.

where $f(x; \mu)$ is the normal density function of $N(\mu, 1)$. Figure 3.4 shows symbolically the joint densities given by (3.9) for several choices of μ. Note that the peaks are at the line $x_1 = x_2$. Again, our goal is to find the value of μ such that the joint density (3.9) is the most likely distribution that could generate our pair of observations (x_1, x_2). Note that the two-dimensional bell-shaped joint densities shown in Figure 3.4 have contour lines in the form of circles with centers at the line $x_1 = x_2$. Smaller circles correspond to higher density.

The smallest circle including point (x_1, x_2) is the one with the center at the point $((x_1 + x_2)/2, (x_1 + x_2)/2)$. Consequently, the joint density with the largest value at point (x_1, x_2) is the one with parameter $\mu = (x_1 + x_2)/2$. This fact can also be demonstrated algebraically by writing (3.9) as

$$\frac{1}{\sqrt{2\pi}}\exp\left[-\frac{(x_1 - \mu)^2}{2}\right] \cdot \frac{1}{\sqrt{2\pi}}\exp\left[-\frac{(x_2 - \mu)^2}{2}\right] = \frac{1}{2\pi}\exp\left[-\frac{(x_1 - \mu)^2 + (x_2 - \mu)^2}{2}\right]$$

$$(3.10)$$

and showing that the minimum of $(x_1 - \mu)^2 + (x_2 - \mu)^2$ with respect to μ is realized when $\mu = (x_1 + x_2)/2$. The estimate $\widehat{\mu} = (x_1 + x_2)/2$ is then called the MLE. For samples with $n > 2$, we will rely on the algebraic approach only. Here, we write the joint density of all n elements of the sample as the product of the normal densities

$$\prod_{i=1}^{n} f(x_i; \mu). \tag{3.11}$$

Again, we want to find the most appropriate value of μ. Hence, we want to think about (3.11) as a function of μ, and we call that function a likelihood function

$$L(\mu; x_1, \ldots, x_n) = \prod_{i=1}^{n} f(x_i; \mu). \tag{3.12}$$

The MLE of μ is that value of μ that gives the largest value of the likelihood, which at the same time is the largest value of the joint density, indicating that this is the most likely model for our observed data. It is often convenient to maximize a natural logarithm of the likelihood (called log-likelihood), which gives the same solution for μ because a logarithm is a strictly increasing function.

In our example of the size n sample from the normal distribution, the log-likelihood is equal to

$$\log L(\mu; x_1, \ldots, x_n) = -\frac{n}{2}\ln(2\pi) - \frac{1}{2}\sum_{i=1}^{n}(x_i - \mu)^2. \tag{3.13}$$

In order to find the maximum, we can equate the derivative of the log-likelihood with respect to μ to zero to obtain the equation

$$\sum_{i=1}^{n}(x_i - \mu) = 0, \tag{3.14}$$

which has the solution $\widehat{\mu} = \overline{x}$. This is the MLE of μ.

In a more general setting, once we assume the form of the joint distribution for the variables $X_1, X_2, \ldots, X_n$, we can treat it as a function of parameters, say $\theta_1, \ldots, \theta_k$, written in the form

$$L(\theta_1, \ldots, \theta_k; x_1, \ldots, x_n). \tag{3.15}$$

The values $\widehat{\theta}_1, \ldots, \widehat{\theta}_k$ that maximize this likelihood are called the MLEs. For a random sample from the normal distribution $N(\mu, \sigma^2)$ (i.e., a set of i.i.d. variables) with both parameters unknown, the MLEs turn out to be

$$\widehat{\mu} = \overline{x} \quad \text{and} \quad \widehat{\sigma}_n^2 = \frac{1}{n}\sum_{i=1}^{n}(x_i - \overline{x})^2. \tag{3.16}$$

As discussed in Section 3.2.2, $\widehat{\sigma}_n^2$ is a biased estimator of σ^2, which demonstrates that the MLEs are not always the best estimators.

In this case, it was possible to obtain explicit formulas for the MLEs, but in most other cases, the maximization can only be obtained by numerical methods, and no explicit formulas exist. This makes it difficult to study properties of the MLEs, especially for small sample sizes. However, it can be proven (under very general conditions) that for a large sample (when n tends to infinity), an MLE is asymptotically unbiased and has nearly the smallest possible variance.

3.3 INTERVAL ESTIMATION

A disadvantage of point estimators is that they give us only a single value as the best guess of the unknown parameter. Of course, knowing the standard error of the point estimator tells us how far the estimate might be from the true value of the parameter, especially for the unbiased estimators. However, this issue can be addressed more directly by constructing an interval that would contain the unknown parameter with a given confidence. This is often done by considering the difference between the estimator $\widehat{\theta}$ and the parameter θ, and then standardizing it by dividing by the known standard error $(SE(\widehat{\theta}))$ or the estimated standard error as follows:

$$Q = \frac{\widehat{\theta} - \theta}{\widehat{SE}(\widehat{\theta})}. \tag{3.17}$$

The random variable Q (called *a pivot*) measures how far the estimate is from the true value in standard error units. If Q has a distribution, say G, which does not depend on any unknown parameters, it can be used to construct a confidence interval (CI). We can write that

$$P\left(\eta_{\alpha/2} < Q < \eta_{1-\alpha/2}\right) = 1 - \alpha, \tag{3.18}$$

where η_{α} is the (100α)th percentile of the distribution G, that is, $G(\eta_{\alpha}) = \alpha$ (where G also denotes the CDF of that distribution). We can express (3.18) equivalently as

$$P\left(\eta_{\alpha/2} \cdot \widehat{SE}(\widehat{\theta}) < \widehat{\theta} - \theta < \eta_{1-\alpha/2} \cdot \widehat{SE}(\widehat{\theta})\right) = 1 - \alpha \tag{3.19}$$

or

$$P\left(\widehat{\theta} - \eta_{1-\alpha/2} \cdot \widehat{SE}(\widehat{\theta}) < \theta < \widehat{\theta} - \eta_{\alpha/2} \cdot \widehat{SE}(\widehat{\theta})\right) = 1 - \alpha. \tag{3.20}$$

The random interval

$$\left(\widehat{\theta} - \eta_{1-\alpha/2} \cdot \widehat{SE}(\widehat{\theta}), \widehat{\theta} - \eta_{\alpha/2} \cdot \widehat{SE}(\widehat{\theta})\right) \tag{3.21}$$

from the last equation is called a *confidence interval* at the $(100(1 - \alpha))\%$ confidence level because it captures the unknown parameter with probability $1 - \alpha$. When the distribution G is symmetric around zero, we have $\eta_{\alpha/2} = -\eta_{1-\alpha/2}$, and the confidence interval can be written in the form

$$\widehat{\theta} \pm \eta_{1-\alpha/2} \cdot \widehat{SE}(\widehat{\theta}). \tag{3.22}$$

The point estimator $\widehat{\theta}$ is the random center of the interval, and its length is equal to $2\eta_{1-\alpha/2} \cdot \widehat{SE}(\widehat{\theta})$. If the standard error is estimated (as in (3.22)), rather than known, the length of the interval is also random.

When the sample values are observed (taking on specific values), this random interval becomes the observed confidence interval, which either contains or does not contain the estimated parameter. Hence, no probability statement can be associated with a single observed confidence interval. Instead, we say that we have a $(100(1 - \alpha))\%$ confidence in the obtained interval in the sense that when repeating the same procedure in repeated samples from the same distribution, we would cover the estimated parameter $(100(1 - \alpha))\%$ of the time. Figure 3.5 shows an example of 20 observed confidence intervals of the parameter θ generated from 20 samples from the normal distribution with the mean θ. Only 1 out of the 20 intervals does not cover the parameter, which matches the 95% confidence level used here. In practice, the value of θ is unknown, and we do not know whether or not a given interval covers the parameter.

Consider now a sample consisting of independent measurements described by random variables X_i, $i = 1, \ldots, n$, all having the same normal distribution $N(\mu, \sigma^2)$. In order to construct a confidence interval for μ, we can use the pivot L of the form shown in formula (3.17), that is,

$$Q = \frac{\overline{X} - \mu}{\widehat{SE}(\overline{X})}, \tag{3.23}$$

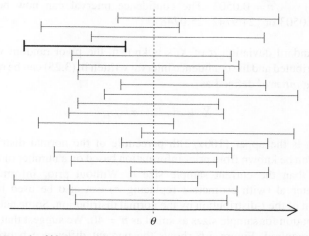

θ

Figure 3.5 An example of 20 observed confidence intervals (at the 95% confidence level) of the parameter θ generated from 20 samples from the normal distribution with the mean θ.

where $\widehat{SE}(\overline{X}) = s/\sqrt{n}$. It is known that the pivot Q follows the t-distribution with $(n-1)$ degrees of freedom when the observations follow the normal distribution, as is the case here. Since the t-distribution is symmetric around zero, we can use property (3.18) written in the form

$$P(-t_{n-1}(\alpha/2) < Q < t_{n-1}(\alpha/2)) = 1 - \alpha, \tag{3.24}$$

where $t_{n-1}(\alpha/2)$ is the *upper*$(100\alpha/2)$th percentile of the t-distribution with $(n-1)$ degrees of freedom. Note that here we use upper percentiles, while in (3.18) we used more general notation with the classic percentile. The resulting confidence interval, called the *one-sample t confidence interval*, is of the form analogous to (3.22), that is, it can be written as

$$\overline{X} \pm t_{n-1}(\alpha/2) \cdot s/\sqrt{n}. \tag{3.25}$$

An implementation of this confidence interval is demonstrated in the following example.

Example 3.1   Consider Small Image data describing an 8 by 13 pixel image of a monochromatic, highly uniform tile in three wide spectral bands (in reflectance units). Our goal is to estimate the "true" tile reflectances μ_1, μ_2, and μ_3 in the three spectral bands, respectively. First, we are going to concentrate on the reflectances in Band 1. Since the tile surface is highly uniform, it makes sense to assume that reflectances in Band 1 for all pixels are independent random variables X_i, $i = 1, \ldots, n = 104$, all having the same distribution with the mean $E(X_i) = \mu_1$ (assuming that the measurements are unbiased). We expect the data to follow the normal distribution because the variability is largely due to the measurement error. For $\alpha = 0.05$, we obtain $t_{n-1}(\alpha/2) = 1.98$. For Band 1 data, we have $\overline{x} = 25.0245$, $s = 0.2586$, and the resulting half of the length of the confidence interval is equal to $h = t_{n-1}(\alpha/2) \cdot s/\sqrt{n} = 0.0503$. The confidence interval can now be written as 25.0245 ± 0.0503 or $(24.9742, 25.0748)$. $\qquad\qquad\square$

When the standard deviation σ of X_i's is known, the pivot random variable L is normally distributed and the confidence interval written in (3.25) can be replaced with a *z confidence interval* defined as

$$\overline{X} \pm z(\alpha/2) \cdot \sigma/\sqrt{n}, \tag{3.26}$$

where $z(\alpha/2)$ is the upper $(100\alpha/2)$th percentile of the normal distribution. The parameter σ can be known from prior information based on a number of observations much larger than the current sample size n. Without prior information, the z confidence interval (with estimate s replacing σ) can also be used based on the approximation of the t-distribution by the normal distribution. Some authors suggest this approximation for sample sizes as small as $n = 40$. We suggest that much larger samples are required. Figure 3.6 shows the percent difference between the two percentiles calculated as $100(t(\alpha) - z(\alpha))/z(\alpha)$. For $n = 200$, the percent difference

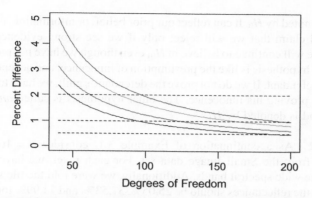

Figure 3.6 Percent difference between the t-distribution and normal distribution kth upper percentile points for a range of degrees of freedom. The four solid lines are for $k = 0.5$, 1, 2.5, and 5 from the top to the bottom, respectively.

ranges between 0.5% and 1%. This means that the t confidence interval is wider than its normal-based approximation by the same percentage.

The above considerations assume normality of observations. It is often argued that without the normality assumption, a z confidence interval can be used for large samples since the numerator of the pivot L is approximately normal based on the central limit theorem and s is close to σ by the law of large numbers (see Section 2.7). We argue that it is a better practice to use the t confidence intervals, even for non-normal samples, unless the sample size is so large that the difference between the two results is not important, or if the resulting simplification is sufficiently beneficial. Note that a z confidence interval is always shorter than the respective t confidence interval. This means that by having less information (potential non-normality), we would be "rewarded" with a shorter interval, which does not seem logical.

3.4 HYPOTHESIS TESTING

Another tool for statistical inference is hypothesis testing. In many instances, it is important to know if a certain statement or hypothesis about a population, or a model, is true or not. Here are some examples of statistical hypotheses:

- The population mean value is equal to 20.
- The sample comes from a normal distribution.
- The population standard deviation is not larger than 5.

Clearly, a hypothesis may refer to a parameter such as the population mean, or it may be a statement about the correct model for the sample (for instance, that the observations come from the normal distribution). In any hypothesis testing problem, we consider two contradictory hypotheses, of which only one can be true. One of the two hypotheses is often a claim that is initially assumed to be true, and it is called *a null hypothesis* denoted by H_0. It can reflect our prior belief, or information, or simply the chosen initial claim that we will reject only if we see strong evidence against it. Otherwise, we will continue to believe in H_0, even though we have not proven it to be true. The null hypothesis is like the presumption of innocence, and accepting it is like acquitting a defendant. If we do not prove the defendant's guilt, we have to acquit him, even without proving his innocence. The other hypothesis is called *an alternative hypothesis* and is denoted by H_1.

Example 3.2   As a continuation of Example 3.1, consider $n = 104$ pixels or observations from the Small Image data set. For each pixel, we have reflectance values in three wide spectral bands. Additionally, we were told the tile specification claiming that the reflectances should be 25.05%, 37.53%, and 74.99% for Bands 1, 2, and 3, respectively. We will first concentrate on Band 1 and consider the specification value $\mu_0 = 25.05$. As in Example 3.1, we assume that the reflectances in Band 1 denoted as X_i, $i = 1, \ldots, n$, are independent and follow the same normal distribution with $E(X_i) = \mu$, where μ is interpreted as the true value. We would like to check if the tile indeed conforms to the specification, which can be written as the null hypothesis $H_0 : \mu = \mu_0$ stating that the true value is equal to μ_0. The alternative hypothesis is that the tile does not conform to the specification, that is, $H_1 : \mu \neq \mu_0$. We will not be able to prove $H_0 : \mu = \mu_0$, because even if the sample mean $\overline{x}$ was equal to $\mu_0 = 25.05$, it could be that the true mean is very close to 25.05, for example, $\mu = 25.0501$. This would explain why $\overline{x} = 25.05$, even if $\mu \neq \mu_0$. On the other hand, if $\overline{x}$ is far from $\mu_0 = 25.05$, then we can easily claim that $H_0 : \mu = \mu_0$ is not true (but only with limited confidence due to a possible sampling error). $\square$

In a general setting, when testing the null hypothesis $H_0 : \mu = \mu_0$ versus the alternative $H_1 : \mu \neq \mu_0$, we want to know how far $\overline{x}$ is from μ_0. Hence, we can use the pivot L given by (3.23) with $\mu = \mu_0$. The resulting random variable given by

$$t = \frac{\overline{X} - \mu_0}{\widehat{SE}(\overline{X})} \tag{3.27}$$

is called the *t-statistic*. It is called a statistic because its value does not depend on any unknown parameters. On the other hand, the pivot L given by (3.23) is not a statistic because it depends on the unknown parameter μ. When the sample comes for a normal distribution, t has *t*-distribution with $n - 1$ degrees of freedom.

Since the departure in any direction from μ_0 is an evidence against $H_0 : \mu = \mu_0$, we decide to reject H_0 when $|t|$ is large enough. Our decision about H_0 versus H_1 will typically be based on a sample that is a representative, though imperfect, reflection of the population. Hence, we might make an incorrect decision.

Table 3.1 shows four possible events. If H_0 is true, and we accept it (do not reject), we are OK. On the other hand, if we reject a true H_0, we make an error, which is traditionally called a *Type I error* or *false alarm*. The probability of this error is often called the *alpha risk* or *false alarm rate* and can be written as $P_{H_0}\{\text{Reject } H_0\}$. If H_0 is not true, and we reject it, we are OK again. However, if H_0 is not true, and we do not

Table 3.1 The Four Scenarios when Testing a Null Hypothesis

Decision H_0	H_0 Is True	H_0 Is *Not* True
Do not reject H_0	OK	Type II error (probability = beta risk)
Reject H_0	Type I error (probability = alpha risk or false alarm rate)	OK (probability = power)

reject it, we again make an error, which is traditionally called a *Type II error*. The probability of this error is often called the *beta risk* and can be written as $P_{H_a}\{\text{Do not reject } H_0\}$. Clearly, it would be best to minimize both alpha and beta risks. We will discuss later on how this can be done.

Definition 3.6 A *test procedure* is a rule, usually based on sample data, for deciding whether to reject the null hypothesis H_0.

Definition 3.7 A *test statistic* is a function of sample data, which is used to decide whether to reject the null hypothesis H_0.

Definition 3.8 A *rejection region* is the set of all values of the test statistic for which the null hypothesis H_0 will be rejected.

We earlier decided to use large $|t|$ as evidence against $H_0 : \mu = \mu_0$. We can now formalize it by selecting a threshold c_{critical}, called a critical value, and defining a test procedure, which rejects H_0, when $|t| \geq c_{\text{critical}}$. The resulting test is called *two-sided* because it tests against a *two-sided alternative* $H_1 : \mu \neq \mu_0$ located on both sides of μ_0. One could also test against a *one-sided alternative*, for example, $H_1 : \mu > \mu_0$, when the other side ($\mu < \mu_0$) is either impossible or consistent with our initial claim. In the latter case, we would consider the null hypothesis $H_0 : \mu \leq \mu_0$ versus the alternative $H_1 : \mu > \mu_0$. For example, if we measure a quality characteristic of a tile produced by a new method, with larger μ indicative of higher quality, the alternative $H_1 : \mu > \mu_0$ means that the new method is better than the current method with $\mu = \mu_0$. If the new method is as good as the current method ($\mu = \mu_0$) or is worse ($\mu < \mu_0$), we would not want to implement the new method. Only when the null hypothesis $H_0 : \mu \leq \mu_0$ is rejected in favor of the one-sided alternative $H_1 : \mu > \mu_0$ would we be interested in implementing the new method.

We will now concentrate on two-sided tests and discuss how to select the threshold c_2 indexed with 2 to emphasize the two-sided alternative. Before selecting the threshold, we need to consider the consequences of picking a specific value. In order to calculate the alpha risk written as $P_{\mu=\mu_0}\{|t| \geq c_2\}$, we will assume that the observations follow the normal distribution. Consequently, the statistic t given by (3.27) follows the t-distribution when $\mu = \mu_0$. Let G be a cumulative distribution function of the t-distribution. From the symmetry of the t-distribution around zero, we obtain $P_{\mu=\mu_0}\{|t| \geq c_2\} = 1 - G(c_2) + G(-c_2) = 2[1 - G(c_2)]$. It is now clear that by varying the critical value c_2, we can obtain any value $\alpha \in (0, 1)$ for the alpha risk equal to $2[1 - G(c_2)]$. More specifically, we should take $c_2 = t_{n-1}(\alpha/2)$

(upper $(100\alpha/2)$th percentile) for a two-sided alternative. Similar calculations for the one-sided alternative $H_1 : \mu > \mu_0$ lead to a test procedure rejecting $H_0 : \mu \leq \mu_0$, when $t \geq c_1$, where $c_1 = t_{n-1}(\alpha)$.

In general, once the α value is determined, the test procedure is defined through the appropriate choice of the threshold c_{critical}. We say that the test is at the significance level α, or we call it "α-level test." It is up to the data analyst to choose the suitable value of α. Traditionally, the largest value that is usually used is $\alpha = 0.05$, but it can also be much smaller if it is important to avoid rejecting H_0 that is in fact true. For example, if investigating the toxicity of a potential pharmaceutical, we should set up H_0 as "the substance is toxic" and use a very small value of α.

Example 3.2 (cont.) For checking the tile specification, as discussed earlier in Example 3.2, we can use $\alpha = 0.05$. With the sample size of $n = 104$, we obtain $c_2 = t_{n-1}(\alpha/2) = t_{103}(0.025) = 1.98$. For Band 1 data, we have $\bar{x} = 25.0245$, $\mu_0 = 25.05$, and $s = 0.2586$. Hence, the resulting t-test statistic calculated from the formula (3.27) is equal to $t = -1.006$. Since $|t| = 1.006 < 1.98 = c_2$, we do not reject the null hypothesis $H_0 : \mu = \mu_0$. There is no evidence against the tile conforming to the specification. □

We can now address the Type II error and its probability. The alternative hypothesis is typically composite, that is, it consists of many possible choices for the parameter, such as μ in this case. This means that the beta risk is in fact a function of μ, which can be written as $\beta = P_\mu\{|t| < c_2\}$ for the two-sided alternative. The beta risk is shown in Figure 3.7 as the area shaded by the slanted lines. The two tails colored in gray represent the alpha risk in that figure. Note that the mean of the t-statistic depends on n. It is convenient to define the *power* of the test as $P_\mu\{|t| \geq c_2\}$ equal to one minus beta risk. Hence, the test power is the probability that we will detect the unknown μ as being different from μ_0. The probabilities of the four events in hypothesis testing are summarized in Table 3.2. Note that the probabilities in columns always sum up to 1 because they represent the only two possible events that can happen. When rejecting $H_0 : \mu = \mu_0$, we can say that we identified the difference to be statistically significant. For a given value of α, a test with smaller β, that is, larger power, is better.

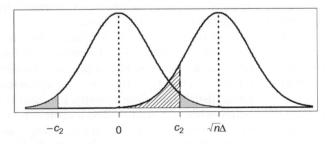

Figure 3.7 The approximate distribution of the t-statistic under the null hypothesis (on the left) and under the alternative (on the right) $(\Delta = (\mu - \mu_0)/\sigma)$. The two tails colored in gray represent the probability of Type I error, and the area shaded by the slanted lines represents the probability of Type II error.

Table 3.2   The Probabilities of the Four Scenarios When Testing a Null Hypothesis

Decision H_0	H_0 Is True	H_0 Is *Not* True
Do not reject H_0	$1 - \alpha$	β
Reject H_0	α	Power $= 1 - \beta$

In order to calculate the power of the two-sided t-test, we are going to make a simplified assumption that n is large enough to justify the normal approximation for the resulting t-distribution. We will also assume a related approximation $s \approx \sigma$. We can write

$$\text{Power}(\mu) = P_\mu\{|t| \geq c_2\} = P_\mu\{t \geq c_2\} + P_\mu\{t \leq -c_2\} \tag{3.28}$$

and calculate the first term as follows:

$$P_\mu\{t \geq c_2\} = P_\mu\left\{\frac{\overline{X} - \mu_0}{\text{SE}(\overline{X})} \geq c_2\right\} = P_\mu\left\{\frac{\overline{X} - \mu + \mu - \mu_0}{\text{SE}(\overline{X})} \geq c_2\right\}$$

$$= P_\mu\left\{\frac{\overline{X} - \mu}{\text{SE}(\overline{X})} \geq c_2 - \frac{\mu - \mu_0}{\text{SE}(\overline{X})}\right\} \approx P\left\{Z \geq c_2 - \frac{\mu - \mu_0}{\sigma/\sqrt{n}}\right\}$$

$$= 1 - \Phi(c_2 - \sqrt{n}\Delta), \tag{3.29}$$

where Z follows the standard normal distribution $N(0, 1)$ and $\Delta = (\mu - \mu_0)/\sigma$ measures (in standard deviation units) how far the alternative μ is from the null hypothesis μ_0 (the sign of Δ indicates the direction). The second term in (3.28) can be calculated the same way, leading to the following formula for power:

$$\text{Power}(\mu) = 1 - \Phi(c_2 - \sqrt{n}\Delta) + \Phi(-c_2 - \sqrt{n}\Delta). \tag{3.30}$$

We can see both algebraically and from Figure 3.7 that larger values of $\sqrt{n}\Delta$ lead to smaller beta risk and larger power. Formula (3.30) can be interpreted as an approximate formula for the t-test or an exact formula for a z-test defined by the z-statistic

$$z = \frac{\overline{X} - \mu_0}{\sigma/\sqrt{n}}. \tag{3.31}$$

The z-test can be used when the σ value is known very precisely from prior experience rather than being estimated from the current sample. A two-sided z-test at level α will reject $H_0 : \mu = \mu_0$, when $|z| \geq c_2 = z(\alpha/2)$.

From formula (3.30), we can see that the test power depends on three quantities—α (through c_2), n, and Δ. We can fix values of two of those three quantities and plot the test power as a function of the third one. When fixing n and Δ, we obtain a receiver operating characteristic (ROC) curve. This is a plot of the test power as a function of the level α. Figure 3.8 shows the ROC curves for the two-sided z-test for various

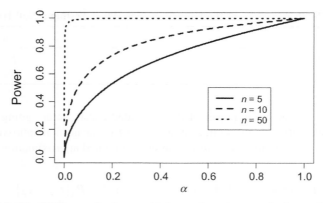

Figure 3.8 The ROC curves for the two-sided z-test for various sample sizes and $\Delta = 0.6$.

sample sizes and $\Delta = 0.6$. We can see that a larger sample size gives a larger power for all $0 < \alpha < 1$ levels. One disadvantage of the ROC curve is that the value of Δ is fixed, so that the power can be shown as a function of α. In practice, we are typically interested in small values of α, especially $\alpha \leq 0.05$. This is why we often plot the ROC curve only in the range of $0 \leq \alpha \leq 0.05$ as shown in Figure 3.9.

It is often more informative to fix the α level and plot power as a function of Δ. The resulting curve is called a power curve. Figure 3.10 shows power curves of the level $\alpha = 0.05$ two-sided z-test as a function of Δ for various sample sizes. Note that $\Delta = 0$ corresponds to the null hypothesis $H_0 : \mu = \mu_0$, and the resulting power Power(μ_0) is equal to α. This is why all functions in Figure 3.10 go down to $\alpha = 0.05$ at $\Delta = 0$. The steeper the function rises from that point, the better the test because of the higher probability (power) of detecting μ as different from μ_0. For the case of the sample size $n = 5$ (the solid line in Figure 3.10), the value $\Delta = 1$ will be detected as significant with probability of only about 0.6. The situation is better for $n = 10$ (dashed line) with

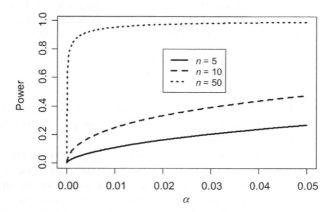

Figure 3.9 A small section of the ROC curve from Figure 3.8 for $0 \leq \alpha \leq 0.05$.

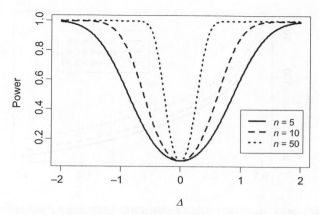

Figure 3.10   The power of the level $\alpha = 0.05$ two-sided z-test given by formula (3.30) as a function of Δ for various sample sizes.

the probability going way above 0.8, and even better for $n = 50$ (dotted line) with the probability of detection equal to almost 1.

Table 3.3 shows some numerical values of the test power, in percent, calculated from formula (3.30). Again, we can clearly see the benefits of larger sample sizes. We can also see that it is easier to detect significance of a larger Δ. The power of the z-test never gets to 1 exactly (the numbers were rounded off to the nearest 0.1% in the table, which explains the resulting values of 100%).

Since we are usually interested in large values of power, we can also fix the power value (together with α) and investigate the relationship between the sample size n and the value of Δ. This approach is especially valuable at the stage of planning a study, so that we can plan how large a sample size n we need in order to achieve the goal of detecting the difference Δ with sufficient probability (power). In order to solve equation (3.30) for n, we assume for simplicity of presentation that $\Delta > 0$ and we eliminate the last term on the right-hand side, which is usually very small (for $\alpha \leq 0.05, \Delta > 0$, and Power ≥ 0.8, the last term is smaller than 10^{-6}). We then obtain the formula for the minimum required sample size as

$$n = \left(\frac{z(\alpha/2) + z(1 - \text{Power})}{\Delta} \right)^2 \qquad (3.32)$$

Table 3.3   Some Numerical Values (Rounded Off to the Nearest 0.1%) of the Power (in Percent) of the Level $a = 0.05$ Two-Sided z-Test

Sample Size	$\Delta = 0.2$	$\Delta = 0.4$	$\Delta = 0.6$	$\Delta = 0.8$	$\Delta = 1$
5	7.3	14.5	26.9	43.2	60.9
10	9.7	24.4	47.5	71.6	88.5
50	29.3	80.7	98.9	100.0	100.0

The values were calculated from formula (3.30).

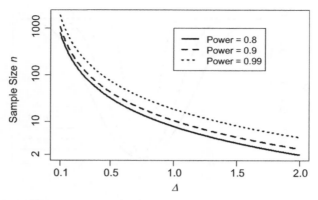

Figure 3.11   The required sample size (on the logarithmic scale) of the level $\alpha = 0.05$ two-sided z-test as a function of Δ for various power levels.

rounded up to the nearest integer. Similar reasoning gives the same formula for $\Delta < 0$. Figure 3.11 shows the required sample size (on the logarithmic scale) of the level $\alpha = 0.05$ two-sided z-test as a function of Δ for various power levels.

Example 3.2 (cont.)   When rejecting the null hypothesis $H_0 : \mu = \mu_0$ that the tile conforms to the specification, we may call the tile a defect. Calling a tile incorrectly as a defect is a loss for the tile producer. Hence, the alpha risk is sometimes called the *producer's risk*. On the other hand, incorrectly accepting a nonconforming tile as a good one creates a potential loss for a consumer. Hence, the beta risk is sometimes called the *consumer's risk*. By increasing the threshold c_2, we reduce the producer's risk but increase the consumer's risk. On the other hand, reducing the threshold has the opposite effect of increasing the producer's risk and reducing the consumer's risk. This means that we need to balance the two types of risk and identify their best combination in a given context. In order to reduce both risks (i.e., get smaller α and larger power), we would need to increase the sample size n, as can be clearly seen from Figures 3.8–3.10.   □

When a value of the test statistic is calculated for a given sample, it would be good to know how extreme or unusual the value is, given the null hypothesis is true. This can be assessed by calculating the probability of such a result or a more extreme result. For the two-sided t-test, we would calculate

$$P_{\mu=\mu_0}\{|t| \geq t^*\}, \tag{3.33}$$

where t^* is the specific value of the t-statistic calculated from the sample. This probability is called a *p-value*. A very small p-value indicates that the t^* value is unusual in the light of the null hypothesis H_0, which is evidence against H_0. This would prompt us to reject H_0. In order to test H_0 at the α level, we would reject H_0 when the p-value is smaller than or equal to α.

For many statistical tests, one can establish a relationship between the test and the confidence interval. For a t confidence interval given as $\overline{X} \pm t_{n-1}(\alpha/2) \cdot s/\sqrt{n}$, a value μ_0 is within the confidence interval if and only if

$$|\overline{X} - \mu_0| < t_{n-1}(\alpha/2) \cdot s/\sqrt{n} \qquad (3.34)$$

or

$$\frac{|\overline{X} - \mu_0|}{s/\sqrt{n}} < t_{n-1}(\alpha/2). \qquad (3.35)$$

The left-hand side of (3.35) is the absolute value of the t-statistic used by the two-sided α-level t-test for testing $H_0 : \mu = \mu_0$. This means that condition (3.35) is equivalent to acceptance of H_0. We have shown that the t confidence interval is the set of values μ_0 of the parameter that would be accepted when tested as $H_0 : \mu = \mu_0$. At the same time, the values outside of the confidence interval would be rejected as the true values of the parameter.

3.5　SAMPLES FROM TWO POPULATIONS

In previous sections, we introduced the basic concepts of statistical inference using examples based on one-sample problems, where the whole sample is drawn from one population. There are many other types of statistical problems, where various types of assumptions are made depending on the type of data being analyzed. A broader treatment of those cases is beyond the scope of this chapter, and the reader can find details in other statistical books. Our goal here is to provide a solid foundation for understanding those other types of statistical problems. In this section, we provide a very brief example of another type of statistical problem. Let us first show an example of data, where such a different approach is needed.

Example 3.3　Consider infrared astronomy data on two types of objects (stars), carbon-rich asymptotic giant branch (C AGB) stars and the so-called "H II regions." The former are dying, sun-like stars (red giants), while the latter are plasmas ionized by hot, massive young stars that are still deeply embedded in the molecular clouds out of which they were formed. Here, we use a subset of data used in Kastner et al. (2008), consisting of 126 such objects located in the Large Magellanic Cloud. Sixty-seven of those objects are C AGB stars and 59 are H II (regions). For each star, we have J (1.25 μm), H (1.65 μm), and K (2.17 μm) band magnitudes from the Two-Micron All-Sky Survey. See Appendix B for more details. Let's concentrate on the J band magnitudes. Our goal is to see if there is any difference between the two types of stars in terms of their J band magnitudes. Statistically, we can describe it as two samples coming from two different populations. Each population is characterized by the distribution of its J band magnitudes. We want to test if the two distributions can be assumed to be the same, or they are different. Figure 3.12 shows a box plot of the two

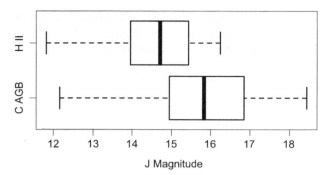

Figure 3.12 A box plot of the two samples of J magnitudes from the two types of stars—"C AGB" and "H II" as discussed in Example 3.3.

samples. There is a substantial overlap of values in the two samples, but there are also some differences. We need to test if the differences are statistically significant or if they could have happened by chance. □

The scenario described in Example 3.3 is called a two-sample problem. In order to show statistical inference in this scenario, we need to start with introducing mathematical notation and formulating our assumptions. We assume that the first sample $X_1, X_2, \ldots, X_n$ comes from the normal distribution $N(\mu_1, \sigma_1^2)$ and the second sample $Y_1, Y_2, \ldots, Y_m$ comes from a potentially different normal distribution $N(\mu_2, \sigma_2^2)$. The sample sizes n and m might also be different. We first want to test if the variances in the two populations are the same. The null hypothesis can be written as $H_0 : \sigma_1^2 = \sigma_2^2$, which will be tested against the alternative $H_1 : \sigma_1^2 \neq \sigma_2^2$. It is clear that the test should be based on a comparison of the sample variances from the two samples. However, instead of considering a difference between them, we should be using their ratio called an F-statistic defined as

$$F = \frac{s_1^2}{s_2^2}. \tag{3.36}$$

It is known that $(\sigma_2^2/\sigma_1^2)F$ follows the F-distribution with $(n-1)$ and $(m-1)$ degrees of freedom (see Appendix A). Under the null hypothesis H_0, we have $\sigma_1^2 = \sigma_2^2$; hence, the F-statistic follows the same F-distribution. The null hypothesis is rejected when either $F \geq F_{n-1,m-1}(\alpha/2)$ or $F \leq F_{n-1,m-1}(1-\alpha/2)$, where $F_{n-1,m-1}(\alpha/2)$ is the (100α)th upper percentile from the F-distribution with $(n-1)$ and $(m-1)$ degrees of freedom.

In order to test the equality of the means, we formulate the null hypothesis $H_0 : \mu_1 = \mu_2$, which will be tested against the alternative $H_1 : \mu_1 \neq \mu_2$. We can say that we are testing $H_0 : \theta = 0$, where $\theta = \mu_1 - \mu_2$, and we can use the pivot Q of the form given in equation (3.17), where $\hat{\theta} = \overline{X} - \overline{Y}$ and

$$\widehat{SE}(\hat{\theta}) = \sqrt{\frac{s_1^2}{n} + \frac{s_2^2}{m}}. \tag{3.37}$$

The pivot Q has approximately a t-distribution with v degrees of freedom estimated from the data by

$$v = \frac{(se_1)^2 + (se_2)^2}{\left[(se_1)^4/(n-1)\right] + \left[(se_2)^4/(m-1)\right]}, \quad \text{where} \quad (se_1)^2 = \frac{s_1^2}{n} \quad \text{and} \quad (se_2)^2 = \frac{s_2^2}{m}.$$

(3.38)

The t-statistic for testing a difference Δ_0 between the two samples, that is, for testing $H_0 : \mu_1 - \mu_2 = \Delta_0$, is defined as

$$t = \frac{\overline{X} - \overline{Y} - \Delta_0}{\sqrt{(s_1^2/n) + (s_2^2/m)}}.$$

(3.39)

The null hypothesis $H_0 : \mu_1 - \mu_2 = \Delta_0$ against the alternative $H_1 : \mu_1 - \mu_2 \neq \Delta_0$ is rejected when $|t| \geq t_v(\alpha/2)$, where $t_v(\alpha/2)$ is the *upper*$(100\alpha/2)$th percentile of the t-distribution with v degrees of freedom calculated from equation (3.38).

There is also a simpler test that could be used under the assumption that the variances in the two populations are the same. However, that test is sensitive to the departures from the model assumptions, including outliers, and we recommend the test given here.

The two-sample t confidence interval for $\mu_1 - \mu_2$ at the confidence level $(1 - \alpha)$ can be found based on the pivot Q in a way analogous to our previous considerations leading to equation (3.22). The interval is given by

$$\overline{X} - \overline{Y} \pm t_v(\alpha/2) \sqrt{\frac{s_1^2}{n} + \frac{s_2^2}{m}}.$$

(3.40)

The numerical calculations on the Example 3.3 data are relegated to the problems at the end of this chapter.

3.6 PROBABILITY PLOTS AND TESTING FOR POPULATION DISTRIBUTIONS

Many statistical methods make some assumptions regarding the population distribution from which the data are drawn. This is why it is important to find a suitable distribution or a family of distributions for our model and then to be able to verify the distributional assumption. In some situations, we might be able to identify the proper distribution based on the physical nature of the data. For example, when modeling a measurement error, we find that the normal distribution is the most likely candidate. In other situations, we may want to start with a probability plot as an exploratory tool. Once a proper distribution is identified, it can be further confirmed by formal hypothesis testing.

3.6.1 Probability Plots

Let us start with discussing probability plots as an informal, but efficient, way to check the population distribution from which a given data set was obtained. The idea is to sort the data, and then compare them to what we would expect to see if the data were coming from a given distribution. For example, for a data set of $n = 10$ observations, we would expect the smallest observation to represent the smallest 10% of the population distribution, the second smallest observation to represent the next 10% of the population, and so on. In order to represent the smallest 10% of the population, we might use the 5th percentile of the hypothesized population distribution, which could be thought of as the median of the lowest 10% of the population. This means that the 10 observations would be compared to the 5th, 15th, and so on until 95th percentiles of the hypothesized population distribution. In general, the percentiles can be calculated as $F^{-1}((i - 0.5)/n)$, $i = 1, \dots, n$, where F^{-1} is the inverse function of the cumulative distribution function of the hypothesized distribution. We can do this comparison by creating a scatter plot with percentiles on one axis and the observations on the other axis. If the observations were to conform ideally to our expectations, the scatter plot points would line up along the straight line $y = x$. In practice, the points will not line up exactly due to natural variability, but as long as they line up approximately along that line, we have reasons to believe that the hypothesized distribution is the true population distribution or at least its reasonable approximation. If the pattern of points is nonlinear, it may give us a hint about a more appropriate distribution as the model for the data.

The above method works when we hypothesize one specific population distribution. More typically, we are dealing with a whole family of distributions with some unknown parameters. In that case, we first need to estimate the parameters and then create a probability plot of the distribution with estimated parameters. For the family of normal distributions, we do not need to estimate the parameters but can simply plot the data versus the percentiles from the standard normal distribution. In order to see why this would work, let us assume that the horizontal axis x represents percentiles from the normal distribution $N(\mu, \sigma)$. Define $z = (x - \mu)/\sigma$, which represents percentiles from the standard normal distribution $N(0, 1)$. When plotting the sorted normal data on y versus x, the percentiles from $N(\mu, \sigma)$, we would expect to see the points to be close to the straight line $y = x$. That line is equivalent to $y = \sigma z + \mu$ when plotting versus z (we substituted $x = \sigma z + \mu$). This means that when plotting the sorted data versus percentiles from the standard normal distribution, we would still expect to see a straight line, but this time, the line would have slope σ and the intercept μ. The percentiles from the standard normal distribution can be calculated based on the previously discussed formula $\Phi^{-1}((i - 0.5)/n)$, $i = 1, \dots, n$, where Φ is the cumulative distribution function of the standard normal distribution. A more precise method is to modify the previous percentiles and use the expected values of the ordered statistics that can be approximated by the following percentiles:

$$z_i = \Phi^{-1} \left[\frac{i - (3/8)}{n + (1/4)} \right], \quad i = 1, \dots, n. \tag{3.41}$$

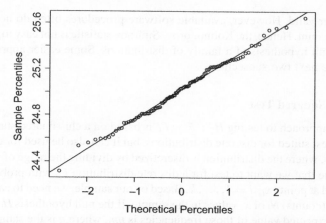

Figure 3.13   A normal probability plot for Band 1 data used in Example 3.4.

In the normal probability plot, we then plot the sorted data (in the increasing order) denoted as $y_1, y_2, \ldots, y_n$ versus the theoretical percentiles $z_1, z_2, \ldots, z_n$. This concept was used in the following example.

Example 3.4   In Examples 3.1 and 3.2, we assumed that the data in each band were normally distributed. This is confirmed in Band 1 by the normal probability plot shown in Figure 3.13 (with percentiles defined in (3.41)), where the points line up along a straight line. Similar plots for the other two bands also confirm their normality. □

Many statistical methods using the assumption of normality are not very sensitive to that assumption. This is why we can often rely on the normal probability plot as a crude assessment of approximate normality. In other cases, we may need to perform formal statistical inference and test the null hypothesis $H_0 : F = F_0$ that the true distribution F of the sample is equal to a given distribution F_0. One such test is discussed in the next subsection.

3.6.2   Kolmogorov–Smirnov Statistic

A natural way to construct a test of the null hypothesis $H_0 : F = F_0$ is to use the empirical distribution F_n defined in Section 3.1 and check how far it is from a hypothesized CDF F_0. For continuous distributions, the distance between F_n and F_0 can be measured by the Kolmogorov–Smirnov statistic defined by

$$D_n = \underset{x \in \mathbb{R}}{\text{supremum}} |F_n(x) - F_0(x)|. \tag{3.42}$$

The limiting distribution of D_n under the null hypothesis $H_0 : F = F_0$, as n tends to infinity, is known, and it is used to set up the critical value for the test. When testing for a family of distributions, the parameters need to be estimated, and those critical values are no longer valid. However, available software procedures often do not take this issue into account. Hence, the Kolmogorov–Smirnov statistic is not easy to use for the composite null hypothesis of a family of distributions. Some easier approaches are shown in the next two subsections.

3.6.3   Chi-Squared Test

A different approach to testing $H_0 : F = F_0$ is based on a chi-squared statistic. This concept is best suited for discrete distributions, but it can also be used for continuous distributions, where the distribution is discretized by dividing the range of values into bins. Assume that we want to test for a discrete distribution F_0 with probabilities p_i concentrated at points x_i, $i = 1, \ldots, k$. Based on our sample, we need to calculate the frequencies (counts) N_i of x_i values in the sample. If the null hypothesis $H_0 : F = F_0$ is true, the expected value of those frequencies is np_i, where n is the sample size. In order to assess how far the actual counts are from their expected values, we can calculate the chi-square statistic

$$\chi^2 = \sum_{i=1}^{k} \frac{(N_i - np_i)^2}{np_i}. \tag{3.43}$$

Under the null hypothesis $H_0 : F = F_0$, the distribution of χ^2 is approximately the chi-squared distribution with $(k-1)$ degrees of freedom. When testing for a family of distributions, the distribution parameters need to be estimated. Assuming that m different parameters are estimated using the maximum likelihood estimators, the χ^2 statistic is approximated by a chi-squared distribution with $(k - 1 - m)$ degrees of freedom (under some general regularity conditions). We can now construct an approximate α-level test of the null hypothesis that F belongs to the given family of distributions. The test rejects H_0 when $\chi^2 \geq \chi^2_{k-1-m}(\alpha)$, where $\chi^2_{k-1-m}(\alpha)$ is the upper (100α)th percentile from the chi-squared distribution with $(k - 1 - m)$ degrees of freedom. The test can be used only if $np_i \geq 5$ for every i.

3.6.4   Ryan–Joiner Test for Normality

In order to test the composite null hypothesis that the distribution of the sample is normal with unknown mean and standard deviation, we can use the Ryan–Joiner test. Let's denote by $y_1, y_2, \ldots, y_n$ the observations sorted in the increasing order as we did in Section 3.6.1 when creating probability plots. For the normal probability plot, we calculated the theoretical percentiles $z_1, z_2, \ldots, z_n$ according to formula (3.41). The idea of a probability plot was that the points (y_i, z_i), $i = 1, \ldots, n$, are supposed to line up along a straight line. In order to verify how strong this linear relationship is, we can calculate the sample correlation coefficient

$$r = \frac{\sum_{i=1}^{n} (y_i - \bar{y})(z_i - \bar{z})}{\sqrt{\sum_{i=1}^{n} (y_i - \bar{y})^2} \sqrt{\sum_{i=1}^{n} (z_i - \bar{z})^2}}, \tag{3.44}$$

which is a sample equivalent of the population correlation coefficient defined in equation (2.34). The Ryan–Joiner test rejects the composite null hypothesis of the normal distribution when $r \leq c_\alpha$, where the critical value c_α can be found in the table in Appendix C.

3.7 OUTLIER DETECTION

An outlier is an observation that does not fit the model assumed for a given data set. More precisely, an outlier is very unlikely to be a realization from a distribution assumed by the model. Since the model is usually based on the given data set, the outlier is an observation that is different from the majority of the data. There are two important considerations when detecting outliers:

1. The assumed model.
2. The sample size.

The second consideration is often ignored by many authors and many software implementations. For example, a direct use of a two- or three-sigma rule is entirely misleading without considering the sample size. We know from Section 2.6 that under the assumption of the normal distribution, we can expect 1 in 20 observations to be outside of two-sigma limits and 3 out of 1000 observations to be outside of three-sigma limits. This means that in a sample of size $n = 1000$, an observation as far as three standard deviations from the mean cannot be considered unusual. However, these guidelines are only approximate because they do not take into account the fact that the standard deviation is estimated from the data. For more precise results, we should use formal tests such as the Grubbs (1969) test. The test works for one suspected outlier, either minimum or maximum denoted here as M, in normally distributed data. The Grubbs test statistic is

$$G = \frac{|M - \overline{X}|}{s}, \qquad (3.45)$$

where $\overline{X}$ is the sample mean and s is the standard deviation, both calculated from the whole sample including the outlier. The formal null hypothesis here is that all observations in the sample follow the same normal distribution (i.e., there is no outlier). We call M an outlier with the confidence of $(1 - \alpha)$ if

$$G > \frac{n-1}{\sqrt{n}} \sqrt{\frac{t_{n-2}^2(\alpha/(2n))}{n - 2 + t_{n-2}^2(\alpha/(2n))}}, \qquad (3.46)$$

where $t_{n-2}(\alpha/(2n))$ is the upper $(100\alpha/(2n))$ percentile from the t-distribution with $(n - 2)$ degrees of freedom. Figure 3.14 shows the critical values for the Grubbs test

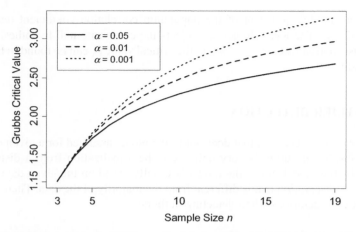

Figure 3.14 Critical values for the Grubbs test (the right-hand side of (3.46)) for small sample sizes n for various levels α.

(the right-hand side of (3.46)) for values of n from 3 to 19. Note that for very small sample sizes, the critical values are below 2. If we used a simple rule of two sigma, we would be missing some of the outliers that could be detected with the Grubbs test.

The critical values for larger sample sizes are shown in Figure 3.15, where n is shown on the horizontal axis on the logarithmic scale. For $n \geq 37$, the critical values are above 3. If we used a simple rule of three sigma, we would be identifying some observations incorrectly as outliers.

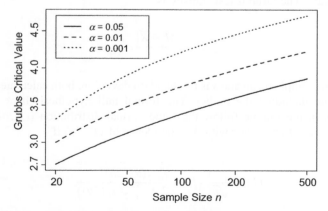

Figure 3.15 Critical values for the Grubbs test (the right-hand side of (3.46)) for large sample sizes n for various levels α.

3.8   MONTE CARLO SIMULATIONS

The method of Monte Carlo simulation is used to approximate probabilities of more complex events that are difficult to calculate analytically through mathematical calculations. In particular, we can approximate distributions of some statistics that would be difficult to calculate otherwise.

Consider a sample $X_1, X_2, \ldots, X_n$ of independent observations from the normal distribution $N(\mu, \sigma^2)$. If we were unsure what the distribution of $s^2 = \sum_{i=1}^{n} (X_i - \overline{X})^2 / (n-1)$ was, we could use the Monte Carlo simulation to approximate it. We know from theory that s^2/σ^2 follows the chi-squared distribution with $(n-1)$ degrees of freedom. For more complex estimators and other distributional assumptions about the sample, we may not know what the sampling distribution of the estimator is.

Let us assume a general context of a sample $X_1, X_2, \ldots, X_n$ drawn from a distribution F. Let's say we want to use a statistic $\widehat{\theta}$, and we want to find out what its distribution is. The statistic $\widehat{\theta}$ depends on the available sample, which we will emphasize by writing $\widehat{\theta} = \widehat{\theta}(X_1, X_2, \ldots, X_n)$. The *Monte Carlo simulation* is performed in the following steps:

1. Select a random number x_1^* from the distribution F. Repeat the process in an independent way in order to obtain the whole sample $x_1^*, x_2^*, \ldots, x_n^*$ of size n from the distribution F.
2. Calculate the value of the estimator $\widehat{\theta}^* = \widehat{\theta}(x_1^*, x_2^*, \ldots, x_n^*)$ for the simulated sample. This value is the simulated value of $\widehat{\theta}$.
3. Repeat N times Steps 1 and 2, and obtain N simulated replications $\widehat{\theta}_1^*, \widehat{\theta}_2^*, \ldots, \widehat{\theta}_N^*$.

The distribution of all numbers $\widehat{\theta}_1^*, \widehat{\theta}_2^*, \ldots, \widehat{\theta}_N^*$ is the empirical distribution of $\widehat{\theta}$ and as such is an approximation of the true distribution of $\widehat{\theta}$. An example of a Monte Carlo simulation study is shown in Section 7.6.2.

3.9   BOOTSTRAP

Earlier in this chapter, we investigated some theoretical properties of estimators and constructed confidence intervals based on the model assumptions and algebra calculations on the estimators. For example, in the context of a one-sample problem, equation (3.2) tells us that the standard error of $\overline{X}$ is equal to $\sigma/\sqrt{n}$, which can then be estimated using $s/\sqrt{n}$. This information can be used to construct confidence intervals. Unfortunately, for some more complex estimators, it might be difficult or impossible to prove similar general properties. This is where bootstrap can be used. This methodology can be used to investigate properties of estimators and to construct confidence intervals based on some computer-intensive calculations.

Let us assume a general context of a sample of numbers $x_1, x_2, \ldots, x_n$ drawn from a distribution F having a parameter θ. Let's say we want to estimate θ by using $\widehat{\theta}$. The estimator $\widehat{\theta}$ depends on the available sample, which we emphasize by writing $\widehat{\theta} = \widehat{\theta}(x_1, x_2, \ldots, x_n)$. The *bootstrap* is performed in the following steps:

1. Select a sample $x_1^*, x_2^*, \ldots, x_n^*$ of size n drawn with *replacement* from the set of values $x_1, x_2, \ldots, x_n$ from our original sample. This is equivalent to sampling from the empirical distribution F_n defined in equation (3.1). Note that some of the original sample values might be repeated and some might not be present at all in the selected sample. The sample $x_1^*, x_2^*, \ldots, x_n^*$ is called a *bootstrap sample*.

2. Calculate the value of the estimator $\widehat{\theta}^* = \widehat{\theta}(x_1^*, x_2^*, \ldots, x_n^*)$ for the bootstrap sample. This value is called a *bootstrap replication* of $\widehat{\theta}$.

3. Repeat N times Steps 1 and 2, and obtain N bootstrap replications $\widehat{\theta}_1^*, \widehat{\theta}_2^*, \ldots, \widehat{\theta}_N^*$.

Note that the process of bootstrapping is very similar to Monte Carlo simulations. The difference is that in simulations we generate random numbers from a theoretical distribution serving as our model. On the other hand, in bootstrap, we generate random numbers from the sample, or more precisely, from the empirical distribution. In Monte Carlo simulations, we need to assume a specific population distribution from which to draw the sample. The bootstrap does not require this assumption. If we assume that the population distribution belongs to a certain family of distributions such as normal, we would still need to assume a specific value of the parameters in order to perform simulations. In bootstrap, we could use the parameter values estimated from the sample and draw the bootstrap samples from that distribution. This approach is called *parametric bootstrap* as opposed to the classic *nonparametric bootstrap* that we introduced earlier.

It can be proven (under some very mild assumptions) that the distribution of the bootstrap replications $\widehat{\theta}_1^*, \widehat{\theta}_2^*, \ldots, \widehat{\theta}_N^*$ approximates the sampling distribution of $\widehat{\theta}$ (under F). This approximation can then be used for finding the estimator's bias or for constructing confidence intervals as will be shown here.

It is important to understand that the bootstrap replications are used to study the properties of the estimator $\widehat{\theta}$, but they are not necessarily any closer to the true value of the parameter θ. In fact, the opposite can be true. For example, assume that $\widehat{\theta}$ has a bias of 10. This means that its value is, on average, 10 units to the right of θ. When creating the bootstrap samples, the resulting bootstrap values $\widehat{\theta}_1^*, \widehat{\theta}_2^*, \ldots, \widehat{\theta}_N^*$ might be the additional 10 units to the right (on average) for a total bias of 20 with respect to the true parameter value (this might not be precisely so, but the bias will be amplified here). This is why we need to use the logic of somewhat counterintuitive reversal as explained below.

For better understanding of bootstrap, it is important to think in terms of the following three concepts: the Population, the Sample, and the Bootstrap Sample. We know that statistical inference is based on the relationship between the Population and the Sample. The main principle of bootstrap is that "the Bootstrap Sample is to the Sample as the Sample is to the Population." A simple application of this principle is the estimation of bias of $\widehat{\theta}$ by using the bootstrap mean $\overline{\theta}^* = \sum_{i=1}^{N} \widehat{\theta}_i^* / N$. Let's say we want to approximate the estimator's bias $E(\widehat{\theta}) - \theta$. The expected value $E(\widehat{\theta})$ can be replaced with $\overline{\theta}^*$, and θ replaced with $\widehat{\theta}$, leading to the *bootstrap estimator of bias* given by $\overline{\theta}^* - \widehat{\theta}$. The *bias-adjusted estimate* can now be defined as

$$\widehat{\theta}_{\text{adj}} = \widehat{\theta} - \left(\overline{\theta}^* - \widehat{\theta} \right) = 2\widehat{\theta} - \overline{\theta}^*. \tag{3.47}$$

This approach works best for the estimators constructed by the plug-in principle discussed in Section 3.2.2. For other estimators, one should make an adjustment for the difference between a given estimator and the plug-in estimator.

We can also use bootstrap for constructing a confidence interval for θ. A popular method is the so-called *bootstrap percentile method* producing a confidence interval given as

$$\left[\widehat{\theta}_{\alpha/2}^*, \widehat{\theta}_{1-\alpha/2}^* \right], \tag{3.48}$$

where the lower and upper limits are the $100(\alpha/2)$ and $100(1 - \alpha/2)$ percentiles from the bootstrap distribution $\widehat{\theta}_1^*, \widehat{\theta}_2^*, \ldots, \widehat{\theta}_N^*$, respectively. The problem with the percentile method is that it is "theoretically backward" as discussed, for example, in Hamilton (1992) with further references there. The method could lead to the approximate bias of 20, as explained earlier. Here we explain a *percentile-reversal method*, also called *hybrid method*, which is based on the main principle of bootstrap. There are also some other, potentially more precise, methods discussed in books such as Efron and Tibshirani (1994), Good (2005), and Davison and Hinkley (1997), where more information about bootstrap can be found. The most popular ones are the *t*-method and the bias-corrected method. However, those methods are even more computationally intensive, and they have their own issues. Having access to statistical software with a bootstrap method already available for a given model would be helpful here. Without such software, people often use the percentile method. We suggest using the percentile-reversal method instead, which does not involve any additional calculations. The percentile method works reasonably well for symmetric and unbiased sampling distributions, but the percentile-reversal method gives very similar results in those cases, and it works better in other cases.

We believe that the method shown here presents a reasonable balance between simplicity and good properties. Another reason for showing it here is that the explanation of the method is a good educational tool for understanding the concept of bootstrap.

In order to find a confidence interval, we want to find L and U (usually dependent on data) such that

$$P\left(L < \theta - \widehat{\theta} < U \right) = 1 - \alpha. \tag{3.49}$$

In the bootstrap context, the distribution of $\theta - \widehat{\theta}$ should be well approximated by the distribution of $\widehat{\theta} - \widehat{\theta}^*$. If we find L^* and U^* such that

$$P_{\text{bootstrap}}\left(L^* < \widehat{\theta} - \widehat{\theta}^* < U^*\right) = 1 - \alpha, \tag{3.50}$$

we should have $P\left(L^* < \theta - \widehat{\theta} < U^*\right) \approx 1 - \alpha$, and the approximate bootstrap CI can be defined as $\left(\widehat{\theta} + L^*, \widehat{\theta} + U^*\right)$. Note that (3.50) can be written as $P_{\text{bootstrap}}\left(\widehat{\theta} - U^* < \widehat{\theta}^* < \widehat{\theta} - L^*\right) = 1 - \alpha$. Hence, we can take $\widehat{\theta} - L^* = \widehat{\theta}^*_{1-\alpha/2}$ and $\widehat{\theta} - U^* = \widehat{\theta}^*_{\alpha/2}$. Finally, the approximate *percentile-reversal* bootstrap CI for θ is given by

$$\left(2\widehat{\theta} - \widehat{\theta}^*_{1-\alpha/2}, 2\widehat{\theta} - \widehat{\theta}^*_{\alpha/2}\right). \tag{3.51}$$

It is worth noting that the percentile-reversal bootstrap CI will adjust for possible lack of symmetry of the sampling distribution and for possible bias in estimation of θ. A disadvantage of the percentile-reversal bootstrap method is that it underestimates the variability and consequently tends to produce too short confidence intervals. An example of bootstrap is shown in Section 7.6.2.

PROBLEMS

3.1. Show that for $\widehat{\sigma}^2_n$ defined in (3.5), we have $E\left(\widehat{\sigma}^2_n\right) = [(n-1)/n]\sigma^2$.

3.2. Show that for a random sample (i.e., a set of i.i.d. variables) from the normal distributions $N(\mu, \sigma^2)$ with both parameters unknown, the MLEs are given as follows:

$$\widehat{\mu} = \overline{x} \quad \text{and} \quad \widehat{\sigma}^2_n = \frac{1}{n}\sum_{i=1}^{n}(x_i - \overline{x})^2.$$

3.3. As a follow-up to Example 3.1, calculate the one-sample t confidence intervals for the means in Bands 2 and 3.

3.4. Verify the results shown in Figure 3.6 by calculating $100(t_n(\alpha) - z(\alpha))/z(\alpha)$ for various values of α and n.

3.5. As a follow-up to Example 3.2, test whether the tile conforms to the specification of 37.53% and 74.99% of reflectance in Bands 2 and 3, respectively.

3.6. As a follow-up to Example 3.4, create normal probability plots of Band 2 and 3 data from the tile image.

3.7. Prove that $\text{MSE}(\hat{\theta}) = \text{Var}(\hat{\theta}) + [\text{Bias}(\hat{\theta})]^2$ (see equation (3.4)). *Hint*: Use definition (3.3) and add and subtract $E(\hat{\theta})$ to obtain $\text{MSE}(\hat{\theta}) = E\left[\left(\hat{\theta} - E(\hat{\theta}) + E(\hat{\theta}) - \theta\right)^2\right]$.

3.8. For the H band magnitude variable introduced in Example 3.3, create a box plot analogous to the one shown in Figure 3.12. Repeat this for the K band magnitude variable. Are there differences between the two groups in the plots?

3.9. Use the Ryan–Joiner test for testing normality of the two samples for "C AGB" and "H II" objects (stars) discussed in Example 3.3. Base your calculations on
   a. the J band magnitudes.
   b. the H band magnitudes.
   c. the K band magnitudes.

3.10. Test the equality of the population *variances* for the two groups (C AGB stars and H II regions) discussed in Example 3.3. Base your calculations on
   a. the J band magnitudes.
   b. the H band magnitudes.
   c. the K band magnitudes.
   d. Which band would be more appropriate for distinguishing between the two samples?

3.11. Explain how the *t*-statistic given in equation (3.39) can be derived based on the pivot Q discussed in Section 3.5. *Hint*: This is analogous to the derivation of the *t*-statistic given in equation (3.27).

3.12. Test the equality of the population *means* for the two groups (C AGB stars and H II regions) discussed in Example 3.3. Base your calculations on
   a. the J band magnitudes.
   b. the H band magnitudes.
   c. the K band magnitudes.
   d. Which band would be more appropriate for distinguishing between the two samples?

3.13. Explain how the confidence interval given in equation (3.40) can be derived based on the pivot Q discussed in Section 3.5. *Hint*: This is analogous to the derivation of the confidence interval given in equation (3.25).

3.14. Construct the two-sample t confidence interval for the difference in the population means for the two groups (C AGB stars and H II regions) discussed in Example 3.3. Use the confidence level $(1 - \alpha) = 0.95$. Base your calculations on

a. the J band magnitudes.

b. the H band magnitudes.

c. the K band magnitudes.

CHAPTER 4

Statistical Models

4.1 INTRODUCTION

In Chapter 3, we discussed statistical inference in the context of simple scenarios, such as a sample of measurements from the same distribution. We considered a sample of independent measurements $X_1, X_2, \ldots, X_n$, where all variables X_i followed the same distribution F. This is the simplest example of a statistical model. A slightly more complex model, considered in Section 3.5, is where one sample of n measurements comes from one population described by a distribution F, and another sample of m measurements comes from a different population described by a distribution G. The usual purpose here is to see if the two distributions are indeed different. In this chapter, we discuss even more complex models describing more intricate structures of the data. In Section 4.2, we introduce regression models for describing relationships among variables and predicting values of one variable based on values of other variables. In Section 4.3, we discuss ways to design and analyze controlled experiments, where values of factors can be deliberately varied in a systematic fashion. Supplements 4A and 4B provide basic properties of vectors and matrices that will be needed throughout this and subsequent chapters.

4.2 REGRESSION MODELS

In regression models, we are trying to establish a relationship between a *response variable y* and one or more *predictors*, also called *x*-variables. We would like to know if the response variable *y responds* to changes in *x*, or if *y* can be *predicted* from the predictor *x*. In some fields, the response variable *y* is called a dependent variable and the predictor *x* is called an independent variable. In statistics, this terminology is avoided so as not to be confused with stochastic independence or independence among multiple predictors.

Statistics for Imaging, Optics, and Photonics, Peter Bajorski.
© 2012 John Wiley & Sons, Inc. Published 2012 by John Wiley & Sons, Inc.

In the following subsection, we will start with the simplest case of a linear regression model with one predictor, and then we will discuss more complex models in subsequent sections.

4.2.1 Simple Linear Regression Model

Example 4.1 The Landsat Program is a series of Earth-observing satellite missions jointly managed by NASA and the U.S. Geological Survey since 1972. Due to the long-term nature of the program, there is a significant interest in the long-term calibration of the results, so that measurements taken at different times can be meaningfully compared. One approach to this calibration problem is discussed by Anderson (2010). As part of the approach, Landsat measurements of a fixed desert site were collected. The desert site was confirmed to be sufficiently stable over time, so that the changes in measurements can be attributed to a drift of the measuring instrument, except for some factors such as the Sun position in the sky. In this example, we consider the surface reflectance measurements of the desert site performed at 76 different times (different days and times of the day). The reflectance measurements are from one spectral band (Band 2) of the instrument. For each time of the measurement, we also know the solar elevation angle. See Appendix B for more details about the data. In order to investigate a relationship between reflectance in Band 2 and the solar elevation angle, we can create a scatter plot of the two variables as shown in Figure 4.1. Based on the pattern in the scatter plot, we expect a linear relationship between the two variables. The line drawn in Figure 4.1 seems to be a reasonable general description of that pattern. The observation points are scattered around the line, reflecting the fact that the linear relationship is not perfect and the reflectance may also depend on factors other than the solar elevation angle. A measurement error may also contribute to this variability. We can say that for a given solar elevation angle, the line shows an average reflectance value, with approximately the same number of observations above and below the line. This example will be continued later on in this subsection. □

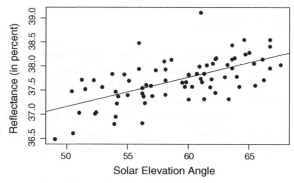

Figure 4.1 A scatter plot of reflectance in Band 2 versus the solar elevation angle for Landsat data discussed in Example 4.1.

In the simplest scenario of a linear relationship between the response Y and a single predictor x, as seen in Figure 4.1, we can describe this relationship using a *population linear regression model* written as

$$Y = \beta_0 + \beta_1 x + \varepsilon, \tag{4.1}$$

where β_0 and β_1 are the unknown coefficients called intercept and slope, respectively. This is called a population model because it establishes a general relationship between the two variables that governs all elements of the population. We can also say that the model (4.1) describes a general property of the output variable (Y) reacting to changes in an input variable (x) of a process under consideration. For instance, the observed reflectance of a given surface (Y) depends on the solar elevation angle (x). If the relationship were perfectly linear, we could use the model (4.1) without the ε term. In practice, there are always some imperfections, which might be due to factors other than x or due to a measurement error. Given the lack of information about those additional sources of variability, we assume that ε is a random variable, and consequently, Y is also a random variable (which is why we use a capital letter to denote the response). The ε term is often called an error term. For mathematical reasons, we assume that x is nonrandom. This means that in practical applications we need to make sure that x is known rather precisely with a small measurement error. If the measurement error in x is large, we need to use more complex models than those discussed here (see Section 15.2 in Montgomery et al. (2006) or Section 4.5 in Kutner et al. (2005)).

We usually assume that $E(\varepsilon) = 0$, which means that $E(Y) = \beta_0 + \beta_1 x$, that is, the population average of Y is a linear function of x. This function is called a *regression function*. The line $y = \beta_0 + \beta_1 x$ is a *regression line*. The regression function $\beta_0 + \beta_1 x$ can be regarded as the deterministic part of the model.

We often make the assumption that the error term ε follows a specific distribution, often a normal distribution with the mean zero. Under this assumption, the distribution of Y is also normal and centered at its expected value $E(Y) = \beta_0 + \beta_1 x$. Figure 4.2 illustrates the normal distribution of Y by drawing a normal density curve

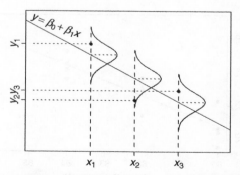

Figure 4.2 Conditional distributions of Y given x are shown here as normal distributions centered at their expected values $E(Y) = \beta_0 + \beta_1 x$, which depend on x in a linear fashion.

in a vertical direction for a fixed value of x. The distribution changes with varying x, but only in terms of its expected value. According to the model, the observations are equally likely to be above or below the line, and most observations should be close to the line where the density function has large values. Only a small fraction of observations is expected to be far away from the line where the density function has very small values (recall that the probability is equal to the area under the curve). The variability around the regression line is assumed here to be the same for all x, which is expressed as $\text{Var}(\varepsilon) = \sigma^2$, where σ^2 is a constant that does not depend on x.

The assumption of normality has been discussed here only for illustration purposes. For the remaining part of Section 4.2, up until Section 4.2.4, we are not going to use this assumption. The normality assumption will be needed in Section 4.2.5 in order to investigate the distributional properties of estimators and perform statistical inference.

We now want to describe another example that will be used throughout this chapter to demonstrate regression methods.

Example 4.2 An experiment was performed in order to find out how much power is lost when sending signals through optical fiber. This was similar to the experiment described in Example 2.1, except that only one piece of optical fiber was tested this time. The input power of a laser light signal sent from one end of the fiber was set at four different levels: 80, 82, 84, and 86 mW, and the corresponding output power was measured at the other end of the fiber. The purpose was to see how the power loss might depend on the power input. The advantages of using only one piece of fiber are that fewer measurements need to be taken and we do not need to deal with fiber-to-fiber variability. An important disadvantage is that we would not know if our findings apply to other pieces of optical fiber as well.

Five repeated runs were performed at each input power level. The resulting 20 runs were done in a random order. Figure 4.3 shows a scatter plot of the output power (Y) versus the input power (x). The straight line in the plot shows the estimated regression line. For the two cases of x equal to 84 and 86 mW, the line goes almost perfectly through the middle of the group of five data points. The other two cases of x are not as perfect, but

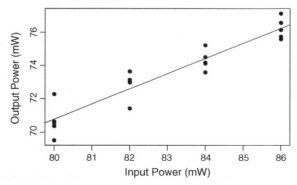

Figure 4.3 The input and output power in a laser light experiment as described in Example 4.2.

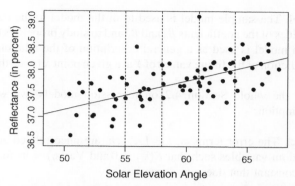

Figure 4.4 The range of x (solar elevation angle) values is divided into smaller sections, so that the distributional assumption is checked separately within each section.

it might be due to the natural variability in the data since we have only five repeats for each x value. The variability of points around the line is similar in all four groups in Figure 4.3, confirming the assumption of constant variance $Var(\varepsilon) = \sigma^2$. It is difficult to tell if the observations are consistent with the normality assumption illustrated in Figure 4.2. However, we can see a general tendency of points to concentrate around the estimated regression line. This example will be further discussed at the end of the subsection, once we learn how to estimate the regression line. □

In controlled experiments, such as the one described in Example 4.2, we can expect repeated values, and it is then easier to observe the distribution of Y for a fixed value of x. For data from observational studies, such as those in Example 4.1, there is usually only one value of Y for a given value of x. In such cases, we can look at a range of x values that captures several data points. The vertical dashed lines in Figure 4.4 identify six of such ranges or sections. Within each section, we can see a balance of points above and below the estimated regression line, which is consistent with the assumption that $E(Y) = \beta_0 + \beta_1 x$. The variability in each section is similar, which is consistent with the assumption of constant variance $Var(\varepsilon) = \sigma^2$.

In order to estimate the model parameters and to check if the model is correct, we need some data on both variables. Let's assume that we have n pairs (x_i, y_i), $i = 1, \ldots, n$, of values of both variables. If these paired observations come from a process following the model (4.1), then each y_i is a realization of a random variable Y_i such that $Y_i = \beta_0 + \beta_1 x_i + \varepsilon_i$, where ε_i is a random variable describing an observation-specific error term. If two x values are the same, for example, $x_1 = x_2$, then the random variables Y_1 and Y_2 are still different random variables, and their realizations y_1 and y_2 will usually not be the same (e.g., as seen in Figure 4.3). This leads to the following *sample regression model*:

$$Y_i = \beta_0 + \beta_1 x_i + \varepsilon_i, \quad i = 1, \ldots, n, \tag{4.2}$$

consisting of n equalities. The sample and the population regression models described by equations (4.1) and (4.2) are really two different ways of looking at the same model. The sample model is used to fit the model to the data, that is, to estimate the values of the coefficients β_0 and β_1 and to study properties of estimators. The population model is used as a general description of the relationship and can be used for prediction of the future values of Y at a given point x (once the coefficients are estimated).

Throughout the whole Section 4.2 on regression models, we will make the following assumption.

Assumption 4.1 The error terms ε_i, $i = 1, \ldots, n$, are uncorrelated and identically distributed random variables such that $E(\varepsilon_i) = 0$ and $\mathrm{Var}(\varepsilon_i) = \sigma^2$ for $i = 1, \ldots, n$, where σ^2 is a constant that does not depend on x.

It is important to understand that even though the error terms ε_i all have the same distribution, the responses Y_i have generally different distributions, each having its own mean $E(Y_i) = \beta_0 + \beta_1 x_i$.

When the variable Y_i takes on a specific value y_i that we observe, the error term variable ε_i also takes on a value, call it e_i^*, that is not directly observed. This means that we obtain a system of n equations

$$y_i = \beta_0 + \beta_1 x_i + e_i^*, \quad i = 1, \ldots, n, \tag{4.3}$$

with $(n + 2)$ unknowns, that is, n values of the error terms and two coefficients. Clearly, the system cannot be solved in general, but it is reasonable to require the error term values to be as small in magnitude as possible. The best way to minimize the error terms depends on their distribution. Minimizing the sum of absolute values $\sum_{i=1}^{n} |e_i^*|$ works best for the Laplace (double-exponential) distribution of ε_i, and it works well for other heavy-tailed distributions. This approach leads to robust regression (see Chapter 12 in Montgomery et al. (2006) or Section 11.3 in Kutner et al. (2005)). Minimizing the sum of squares $\sum_{i=1}^{n} \left(e_i^*\right)^2$ is easier mathematically, and it leads to the *least-squares regression* that we will use in this chapter. The least-squares regression has many desirable properties that will be discussed later on, and it works especially well when ε_i's follow the normal distribution. Since $e_i^* = y_i - \beta_0 - \beta_1 x_i$, we define the *least-squares estimates* b_0 and b_1 of the coefficients β_0 and β_1 as the values minimizing the sum of squares

$$S(\beta_0, \beta_1) = \sum_{i=1}^{n} (y_i - \beta_0 - \beta_1 x_i)^2. \tag{4.4}$$

Once we impose the least-squares minimization, the resulting values of e_i^* are no longer their true values, and consequently, the least-squares estimates b_0 and b_1 are different from the true values β_0 and β_1 of the parameters. Note that the function $S(\beta_0, \beta_1)$ in (4.4) is a second-degree polynomial with respect to β_0 and β_1, and it takes the minimum value at a point with both partial derivatives equal to zero. This means that the least-squares estimates b_0 and b_1 must satisfy the equations

$$\left.\frac{\partial S(\beta_0, \beta_1)}{\partial \beta_0}\right|_{b_0, b_1} = -2\sum_{i=1}^{n}(y_i - b_0 - b_1 x_i) = 0,$$

$$\left.\frac{\partial S(\beta_0, \beta_1)}{\partial \beta_1}\right|_{b_0, b_1} = -2\sum_{i=1}^{n}(y_i - b_0 - b_1 x_i)x_i = 0,$$

(4.5)

which can be simplified to the following system of two equations:

$$nb_0 + b_1\sum_{i=1}^{n}x_i = \sum_{i=1}^{n}y_i,$$

$$b_0\sum_{i=1}^{n}x_i + b_1\sum_{i=1}^{n}x_i^2 = \sum_{i=1}^{n}y_i x_i$$

(4.6)

called the *least-squares normal equations*. The solutions to equations (4.6), called the *least-squares estimates*, are given as

$$b_1 = \frac{S_{xy}}{S_{xx}}, \qquad b_0 = \bar{y} - b_1\bar{x},$$

(4.7)

where $\bar{x}$ and $\bar{y}$ are the sample means of the x and y values and

$$S_{xx} = \sum_{i=1}^{n}(x_i - \bar{x})^2, \qquad S_{xy} = \sum_{i=1}^{n}y_i(x_i - \bar{x}).$$

(4.8)

The *estimated regression line* is given by the equation $y = b_0 + b_1 x$. The regression lines in Figures 4.1, 4.3 and 4.4 were estimated using the least-squares method. The second equation in (4.7) can be written as $\bar{y} = b_0 + b_1\bar{x}$, which tells us that the estimated regression line goes through the center point $(\bar{x}, \bar{y})$ of the data.

For an arbitrary value x, we define the *fitted value* as $\hat{y} = b_0 + b_1 x$, which estimates the expected value $E(Y) = \beta_0 + \beta_1 x$. For the ith observation pair (x_i, y_i), we define a fitted value as $\hat{y}_i = b_0 + b_1 x_i$ and a *residual* $e_i = y_i - \hat{y}_i$. Figure 4.5 shows a fitted value and a residual for the first observation pair (x_1, y_1).

The *residual* e_i approximates the true (but unobserved) realization e_i^* of the error term ε_i. There are some important differences between properties of the error terms and the residuals. For example, the error terms ε_i, $i = 1, \ldots, n$, are uncorrelated and the residuals are correlated. We say that there are n degrees of freedom in the responses y_i (or in the unobserved realization e_i^*), which means that they can be arbitrary numbers without any constraints. On the other hand, there are only $(n-2)$ degrees of freedom in the residuals because they need to conform to two constraints (this will become clearer from the geometric interpretation discussed in Section 4.2.4). One of those constraints is that the sum of all residuals is equal to zero ($\sum_{i=1}^{n} e_i = 0$).

Let us now discuss estimation of the variance σ^2 of the error terms. Since $E(\varepsilon_i) = 0$, we have $\sigma^2 = \text{Var}(\varepsilon_i) = E(\varepsilon_i^2)$. If we knew the true realization e_i^* of the error term ε_i, we would simply use $\sum_{i=1}^{n}(e_i^*)^2/n$ to estimate σ^2. However, in the process of

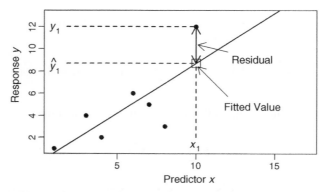

Figure 4.5   The fitted value and the residual for the first observation pair (x_1, y_1).

calculating the residuals (approximating e_i^*'s), we say that we lose two degrees of freedom for estimation of the regression coefficients, and an unbiased estimator of σ^2 turns out to be

$$\widehat{\sigma}^2 = \frac{1}{n-2} \sum_{i=1}^{n} e_i^2 = \frac{1}{n-2} \sum_{i=1}^{n} (y_i - \widehat{y}_i)^2. \tag{4.9}$$

Note that the above formula is very different from the formula for the sample variance of the y-values $\sum_{i=1}^{n} (y_i - \bar{y})^2 / (n-1)$, where the variability is calculated with respect to $\bar{y}$, while in formula (4.9) the variability is with respect to $\widehat{y}_i$ that is different for each data point.

An overall variability in all residuals can be measured by the residual sum of squares $SS_{Res} = \sum_{i=1}^{n} e_i^2 = \sum_{i=1}^{n} (y_i - \widehat{y}_i)^2$, which is the variability of the response values around the regression line. The total variability of the response values (around their mean) can be measured by the total sum of squares $SS_{Total} = \sum_{i=1}^{n} (y_i - \bar{y})^2$. In the so-called *analysis of variance* (ANOVA), we can partition the total sum of squares into SS_{Res} and the regression sum of squares defined as $SS_{Regr} = \sum_{i=1}^{n} (\widehat{y}_i - \bar{y})^2$. That is,

$$SS_{Total} = SS_{Regr} + SS_{Res}. \tag{4.10}$$

The total sum of squares SS_{Total} measures the variability in the response Y without any input from the regression model. Some of this variability is explained by the predictor x since we know that Y changes as x changes. The variability that is not explained by the predictors, that is, by the regression model, is measured by SS_{Res} because it tells us how far the actual observations y_i are from what can be predicted from the model, that is, $\widehat{y}_i$. For example, if all observations were lying exactly on a straight line, then SS_{Res} would be equal to zero, and all variability in Y would be explained by changes in x. Based on equation (4.10), the remaining variability, that is, the regression sum of squares $SS_{Regr} = SS_{Total} - SS_{Res}$, is the amount of variability explained by the model. The fraction of the total variability explained by the model is measured by the *coefficient of determination* defined as

$$R^2 = \frac{SS_{Regr}}{SS_{Total}} = 1 - \frac{SS_{Res}}{SS_{Total}}. \qquad (4.11)$$

We always have $0 \leq R^2 \leq 1$. The R^2 coefficient may serve as a general indicator by how much a given model can be potentially improved. For example, if $R^2 = 0.7$, we may try to find additional predictors that would explain the remaining 30% of variability. On the other hand, when $R^2 = 0.95$, we know that almost all variability has been explained, and not much more can be explained by finding a better model. At the same time, explaining an additional 3% of variability might be important in some applications.

Example 4.1 (cont.). As a continuation of the Landsat data example, we find the estimated least-squares regression line as $y = 0.3412 + 0.00061x$. The intercept is the value of y for $x = 0$, but the solar elevation angle never gets close to zero in our data set, and it would not be reasonable to extrapolate our model to such values. Hence, the intercept has no particular interpretation in this case. The slope of 0.00061 means that for each degree of the solar elevation angle, the average reflectance increases by 0.061% of reflectance. The variance σ^2 was estimated as $\widehat{\sigma}^2 = 0.0000132$. It is easier to interpret the estimated standard deviation $\widehat{\sigma} = \sqrt{\widehat{\sigma}^2} = 0.00363$ or 0.363% of reflectance. As an approximate calculation assuming the normal distribution of the error term, we can use the rule of two sigma from Section 2.6 and conclude that 95% of reflectance values in Band 2 will be within $\pm 2 \times 0.363 = \pm 0.726\%$ of reflectance from the regression line $y = 0.3412 + 0.00061x$ drawn in Figure 4.1. This calculation does not take into account the uncertainty in the parameters that were estimated. More precise calculations will be performed in Section 4.2.6.

The sums of squares were calculated as $SS_{Regr} = 0.000630$ and $SS_{Res} = 0.000977$ for the total of $SS_{Total} = 0.001607$. Hence, the fraction of variability explained by the model is $R^2 = 0.392$ or 39.2%. From a statistical point of view, there is still room for model improvement (by using other predictors), although it might be difficult or impossible in practice. □

Example 4.2 (cont.). For the optical fiber data, we can fit the least-squares regression line $y = -1.99 + 0.91x$. Based on the physical interpretation, the input power of zero $(x = 0)$ should result in the output power of zero as well $(y = 0)$. However, such considerations are not relevant when the range of values for x is far away from zero, which is the case here. In order to explain this point, assume for a moment that the intercept was even more negative, let's say, equal to (-37). The resulting regression line $y = -37 + 0.91x$ is shown in Figure 4.6 as a solid line. The true relationship between y and x could in fact be a quadratic function, shown in Figure 4.6 as a dashed line, going through the origin (to satisfy the above physical requirement). Note that within our range of x values between 80 and 86 (shown by two vertical dotted lines), the two lines almost overlap. This means that a linear model can

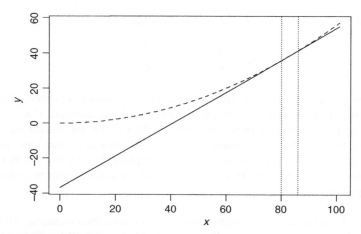

Figure 4.6 A hypothetical example of a true quadratic relationship (dashed line) approximated by a linear function (solid line) within the narrow range of data shown by two vertical dotted lines.

be a satisfactory approximation of a true quadratic relationship, if the range of x values is sufficiently narrow. It is clear from Figure 4.6 that the intercept value of (-37) has no practical interpretation.

In our fitted regression line, the intercept of (-1.99) is fairly close to zero, and we may suspect that it might not be significantly different from zero. We will check this in Section 4.2.5 by testing the null hypothesis $H_0 : \beta_0 = 0$. If we can accept that $\beta_0 = 0$, then we could fit a no-intercept regression line $y = b_1 x$, where b_1 would be interpreted as the fraction of the input power that is successfully transmitted (on average) through the whole length of the optical fiber. The power loss in decibels could then be calculated as $10 \log_{10}(b_1)$. According to the no-intercept model, the calculated power loss would apply to any level of input power within the range of the data used for fitting the model.

The fraction of variability explained by the model is $R^2 = 0.89$ or 89%. When interpreting this number, we need to understand that the R^2 value depends on the range of x values in data. Assuming that the same relationship would hold for a much larger range of x values, we could make the R^2 value larger by increasing the range of x values, and we could also make it very small by considering a very narrow range of x values. This means that the R^2 statistic has a limited value when comparing models across different data sets with varying ranges of x values. □

4.2.2 Residual Analysis

The most important part of Assumption 4.1 introduced in the previous subsection was that $E(\varepsilon_i) = 0$ or equivalently $E(Y_i) = \beta_0 + \beta_1 x_i$, that is, the relationship between the two variables is linear. In Figure 4.3, we were checking this assumption by observing the distribution of points around the estimated regression line for a fixed value x. This was made possible by the presence of repeated observations. In Figure 4.4, we considered a different example that did not have repeats. In that case, we identified

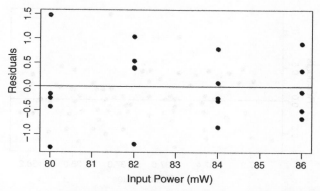

Figure 4.7   The residuals plotted versus Input Power (the x predictor) for the model fitted in Figure 4.3.

vertical ranges, so that we can look at a pattern of responses for predictor values close to each other. A more precise method is based on residuals $e_i = y_i - \widehat{y}_i$ because each observation y_i has its own adjustment by the fitted value $\widehat{y}_i$. We also know that residuals estimate the realizations of the error terms ε_i. So, observing residuals will give us hints about properties of the error terms.

In residual analysis, we create plots of residuals in order to check various assumptions. Figure 4.7 shows residuals plotted versus Input Power (the x predictor) for the model fitted in Figure 4.3. The horizontal line at the zero level can be regarded as equivalent to the level of the estimated regression line. The pattern of points is almost the same as the one shown in Figure 4.3, except that for all four levels of Input Power, the points are shifted vertically to the same common level. Again, the pattern is consistent with the assumption that $E(\varepsilon_i) = 0$ and $\mathrm{Var}(\varepsilon_i) = \sigma^2$ for all four levels of x.

Figure 4.8 shows residuals plotted versus the solar elevation angle (the x predictor) for the model fitted in Figure 4.4. Here each point has its own vertical adjustment by its fitted value, so that the zero level (the horizontal line) in Figure 4.8 represents the fitted values or the level of the estimated regression line. We again want to observe the distribution of residuals within each vertical section. The plot confirms the

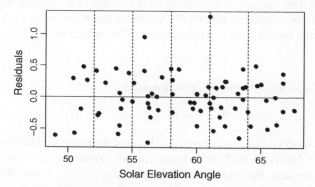

Figure 4.8   The residuals plotted versus the solar elevation angle (the x predictor) for the model fitted in Figure 4.4.

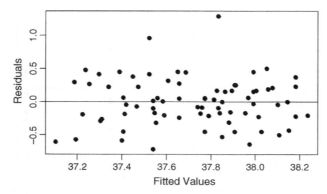

Figure 4.9 The residuals plotted versus fitted values for the model fitted in Figure 4.4.

assumption that $E(\varepsilon_i) = 0$ (there is an approximately even balance of points above and below the line in each segment) and $\text{Var}(\varepsilon_i) = \sigma^2$ (the variability of values is approximately the same in each section). The only concern might be one or two exceptionally large residuals. This issue will be further discussed in Section 4.2.7. The vertical lines are plotted here in order to convey the idea of the vertical sections, but they are usually not plotted in residual plots.

The residuals can also be plotted against the expected magnitude of the response, that is, the fitted value as shown in Figure 4.9. The positions of points in Figures 4.8 and 4.9 are identical because the only difference between the two plots is a linear transformation on the horizontal axis (the fitted value $\widehat{y} = b_0 + b_1 x$ is a linear function of x). The two plots will no longer be identical in the presence of multiple predictors (discussed in the next subsection). More on residual analysis can be found in Section 4.2.7.

4.2.3   Multiple Linear Regression and Matrix Notation

In practical applications, we usually have multiple predictors that may potentially impact the response variable Y. In the context of Example 4.1, the solar azimuth angle is another variable that may potentially influence the reflectance Y. A population linear regression model with two predictors can be written as

$$Y = \beta_0 + \beta_1 x_1 + \beta_2 x_2 + \varepsilon, \tag{4.12}$$

and the equivalent sample linear regression model is

$$Y_i = \beta_0 + \beta_1 x_{i,1} + \beta_2 x_{i,2} + \varepsilon_i, \quad i = 1, \ldots, n. \tag{4.13}$$

The above model is called a *linear* model because the regression function is a linear function of the β parameters. This is a broad class of models that also covers nonlinear relationships between the response and the predictors. For example, the model

$$Y = \beta_0 + \beta_1 \log(x_1) + \beta_2 \sqrt{x_2} + \varepsilon \tag{4.14}$$

is also a linear regression model because we can define transformed variables $z_1 = \log(x_1)$ and $z_2 = \sqrt{x_2}$, and the model can be written as

$$Y = \beta_0 + \beta_1 z_1 + \beta_2 z_2 + \varepsilon. \tag{4.15}$$

With multiple predictors, the formulas for regression analysis become more complex and more difficult to interpret. For more clarity and more thorough understanding of regression, it is convenient to use matrix notation. See Supplement 4A for background information about matrix algebra. The sample model (4.13) is really a set of n equations that can be written in n rows:

$$Y_1 = \beta_0 + \beta_1 x_{1,1} + \beta_2 x_{1,2} + \varepsilon_1,$$

$$Y_2 = \beta_0 + \beta_1 x_{2,1} + \beta_2 x_{2,2} + \varepsilon_2,$$

$$\vdots \tag{4.16}$$

$$Y_n = \beta_0 + \beta_1 x_{n,1} + \beta_2 x_{n,2} + \varepsilon_n.$$

The columns in (4.16) can be written as n-dimensional vectors $\mathbf{Y} = [Y_1, Y_2, \ldots, Y_n]^T$ (for responses), $\mathbf{x}_1$ (for values of the first predictor), $\mathbf{x}_2$ (for values of the second predictor), and $\boldsymbol{\varepsilon}$ for the n error terms. The column of intercept terms can be captured as $\beta_0 \mathbf{1}_n$, where $\mathbf{1}_n$ is an n-dimensional vector with all coordinates equal to 1. Hence, (4.16) can be written in vector notation as

$$\mathbf{Y} = \beta_0 \mathbf{1}_n + \beta_1 \mathbf{x}_1 + \beta_2 \mathbf{x}_2 + \boldsymbol{\varepsilon}. \tag{4.17}$$

The vectors $\mathbf{1}_n$, $\mathbf{x}_1$, and $\mathbf{x}_2$ can be placed as three columns in a matrix $\mathbf{X}$, and the three beta parameters can be written as a vector $\boldsymbol{\beta} = [\beta_0, \beta_1, \beta_2]^T$. This leads to the following matrix notation for the sample linear regression model:

$$\mathbf{Y} = \mathbf{X}\boldsymbol{\beta} + \boldsymbol{\varepsilon}. \tag{4.18}$$

One advantage of the matrix notation in (4.18) is that it works the same way for any number of predictors. From now on, we will be discussing a general case of k predictors, which means that $\mathbf{X}$ is an n by $(k + 1)$ matrix and $\boldsymbol{\beta}$ is a $(k + 1)$-dimensional vector of regression coefficients. The matrix $\mathbf{X}$ is called a design matrix, with this name being especially relevant in designed studies when the values of the predictors are selected by the investigator. We can use the matrix notation in order to obtain an equivalent formulation of Assumption 4.1 from Section 4.2.1 as follows (see Supplement 4B for random vector notation).

Assumption 4.1 The error term in (4.18) is a random vector such that $E(\boldsymbol{\varepsilon}) = \mathbf{0}$ and $\text{Var}(\boldsymbol{\varepsilon}) = \sigma^2 \mathbf{I}$, where $\mathbf{I}$ is an n by n identity matrix.

Based on the above assumption, we have $E(\mathbf{Y}) = \mathbf{X}\boldsymbol{\beta}$ and $\text{Var}(\mathbf{Y}) = \sigma^2\mathbf{I}$. In this general setup, we can write many formulas in a very concise way. For example, the least-squares normal equations (see (4.6)) have the form

$$\mathbf{X}^\mathrm{T}\mathbf{X}\mathbf{b} = \mathbf{X}^\mathrm{T}\mathbf{Y}, \tag{4.19}$$

where the vector $\mathbf{b} = [b_0, b_1, \ldots, b_k]^\mathrm{T}$ is the least-squares estimator of $\boldsymbol{\beta}$. The matrix equation (4.19) (which is really a system of linear equations) has a unique solution for $\mathbf{b}$ when the matrix $\mathbf{X}^\mathrm{T}\mathbf{X}$ is nonsingular. The solution gives the least-squares estimators of the regression coefficients

$$\mathbf{b} = \left(\mathbf{X}^\mathrm{T}\mathbf{X}\right)^{-1}\mathbf{X}^\mathrm{T}\mathbf{Y}. \tag{4.20}$$

Note that $\mathbf{b}$ is a *linear estimator* of the responses because it has the form of a matrix, here $\left(\mathbf{X}^\mathrm{T}\mathbf{X}\right)^{-1}\mathbf{X}^\mathrm{T}$, multiplied by the vector $\mathbf{Y}$ of responses. The matrix notation allows a straightforward derivation (see Problems 4.2 and 4.3) of the following property of the least-squares estimators.

Property 4.1
(a) The least-squares estimator $\mathbf{b}$ given by (4.20) is an unbiased estimator of $\boldsymbol{\beta}$, that is, $E(\mathbf{b}) = \boldsymbol{\beta}$.
(b) The variance–covariance matrix of the least-squares estimator $\mathbf{b}$ is equal to $\text{Var}(\mathbf{b}) = \sigma^2(\mathbf{X}^\mathrm{T}\mathbf{X})^{-1}$.

This tells us that $\mathbf{b}$ is a *linear unbiased estimator*. An estimator is called the *best linear unbiased estimator* (or BLUE) if it has the minimum variance among all linear unbiased estimators. Clearly, this is a desirable property of an estimator.

Property 4.2 The least-squares estimator $\mathbf{b}$ given by (4.20) is the best linear unbiased estimator (or BLUE) of $\boldsymbol{\beta}$.

The proof can be found in Montgomery et al. (2006) (in their Appendix C.4). It is worth noting that the above properties hold without any specific distributional assumption about the model other than Assumption 4.1*.

We note that the n-dimensional vector $\widehat{\mathbf{Y}}$ of fitted values is equal to $\mathbf{X}\mathbf{b}$. This means that

$$\widehat{\mathbf{Y}} = \mathbf{X}\mathbf{b} = \mathbf{X}\left(\mathbf{X}^\mathrm{T}\mathbf{X}\right)^{-1}\mathbf{X}^\mathrm{T}\mathbf{Y} \tag{4.21}$$

or equivalently

$$\widehat{\mathbf{Y}} = \mathbf{H}\mathbf{Y}, \tag{4.22}$$

where

$$\mathbf{H} = \mathbf{X}\left(\mathbf{X}^\mathrm{T}\mathbf{X}\right)^{-1}\mathbf{X}^\mathrm{T} \tag{4.23}$$

is called the hat matrix. The hat matrix $\mathbf{H}$ is symmetric and idempotent, that is, $\mathbf{HH} = \mathbf{H}$. The vector of residual can be written as

$$\mathbf{e} = \mathbf{Y} - \widehat{\mathbf{Y}} = \mathbf{Y} - \mathbf{HY} = (\mathbf{I} - \mathbf{H})\mathbf{Y}, \tag{4.24}$$

where $(\mathbf{I} - \mathbf{H})$ is also symmetric and idempotent. It is now easy to show (see Problem 4.4) that

$$\mathrm{Var}(\mathbf{e}) = \sigma^2(\mathbf{I} - \mathbf{H}). \tag{4.25}$$

In Section 4.2.1, we estimated sigma in a simple regression model using formula (4.9). In multiple regression, we need to estimate $(k + 1)$ regression coefficients. Consequently, the residuals have $(n-k-1)$ degrees of freedom, and an unbiased estimator of σ^2 is

$$\widehat{\sigma}^2 = \frac{1}{n-k-1} \sum_{i=1}^{n} e_i^2 = \frac{1}{n-k-1} \sum_{i=1}^{n} (y_i - \widehat{y}_i)^2. \tag{4.26}$$

The analysis of variance partitioning of the total variability explained in Section 4.2.1 (see (4.10)) works in the same way for multiple regression, and the coefficient of determination R^2 is also defined by formula (4.11).

4.2.4 Geometric Interpretation in an n-Dimensional Space

When introducing the matrix notation in the previous section, we used vector notation as an intermediate step shown in equation (4.17). The vector $\mathbf{Y}$ of n responses is an n-dimensional vector in $\mathbb{R}^n$. For k predictors, the deterministic part $\mathbf{X}\boldsymbol{\beta}$ of the model (4.18) can be written as

$$\mathbf{X}\boldsymbol{\beta} = \beta_0 \mathbf{1}_n + \beta_1 \mathbf{x}_1 + \cdots + \beta_k \mathbf{x}_k, \tag{4.27}$$

where the vectors on the right-hand side are n-dimensional. Hence, the deterministic part of the model is a linear combination of those vectors, that is, it belongs to a $(k + 1)$-dimensional linear subspace V of $\mathbb{R}^n$ generated (or spanned) by the vectors $\mathbf{1}_n, \mathbf{x}_1, \ldots, \mathbf{x}_k$. The subspace V is called an *estimation space*. The least-squares method of estimation introduced in Section 4.2.1 (see formula (4.4)) minimizes the sum of squares

$$S(\boldsymbol{\beta}) = \sum_{i=1}^{n} (y_i - \beta_0 - \beta_1 x_{1i} - \cdots - \beta_k x_{ki})^2 = \|\mathbf{Y} - \mathbf{X}\boldsymbol{\beta}\|^2, \tag{4.28}$$

which is equal to the squared length of the distance between $\mathbf{Y}$ and $\mathbf{X}\boldsymbol{\beta}$. We can say that the least-squares method tries to find a vector in the estimation space V that is closest to $\mathbf{Y}$. This vector is the vector $\widehat{\mathbf{Y}}$ of the fitted values equal to $\mathbf{Xb}$, where $\mathbf{b}$ is the vector of the least-squares estimates. A vector from V that is closest to $\mathbf{Y}$ is obtained by an orthogonal projection of $\mathbf{Y}$ on V, and the hat matrix $\mathbf{H}$ defined in (4.23) is the matrix of that projection. This is consistent with formula (4.22) saying that $\widehat{\mathbf{Y}} = \mathbf{HY}$. Figure 4.10 shows the vector $\mathbf{Y}$ projected on the estimation space. The vector $\mathbf{e} = \mathbf{Y} - \mathbf{Xb}$

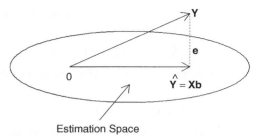

Figure 4.10 The vector **Y** of responses is projected on the estimation space resulting in the vector of fitted values.

of residuals is a vector connecting the point of fitted values $\widehat{\mathbf{Y}} = \mathbf{Xb}$ with **Y**. The residual vector **e** is orthogonal to $\widehat{\mathbf{Y}}$ based on the property of an orthogonal projection, but this fact can also be verified algebraically.

Formula (4.20) for the least-squares estimates can be applied when the inverse matrix $(\mathbf{X}^T\mathbf{X})^{-1}$ exists. We may wonder if the least-squares estimates exist when the matrix $\mathbf{X}^T\mathbf{X}$ is singular and $(\mathbf{X}^T\mathbf{X})^{-1}$ does not exist. The orthogonal projection of **Y** on the estimation space, which is a linear subspace, is obtained by dropping a perpendicular onto the estimation space, which always gives us a unique point $\widehat{\mathbf{Y}}$. From this geometric interpretation, we realize that the least-squares solution always exists and is unique in the sense of fitted values. The matrix $\mathbf{X}^T\mathbf{X}$ is singular in the presence of multicollinearity in the vectors $\mathbf{1}_n, \mathbf{x}_1, \ldots, \mathbf{x}_k$, that is, when they span a subspace of dimension lower than $(k + 1)$. In such cases, some of the predictors can be dropped, resulting in a new matrix $\mathbf{X}^*$ such that $(\mathbf{X}^*)^T\mathbf{X}^*$ is nonsingular. We then obtain the same vector of fitted values $\widehat{\mathbf{Y}}$ as $\mathbf{X}^*\left((\mathbf{X}^*)^T\mathbf{X}^*\right)^{-1}(\mathbf{X}^*)^T\mathbf{Y}$, which is analogous to formula (4.21). There are several choices for the predictors to be dropped (each leading to the same vector of fitted values $\widehat{\mathbf{Y}}$), which means that the least-squares solution is not unique in the sense of being represented as a linear combination of different sets of predictors. However, it is unique in the sense of a unique vector $\widehat{\mathbf{Y}}$ of fitted values.

The geometric interpretation in an n-dimensional space also helps with explaining the concept of degrees of freedom. The vector **Y** of responses has n degrees of freedom because it can be anywhere in the n-dimensional space $\mathbb{R}^n$. On the other hand, the vector $\widehat{\mathbf{Y}}$ of fitted values has $(k + 1)$ degrees of freedom (assuming no multicollinearity) because it lies in a $(k + 1)$-dimensional estimation space V. The residuals are associated with $(n-k-1)$ degrees of freedom because the residual vector **e** lies in a $(n-k-1)$-dimensional subspace orthogonal to V.

4.2.5 Statistical Inference in Multiple Linear Regression

The multiple regression model $\mathbf{Y} = \mathbf{X}\boldsymbol{\beta} + \boldsymbol{\varepsilon}$ was earlier introduced in formula (4.18), and now we write the model as

$$Y_i = \beta_0 + \beta_1 x_{i,1} + \cdots + \beta_k x_{i,k} + \varepsilon_i, \quad i = 1, \ldots, n. \tag{4.29}$$

In Section 4.2.3, we introduced Assumption 4.1* about the error term ε. In order to perform statistical inference, we will make the following additional assumption for the remaining part of Section 4.2.

Assumption 4.2 The error terms ε_i, $i = 1, \ldots, n$, follow the normal distribution $N(0, \sigma^2)$.

An important question in regression is whether the predictors indeed have any impact on the response. We want to test the null hypothesis of no impact of the predictors, that is,

$$H_0 : \beta_1 = \beta_2 = \cdots = \beta_k = 0 \tag{4.30}$$

versus an alternative that at least one of the predictors is significant, that is,

$$H_a : \beta_j \neq 0 \text{ for at least one } j. \tag{4.31}$$

If the null hypothesis H_0 is true, one might expect that no variability is explained by the predictors. Indeed, if the regression line were flat, all fitted values $\widehat{y}_i$ would be equal to the mean $\bar{y}$ and the regression sum of squares $SS_{Regr} = \sum_{i=1}^{n} (\widehat{y}_i - \bar{y})^2$ would be equal to zero. However, due to the sampling variability, the estimated coefficients are different from their true β-values, and the resulting fitted values will be on a line (or a surface for multiple predictors) that is not exactly flat. Therefore, SS_{Regr} will "accidentally" show some variability due to the natural sampling variability. This means that SS_{Regr} will typically be different from zero, but its small value will indicate "accidentally" explained variability. Hence, we will reject the null hypothesis H_0 only for SS_{Regr} large enough. In order to find the appropriate threshold, we need to follow up on the analysis of variance shown in equation (4.10) and create the so-called *ANOVA table* shown in Table 4.1.

The first column in Table 4.1 specifies the two sources of variability—the regression model (i.e., the predictors as the sources of variability) and the remaining variability associated with the error term as described by residuals. The second column in the ANOVA table gives the sums of squares for the two sources of variability and the sum of the two as the total sum of squares in accordance with equation (4.10). The concept of degrees of freedom was explained in Section 4.2.4

Table 4.1   The ANOVA Table for Multiple Regression with k Predictors

Source of Variability	Sum of Squares	Degrees of Freedom	Mean Square	F-Statistic
Regression	SS_{Regr}	k	$MS_{Regr} = \dfrac{SS_{Regr}}{k}$	$F_0 = \dfrac{MS_{Regr}}{MS_{Res}}$
Residual (error)	SS_{Res}	$n-k-1$	$MS_{Res} = \dfrac{SS_{Res}}{n-k-1}$	
Total	SS_{Total}	$n-1$		

using the geometric interpretation. We associated n degrees of freedom with the vector $\mathbf{Y}$ of responses. However, the total variability is measured with respect to the estimated mean $\bar{y}$, so we lose one degree of freedom for this estimation, and we are left with $n-1$ degrees of freedom for the total variability. As explained in Section 4.2.4, each predictor is associated with one degree of freedom resulting in k degrees of freedom for the whole regression. Note that the intercept does not count here because the presence of the intercept was already taken into account when subtracting one degree of freedom in the calculation of the $n-1$ degrees of freedom for the total variability. The residual vector $\mathbf{e}$ has the remaining $(n-k-1)$ degrees of freedom (see Section 4.2.4 for a geometric justification) associated with SS_{Res}. The regression and the residual degrees of freedom sum up to the degrees of freedom for the total variability. Recall that $\widehat{\sigma}^2$ defined by formula (4.26) is an unbiased estimator of σ^2. That estimator can also be written as $\widehat{\sigma}^2 = SS_{\text{Res}}/(n-k-1)$, which is the same as the *residual mean square* MS_{Res} defined in Table 4.1. Hence, $MS_{\text{Res}} = \widehat{\sigma}^2$ is an unbiased estimator of σ^2, that is, $E(MS_{\text{Res}}) = \sigma^2$. The *regression mean square* MS_{Regr} is defined in an analogous way by dividing the regression sum of squares by its degrees of freedom. It turns out that $E(MS_{\text{Regr}}) = \sigma^2$ under the null hypothesis H_0, which means that the average amount of variability in SS_{Regr} with k insignificant predictors is $k\sigma^2$. We can now define an F-statistic as

$$F_0 = \frac{MS_{\text{Regr}}}{MS_{\text{Res}}}. \tag{4.32}$$

Under the null hypothesis H_0, both the numerator and the denominator of F_0 are expected to be around σ^2. Hence, F_0 is expected to be close to 1. More variability explained by the model will lead to larger MS_{Regr} and hence larger F_0. For F_0 above a certain threshold, we can no longer explain the regression variability as merely due to the random sampling variability, and we will reject the null hypothesis H_0.

In order to find a suitable threshold for F_0, we use a property that its distribution under the null hypothesis H_0 is known to be the F-distribution with k and $(n-k-1)$ degrees of freedom. The α-level F-test of *significance of regression* rejects the null hypothesis H_0 when $F_0 \geq F_{k,n-k-1}(\alpha)$, where $F_{k,n-k-1}(\alpha)$ is the upper (100α)th percentile from the F-distribution with k and $(n-k-1)$ degrees of freedom.

Our earlier considerations about the variability expressed by SS_{Regr} could also be translated to the percent of explained variability measured by R^2, which could also be used as a test statistic. As expected, F_0 is closely related to R^2 by the following formula:

$$F_0 = \frac{(n-k-1)}{k} \frac{R^2}{1-R^2} = \frac{(n-k-1)}{k} \frac{1}{(1/R^2)-1}. \tag{4.33}$$

That is, F_0 is a strictly increasing function of R^2. Hence, one could use formula (4.33) to obtain an equivalent threshold for R^2 based on the earlier critical value $F_{k,n-k-1}(\alpha)$ for the F-test. Clearly, it is easier to use the F-test directly.

Once we find out that the whole regression model is significant, we may want to test which predictors are significant. If some of them are not significant, we might be able to remove them from the model in order to simplify it. The jth predictor x_j is significant, that is, it impacts the response, when $\beta_j \neq 0$. We may also want to test the significance of the intercept β_0. For any $j = 0, 1, 2, \ldots, k$, we test the null hypothesis

$$H_0 : \beta_j = 0 \text{ versus the alternative } H_a : \beta_j \neq 0 \quad (4.34)$$

using a t-test defined by the t-statistic

$$t_0 = \frac{b_j}{\widehat{SE}(b_j)}, \quad (4.35)$$

where $\widehat{SE}(b_j)$ is the estimated standard error of b_j calculated as the jth diagonal element of the matrix $\widehat{Var}(\mathbf{b}) = \hat{\sigma}^2(\mathbf{X}^T\mathbf{X})^{-1}$ (see Property 4.1b). We reject $H_0 : \beta_j = 0$ when $|t_0| \geq t_{n-k-1}(\alpha/2)$, where $t_{n-k-1}(\alpha/2)$ is the *upper* $(100\alpha/2)$th percentile of the t-distribution with $(n-1)$ degrees of freedom.

Example 4.3 As a continuation of Example 4.1, an additional predictor, the solar azimuth angle (x_2), was added in an attempt to explain more variability in Band 2 reflectance. The regression function was estimated as $y = 0.3370 + 0.00065x_1 + 0.000018x_2$. Note some changes in the intercept and the slope by x_1 in relation to the previous model $y = 0.3412 + 0.00061x_1$ without x_2. Such changes will occur when the predictors are correlated, that is, some of the information in x_1 is also contained in x_2.

The standard deviation was estimated as $\hat{\sigma} = 0.00365$. As can be seen from Table 4.2, the regression sums of squares increased slightly to $SS_{Regr} = 0.000633$ at the expense of the decreasing $SS_{Res} = 0.000974$. As expected, the total sum of squares stayed the same because it does not depend on the model. The fraction of explained variability increased slightly to $R^2 = 0.394$. We may wonder if the small increase is worth the added complexity created by an extra predictor. This question can be addressed more formally by testing the significance of individual predictors.

First, we test the significance of the whole regression by calculating the value of the F-statistic $F_0 = 23.7$, which is highly significant because the critical value at the $\alpha = 0.05$ level is $F_{2,74}(0.05) = 3.12$.

Table 4.3 shows the calculations needed for checking the significance of all three regression coefficients. The critical value for each t-statistic is $t_{n-k-1}(\alpha/2) = t_{74}(0.025) = 1.99$. The p-values in the last column of Table 4.3 are calculated based on formula (3.33), that is, as $2(1-G(t_0))$, where t_0 is the value of

Table 4.2 The ANOVA Table for Example 4.3 Data with Two Predictors

Source of Variability	Sum of Squares	Degrees of Freedom	Mean Square	F-Statistic
Regression	0.000633	2	0.000316	23.70
Residual (error)	0.000974	74	0.000013	
Total	0.001607	75		

Table 4.3   Calculations Needed for Checking the Significance of All Three Regression Coefficients for Example 4.3 Data with Two Predictors

Predictor	Coefficient	Standard Error	t-Statistic	p-Value
Constant (intercept)	0.3370	0.0114	29.45	1.31×10^{-42}
Solar elevation angle	0.00065	0.00013	5.11	2.43×10^{-6}
Solar azimuth angle	0.000018	0.000044	0.42	0.676

the t-statistic and G is the CDF of the t-distribution with 74 degrees of freedom. We can see that the intercept term and the slope by the solar elevation angle are both highly significant. On the other hand, the slope by the solar azimuth angle is not significant. We conclude that the solar azimuth angle does not contribute significantly to the prediction of the Band 2 reflection in the presence of the solar elevation angle. This means that the previous model with only one predictor (the solar elevation angle) is a better model. $\square$

Example 4.2 (cont.). This is a continuation of Example 4.2 from Section 4.2.1, where we fitted the least-squares regression line $y = -1.99 + 0.91x$ to the output power (y) and the input power (x) data from an experiment on an optical fiber. Table 4.4 shows a high p-value of 0.76 for the significance of the intercept, which means that it is not significant. On the other hand, the slope is highly significant. Since the simple regression model has only one predictor, we do not need to check the significance of the whole regression through the ANOVA table. In such a model, the F_0 statistic is equal to the square of the t-statistic for the slope, and the p-values for both tests are the same.

Since the intercept is not significant, we can fit a model without the intercept. For such a model, the matrix notation formulas still hold, but the design matrix $\mathbf{X}$ no longer contains the column of values of 1. We obtain the least-squares regression line $y = 0.886x$. We can now interpret $b_1 = 0.886$ as the fraction of the input power that is successfully transmitted (on average) through the optical fiber for the range of input power between 80 and 86 mW. According to formula (2.1), the average power loss in decibels is then calculated as $10 \log_{10}(b_1) = -0.5257$. $\square$

4.2.6   Prediction of the Response and Estimation of the Mean Response

An important application of the regression models is the prediction of the response variable Y based on the values of predictors for a new observation that is not within the

Table 4.4   Calculations Needed for Checking the Significance of Regression Coefficients for Example 4.2 Data

Predictor	Coefficient	Standard Error	t-Statistic	p-Value
Constant (intercept)	-1.99	6.26	-0.32	0.76
Input power (x)	0.91	0.075	5.11	4.7×10^{-10}

available data set. In Example 4.1, where the Band 2 reflectance was modeled as a linear function of the solar elevation angle, we may want to predict the reflectance at a specific future time with a known solar elevation angle. In a general context, we will denote by $\mathbf{x}_h^T = [1, x_1, \ldots, x_k]$ the values of the predictors for which we want to predict the response denoted by Y_h. We will start with an easier task of estimating the expected value $E(Y_h)$ of the response. In the context of Example 4.1, $E(Y_h)$ can be interpreted as the average Band 2 reflectance on all occasions with a given solar elevation angle. Since $E(Y_h) = \beta_0 + \beta_1 x_1 + \cdots + \beta_k x_k$, we can estimate it with the fitted value $\widehat{Y}_h = \mathbf{x}_h^T \mathbf{b} = b_0 + b_1 x_1 + \cdots + b_k x_k$. In order to assess precision of this estimation, we can construct a *confidence interval for the mean response* $E(Y_h)$ given by

$$\widehat{Y}_h \pm t_{n-k-1}(\alpha/2) \cdot \widehat{SE}\left(\widehat{Y}_h\right), \tag{4.36}$$

where

$$\widehat{SE}\left(\widehat{Y}_h\right) = \sqrt{\mathbf{x}_h^T \widehat{Var}(\mathbf{b})\mathbf{x}_h} = \sqrt{\widehat{\sigma}^2 \mathbf{x}_h^T (\mathbf{X}^T \mathbf{X})^{-1} \mathbf{x}_h}. \tag{4.37}$$

Example 4.4 This is a continuation of Example 4.1, where the Band 2 reflectance was modeled as a linear function of the solar elevation angle. For a model with one predictor x, we have $\mathbf{x}_h^T = [1, x_h]$ and formula (4.37) takes the form

$$\widehat{SE}\left(\widehat{Y}_h\right) = \sqrt{\widehat{\sigma}^2 \left(\frac{1}{n} + \frac{(x_h - \bar{x})^2}{S_{xx}}\right)}, \tag{4.38}$$

where $S_{xx} = \sum_{i=1}^n (x_i - \bar{x})^2$. Figure 4.11 shows the estimated regression line (solid line) and the 95% confidence interval limits (the two dashed lines)

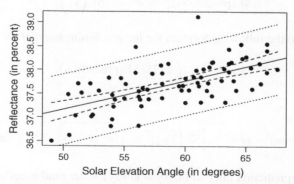

Figure 4.11 The estimated regression line (solid line) and the 95% confidence interval limits (the two dashed lines) together with the 95% prediction interval limits (the two dotted lines) based on Example 4.4 data.

calculated based on formula (4.38) for the whole range of the solar elevation angles as values of x_h. Let's assume we are interested in the mean response $E(Y_h)$ for $x_h = 55$. We could then draw a vertical line at the solar elevation angle value of 55. The points of intersection between the vertical line and the two dashed lines signify the confidence interval for $E(Y_h)$. The dashed lines are fairly close to the solid line, which means that the fitted values $\widehat{Y}_h$ estimate $E(Y_h)$ fairly precisely. The precision of estimation is best when $x_h = \bar{x}$ and then the confidence interval gets wider as $|x_h - \bar{x}|$ increases. The reason is that the uncertainty in the slope has the largest impact at the extremes due to the tilting of the regression line, which always goes through the center point $(\bar{x}, \bar{y})$.

Note that with the larger sample size, both denominators in formula (4.38) will get larger, and the confidence interval will get narrower. This is reasonable because with more data, we should be able to estimate more precisely the regression coefficients, and consequently, the mean $E(Y_h)$. Hence, the dashed lines of confidence limits will be getting very close to the estimated regression line with increased sample sizes. With a sufficient amount of data, we would know the average Band 2 reflectance very precisely based on the solar elevation angle. However, this does not mean that we know the actual Band 2 reflectance (rather than the average) on a given day. In order to predict the actual reflectance, we need to construct prediction intervals discussed next. □

We now want to tackle the more difficult problem of predicting the response Y_h. We know that Y_h varies around its mean $E(Y_h)$ with the variability described by the variance σ^2. It is intuitively appealing to use $E(Y_h)$ as one single number that best predicts the random response Y_h. This intuition is supported by the fact that the mean squared error of prediction $\mathrm{MSE}(c) = E(Y - c)^2$ is minimized by $c = E(Y)$ (see Problem 4.5). Since $E(Y_h)$ is unknown, we will use its estimate, the fitted value $\widehat{Y}_h$, as a predictor of Y_h. Uncertainty in the prediction is due to the variance σ^2 and the uncertainty in the estimation of $E(Y_h)$. We can express it with the variance of the prediction error estimated as

$$\left[\widehat{\mathrm{SE}}(\text{prediction})\right]^2 = \widehat{\sigma}^2 + \left[\widehat{\mathrm{SE}}\left(\widehat{Y}_h\right)\right]^2. \tag{4.39}$$

This leads to the following formula for the *prediction interval* for Y_h:

$$\widehat{Y}_h \pm t_{n-k-1}(\alpha/2) \cdot \widehat{\mathrm{SE}}(\text{prediction}), \tag{4.40}$$

where

$$\widehat{\mathrm{SE}}(\text{prediction}) = \sqrt{\widehat{\sigma}^2 + \left[\widehat{\mathrm{SE}}\left(\widehat{Y}_h\right)\right]^2} = \sqrt{\widehat{\sigma}^2\left[1 + \mathbf{x}_h^{\mathrm{T}}(\mathbf{X}^{\mathrm{T}}\mathbf{X})^{-1}\mathbf{x}_h\right]}. \tag{4.41}$$

Clearly, the prediction interval (4.40) is always wider (and usually much wider) than the equivalent confidence interval (4.36). With increased sample size, the estimated standard error $\widehat{\mathrm{SE}}\left(\widehat{Y}_h\right)$ is getting close to zero, but $\widehat{\mathrm{SE}}(\text{prediction})$ is getting close to σ.

Example 4.4 (cont.). For a model with one predictor x, formula (4.41) takes the form

$$\text{SE}\left(\widehat{Y}_h\right) = \sqrt{\widehat{\sigma}^2\left(1 + \frac{1}{n} + \frac{(x_h - \bar{x})^2}{S_{xx}}\right)}. \tag{4.42}$$

Figure 4.11 shows the 95% prediction interval limits (the two dotted lines) that are much wider than the confidence intervals (dashed lines). We would expect approximately 5% of all $n = 76$ observations to be outside of the prediction limits. Hence, the two observations above the limits and one observation on the boundary line should not be surprising here. We can say that at any future time with a known solar elevation angle, the Band 2 reflectance will be within the plotted prediction limits with 95% confidence. □

4.2.7 More on Checking the Model Assumptions

In Section 4.2.2, we introduced some residual plots for checking model assumptions. For more refined plots, we should take into account that residuals may have different variances, as can be seen from formula (4.25). That is, $\text{Var}(e_i) = \sigma^2(1 - h_{ii})$, where h_{ii} is the ith diagonal element of the hat matrix $\mathbf{H}$. Hence, we can expect different variability from different residuals, and their direct comparison can be questionable. If we knew σ, we could use the standardized residuals $e_i/\left(\sigma\sqrt{1 - h_{ii}}\right)$. In practice, we need to estimate σ, and we define *externally studentized residuals* as

$$r_i = \frac{e_i}{s_{(i)}\sqrt{1 - h_{ii}}}, \tag{4.43}$$

where

$$s_{(i)} = \sqrt{\frac{(n-k)\widehat{\sigma}^2 - e_i^2/(1 - h_{ii})}{n-k-1}} \tag{4.44}$$

is used to estimate σ, and $\widehat{\sigma}^2$ is calculated from formula (4.26). The term "studentized," as a version of "standardized," is used after a statistician, William Gosset, who wrote under the pseudonym Student. The reason for using formula (4.44) is that it estimates σ without the impact of the ith observation y_i, unlike the direct estimator $\widehat{\sigma}$. This is why we call the residual *externally* studentized. This methodology is especially relevant for outlying observations that may significantly impact the $\widehat{\sigma}^2$ estimator.

Since we replaced the true σ with its estimated value $s_{(i)}$, the variance of r_i is only approximately equal to 1. Based on Assumption 4.2 and formula (4.24), the residuals e_i follow a normal distribution. It turns out that the externally studentized residuals follow a t-distribution with $(n-k-1)$ degrees of freedom. Hence, they all have the same variance $\text{Var}(r_i) = (n-k-1)/(n-k-3)$ for $(n-k-1) > 2$ (see Appendix A for properties of the t-distribution). In residual plots, we can now make direct comparisons of the externally studentized residuals.

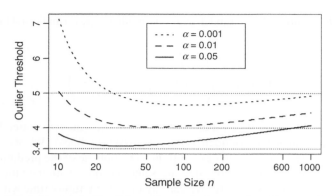

Figure 4.12 Threshold values for identifying outliers in simple regression with one predictor ($k = 1$) as a function of the sample size n for various α levels. Externally studentized residuals larger than the threshold suggest outliers.

It is often of interest to detect unusually large residuals as outliers. To this end, we want to calculate a threshold such that it is unlikely that the most extreme absolute value of the externally studentized residuals would be above the threshold if all observations followed the assumed model. Using a Bonferroni-type approach, we can use $t_{n-k-1}(\alpha/(2n))$ as the *outlier threshold*, where α is the acceptable probability of exceeding the threshold by the most extreme residual. The same threshold can also be used in the presence of multiple outliers. Although one could develop a different threshold for the second largest observation and then subsequent order statistics, it would make little difference and those thresholds would be smaller than our threshold. Hence, our approach can be considered a slightly more conservative approach. An outlier based on our threshold would also be an outlier based on those more precise, but also more complex, thresholds.

We recommend the following terminology. We detect a *likely outlier* when using the outlier threshold with $\alpha = 0.05$, a *definite outlier* when using $\alpha = 0.01$, and an *extreme outlier* for $\alpha = 0.001$. Figure 4.12 shows the threshold values for simple regression with one predictor ($k = 1$) as a function of the sample size n for various α levels. The smallest threshold value is 3.479, so as a practical rule of thumb, we recommend using the likely outlier threshold of at least 3.5 in any situation. For larger k, the thresholds are larger, but they are usually only slightly larger, except for very small sample sizes. However, such small sample sizes are not recommended when dealing with multiple predictors.

Figure 4.13 shows a plot of the externally studentized residuals versus fitted values for the simple regression model fitted to the Example 4.1 data. The likely outlier thresholds are plotted at the levels of $\pm t_{76-1-1}(0.05/(2 \cdot 76)) = \pm 3.56$. There is one observation that is likely to be an outlier.

Residuals can also be plotted in the order in which the observations were collected, so that we can detect a potential impact of other factors that change over time but have not been recorded. For the Example 4.1 data, we can plot the externally studentized residuals versus the recorded day of the year as shown in Figure 4.14. We can see that

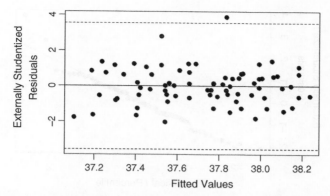

Figure 4.13 The externally studentized residuals plotted versus fitted values for the simple regression model fitted to the Example 4.1 data. The likely outlier thresholds are plotted at the levels of $\pm t_{76-1-1}(0.05/(2 \cdot 76)) = \pm 3.56$.

the residuals for the earlier part of the year (up until Day 120) tend to be somewhat larger than the remaining residuals, with the two largest residuals belonging to that group. This is a minor issue, and it could be ignored for this data set. Another option could be to try to collect more data for the earlier part of the year, and see if a separate model is needed for those data.

In order to verify the normality assumption of the error term in the model (see Assumption 4.2 in Section 4.2.5), the normal probability plots (see Section 3.6) of classic residuals are used. The plots usually work reasonably well. Under the model assumptions, the classic residuals e_i follow the normal distribution, but they have different variances as pointed out earlier (see also formula (4.25)). Hence, using the classic residuals in a normal probability plot could be potentially misleading. A more formal approach is to create a probability plot (see Section 3.6) of the

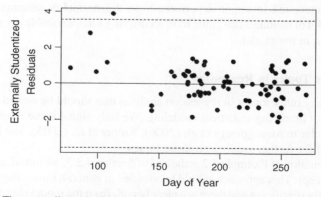

Figure 4.14 The externally studentized residuals plotted versus the day of the year for the simple regression model fitted to the Example 4.1 data. The likely outlier thresholds are plotted at the levels of $\pm t_{76-1-1}(0.05/(2 \cdot 76)) = \pm 3.56$.

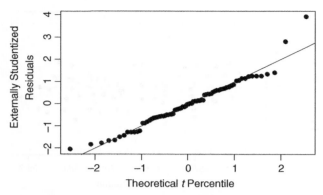

Figure 4.15 A probability plot of the externally studentized residuals versus percentiles from the t-distribution with $(n-k-1)$ degrees of freedom for the Example 4.1 data.

externally studentized residuals versus percentiles from the t-distribution with $(n-k-1)$ degrees of freedom. Figure 4.15 shows such a plot for the Example 4.1 data. The plotted line $y = x$ is the line of equality for the residuals versus the theoretical percentiles. Here we used a simplified method by taking $t_{n-k-1}((i-0.5)/n)$ as the theoretical percentile corresponding to the ith sorted residual. The points line up reasonably close to the line except for the outlier discussed earlier. This means that the normality assumption is violated only mildly, which would not require any action to modify the model.

If simplicity is desired, a normal probability plot of classic residuals can also be used since it will usually look very similar to the plot of the externally studentized residuals proposed here. On the other hand, the calculation of the externally studentized residuals is rather straightforward, and it could easily be implemented in any software.

Formal testing of the distribution of residuals is rarely done because the statistical inference in regression is not particularly sensitive to the assumption of normality. We are mainly interested in major departures from the normality assumption, and a probability plot of residuals can guide us in identifying a more suitable distribution for the error term in the model.

4.2.8 Other Topics in Regression

There are many other topics in regression analysis that should be considered in any serious project involving statistical modeling. We only signal some issues here and direct the reader to Montgomery et al. (2006), Kutner et al. (2005), and Draper and Smith (1998) for further reading on those topics.

In a continuation of Example 4.2 at the end of Section 4.2.5, we introduced a model with no intercept. This approach should be avoided in general, unless the intercept is not statistically significant and there is a clear benefit from the model simplification as was the case in Example 4.2. Such models have some properties different from those discussed here. For example, the sum of residuals is not necessarily equal to zero, and the coefficient of determination R^2 no longer has the usual interpretation of the fraction of explained variability (hence it should not be used).

In the presence of many outlying residuals, we may wish to change the normality assumption and consider the so-called robust regression. We may also encounter a situation where an observation has unusual values of predictors. We would say that we have an outlier in the values of predictors, which results in the so-called high leverage. This often leads to excessive influence of the observation on the estimated regression model, which may not be desirable.

When the assumption of the constant variance does not hold, we might be able to resolve the problem by applying transformations to the predictors, or to the response, or both. Sometimes additional information is available regarding the variability of the error term for each single observation. In that case, it might be helpful to use the weighted least-squares regression.

Sometimes the predictors are highly correlated or multicollinear, that is, they convey equivalent, or almost equivalent, information. This problem can be dealt with by dropping some predictors or by constructing a better set of predictors. Other methods for dealing with multicollinearity include ridge regression, principal component regression, and partial least-squares regression.

When the regression function cannot be written as a linear function of parameters, we are dealing with nonlinear regression. There are typically no closed-form formulas for the least-squares estimates of nonlinear regression coefficients, and the estimates need to be calculated through some numerical optimization methods. Statistical inference is more difficult in such models since the distributional properties of estimators are known only asymptotically.

When the response variable is discrete, we may need to use logistic regression or Poisson regression. Those are examples of a broader class of models called generalized linear models, which allow more flexibility in statistical modeling. In addition to increased computational challenges, the user also needs to deal with more complex interpretations of the resulting models. Specialized references on these topics are Hosmer (2000) and Agresti (2002).

4.3 EXPERIMENTAL DESIGN AND ANALYSIS

When we plan or analyze a study, it is important to distinguish between an observational study and a controlled experiment. The Landsat data study described in Example 4.1 is an observational study because we simply measured reflectance (Y) and recorded the solar elevation angle (X). No attempt was made to move the Sun back and forth across the sky, so that we could see its impact on reflectance. Figure 4.1 shows a clear correlation between Y and X, and the linear relationship between the two variables can be verified to be statistically significant. However, the measurements were taken on various days, and many other factors might have been changing in the meantime as well. Hence, we cannot be sure if the changes in X were the reason for changes in Y, or perhaps there was another unobserved factor causing those changes. In general, it is difficult to establish a causal relationship between the response Y and the predictors. An extreme example is that of elementary school children where for each child we record the number of words the child knows (Y) and the child's shoe size (X). Some people are surprised that the two variables are highly correlated. The confusion is that people often associate correlation with causation. Clearly, there is no direct causal relationship between these variables, but there is a third factor, age, which is causing both variables to change (we are assuming a substantial range of ages among those children). In other situations, the third factor might not be as obvious as in this case.

In order to mitigate the impact of such third factors and be able to prove causation, we should use controlled experiments whenever possible. The optical fiber experiment discussed in Example 4.2 is a controlled experiment because the input power of a laser light signal sent into the optical fiber was being changed and the resulting output power was recorded. One way to perform such an experiment is to make five replicate runs at 80 mW, and then change the laser setup to 82 mW power and perform five replicate runs at that level. We could then continue with the remaining levels of 84 and 86 mW. This way, we would save on some effort of switching between the power levels. Unfortunately, this approach would create an increasing pattern in the input power over time, which could coincide with an unobserved third factor being the real reason for changes in the output power. To avoid this problem, we could perform all 20 runs in a random order. This way, it becomes highly unlikely that a third factor would keep changing in the same way.

Another consideration in a controlled experiment is the choice of values, or levels, for the factor x (we will now use the lowercase x to recognize the fact that the factor is not random). In the context of experiment design, we often use the term "factor" instead of predictor. In the case of a quantitative factor, such as input power, we are left with many choices. First we need to decide on the range of x values we want to investigate. This may call for some preliminary runs to see how extreme we should go with the smallest and the largest value for x. Let us consider an experiment where Y is the surface reflectance and x is the angle, with respect to the surface normal, at which we send a beam of light at the surface. Such an experiment can be performed to estimate the bidirectional reflectance distribution function (BRDF) discussed in Schott (2007). When other factors, such as the viewing angle, the light wavelength, and so on, are kept constant, the BRDF simplifies to Y as a function of x. The angle x could go from $0°$ to $90°$, but the values close to $90°$ (almost parallel to the surface) would cause significant nonlinearity in Y as a function of x, and we should exercise caution in including such grazing angles into the range of x values.

Let us say we decided to experiment on the range of x values from a to b, and we are interested in the relationship of Y as a function of x over this range. Figure 4.16 shows some of the choices we have in terms of specific values for all 20 runs. In Design A, 10 runs are performed at each of the two extreme ends of the range. If we can be 100% certain that the relationship between Y and x is linear, Design A is optimal in the sense of providing the most precise estimates of the model parameters. On the other hand, the resulting data will not tell us if the relationship is indeed linear. Hence, if we are not very certain about the linearity, we need to use more levels of the x values. The more complex the relationship between Y and x we expect, the more levels we need. At the

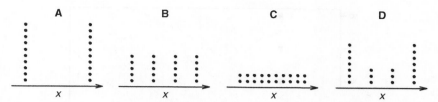

Figure 4.16   Four different designs of an experiment with 20 runs and one quantitative factor x.

same time, there are some benefits of having repeated values of x. Such repeats allow the calculation of variability due to factors other than x that are not observed and recorded in the experiment. The variability observed at these repeated x values is called *pure error*, as opposed to *modeling error* caused by the model imperfections. Hence, Design C with two repeats at each level is a reasonable solution if we expect a high degree of nonlinearity in Y as a function of x. If we expect a simpler relationship, such as a second-degree polynomial, we might also consider Design B with only four levels. Why don't we simply choose Design C, which affords us the most protection? Well, if the relationship actually is linear, then the model's parameters, for example, the slope of the line, are estimated less precisely than if Design A (best) or Design B had been chosen. Based on this reasoning, if we are fairly certain about the linear relationship but we want to make sure that the experiment confirms that, we might select Design D, which not only uses more runs at the extreme values, but also covers some intermediate values.

These considerations assume that we have no prior knowledge about the expected nonlinearity. In the case of estimating the BRDF discussed earlier, we might want to sample the grazing angles more densely, if such angles were of interest, since we would expect more nonlinearity there.

Things become more complex when we use two quantitative factors x_1 and x_2. The purpose of an experiment is often to find the values of x_1 and x_2 giving an optimum value of Y, let's say the largest possible value. If the number of runs was not limited, we could simply use a dense grid of possible values for x_1 and x_2, fit the model of Y as a function of the two factors, and find the maximum of that function. However, in most practical situations, each run costs money and effort of the investigators. Faced with this reality, researchers often invoke a rule of changing one factor at a time. This is a correct scientific rule in general. If you changed both factors at the same time, you would not know if the change in Y was due to a change in x_1 or in x_2. What often goes wrong is that the rule is applied too rigidly in the so-called one-factor-at-a-time experiments, which are performed as follows.

Consider an experiment, where (unknown to the experimenter) the response Y is a second-degree polynomial of x_1 and x_2. The contour lines of such a function are ellipses as shown in Figure 4.17. The function could also be plotted in three dimensions to see the so-called *response surface* directly, but the same type of information appears in Figure 4.17. The maximum value of Y is at the point $(12, 4)$ in this case. Now suppose the experimenter's one-factor-at-a-time strategy was to start with fixing the value of x_2 at 2, and performing runs at multiple levels of x_1 from 9 to

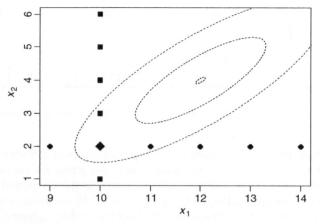

Figure 4.17 An example of a one-factor-at-a-time design where the first phase used $x_2 = 2$ and the second phase concentrated on the "best" value of $x_1 = 10$. The maximum was incorrectly identified around the point (10, 2) instead of the correct point (12, 4).

14 (the horizontal row of dots in Figure 4.17). Based on the resulting fit to the data, the experiment would find the maximum value in the vicinity of $x_1 = 10$. Now the experimenter would fix the value $x_1 = 10$ and would change the values of x_2 (the vertical row of squares). The resulting data would tell us that the largest value of Y is in the vicinity of the point (10, 2). We can see that the experimenter entirely missed the true shape of the relationship between Y and the factors, and he also missed the true maximum at the point (12, 4).

The problem with the previous experiment can be easily fixed by a different implementation of the rule of changing one factor at a time. Consider a *full factorial design* shown in Figure 4.18 consisting of four runs. For each factor, we choose two values, one is called a low (*L*) level and the other is the high (*H*) level. Run I is at a low level of both factors, while Run II is at a low level of x_1 and a high level of x_2, and so on for Runs III and IV. Each set of two runs that are connected by an arrow can be considered a mini round of a one-factor-at-a-time experiment. Here we have four

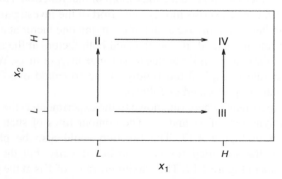

Figure 4.18 The 2^2 factorial design shown as four rounds of one-factor-at-a-time experiments, each consisting of two runs.

rounds of one-factor-at-a-time experiments, each consisting of two runs. However, instead of using a total of eight runs, we only need four runs because each run is a part of two experiments at the same time. This way, we are conforming to the rule of changing one factor at a time, but at the same time, we are investigating two levels of each factor at a time. This design is called the 2^2*factorial design.*

If we choose three levels of each factor in a grid of nine points, we have a 3^2 factorial design. This design consists of six rounds of one-factor-at-a-time experiments, each consisting of three runs.

Factorial experiments are very powerful in discovering various shapes of the response surface such as the one shown in Figure 4.17. However, in order to minimize the number of runs and still be able to estimate a second-degree response surface, we can use *central composite designs*. One such design is shown in Figure 4.19, where five levels of each factor are used. Rather than using all 25 combinations of five levels for each factor, the design uses only nine combinations placed in strategic positions. The extreme points have the coordinates $\sqrt{2} \approx 1.414$, and the four points connected into a square in Figure 4.19 are the 2^2 factorial part of this design. The design is shown here in a coded form, where $z_i = (x_i - m_i)/b_i$, $i = 1, 2$, and m_i and b_i are chosen for the ith variable, depending on the desired experimental region. The central point at $(0, 0)$ has a special purpose, and it is often chosen for repeats (called replicater in experimental design) that are needed to estimate the pure error discussed earlier. One would usually use three to five replicater of the center point. An analogous center point with replicater can also be added to the earlier discussed 2^2 factorial design in order to assess nonlinearity of the response surface.

The so-called *response surface methodology* incorporates multiple rounds of the designs discussed here, so that the optimum value for Y can be found. Ientilucci and Bajorski (2008) show an application of similar designs in the estimation of a response surface in the context of remote sensing.

In the experiments with quantitative factors considered so far, the regression methods discussed in Section 4.2 are used to fit the appropriate model and

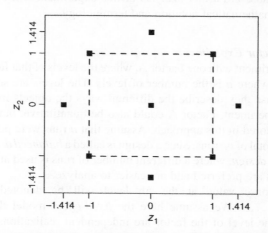

Figure 4.19 A central composite design allowing estimation of a second-degree response surface.

draw conclusions. When the factors are qualitative, the regression models can still be used, but some special approaches are required as discussed in the following subsection.

4.3.1 Analysis of Designs with Qualitative Factors

The following example demonstrates an experiment with qualitative factors.

Example 4.5   An experiment was designed in order to investigate sources of variability in spectrometer readings. Three tiles were chosen for the experiment—a white, a gray, and a black tile coded as 1, 2, and 3, respectively. Two spectrometers of the same type were chosen and were coded as 1 and 2, respectively. Two operators performed the measurements and were also coded. We say that we have three factors here—the tiles (Factor A), the spectrometers (Factor B), and the operators (Factor C). These three factors are the potential sources of variability in spectrometer readings. We would expect Factor A to generate a lot of variability in the measurements because very different tiles are measured here. The other two factors contribute to the measurement error variability that we would like to see as small as possible. There are also many other factors, not recorded here, that might contribute to the measurement error. They may cause some drift or trend in spectrometer readings over time, and we are interested in discovering such trends.

There are three qualitative levels of Factor A and two qualitative levels of each of the remaining two factors. There are no intermediate levels that can be chosen, and the central composite designs would not be possible here, but we can use a full factorial design. In order to cover all possible combinations, we need 12 runs. Since it is useful to have replicates, 24 runs were used with two replicates for each combination. In each run, a 31-dimensional vector of reflectances was recorded, but for simplicity we will consider the reflectance in the middle spectral band with wavelength of 550 nm as the response variable Y. Further analysis of this experiment will be done after we introduce some mathematical notation and terminology.   □

4.3.1.1   One-Factor Experiments

Consider an experiment with one Factor A, where the levels of that factor were coded into 1, 2, ..., a, where a is the number of levels. The levels are sometimes called *treatments* because they describe the different ways the objects in the study were treated in the experiment. Factor A could also be quantitative, but its quantitative nature will be ignored in this approach. Assume that n runs were performed at each factor level for a total of na runs. Such a design is called a *balanced design*, as opposed to an *unbalanced design*, where a different number of runs is used at different levels. Balanced designs are preferred and are easier to analyze.

The ith response value at the jth level will be denoted by $Y_{ji}, j = 1, \dots, a$ and $i = 1, \dots, n$. We assume here the *fixed-effect* model that all response values at the same level of the factor are independent realizations from the same normal distribution $N(\mu_j, \sigma^2)$, where μ_j is the mean at the jth level and σ^2 is the variance that is the same for all observations in the experiment. The impact of Factor A will be evaluated based on the means μ_j. If they are all the same, this means that the factor has no impact on the response Y. Hence, it is common and useful to define an *overall mean* $\mu = \sum_{j=1}^{a} \mu_j/a$ and the *treatment effects* $\tau_j = \mu_j - \mu$. The fixed-effect model can now be written as

$$Y_{ji} = \mu + \tau_j + \varepsilon_{ji}, \quad \text{where } j = 1, \ldots, a \text{ and } i = 1, \ldots, n, \qquad (4.45)$$

and ε_{ji} are independent random variables from the normal distribution $N(0, \sigma^2)$. Note that $\sum_{j=1}^{a} \tau_j = 0$, and the null hypothesis of no impact of Factor A can be written as

$$H_0: \ \tau_1 = \tau_2 = \cdots = \tau_a = 0. \qquad (4.46)$$

In order to test this null hypothesis, we can perform an analysis of variance similar to the one discussed in Section 4.2.5 and summarized in Table 4.1 for regression models. The significance of Factor A can be determined based on the *treatment sum of squares* defined as

$$SS_{\text{Treatment}} = n \sum_{j=1}^{a} \left(\bar{Y}_{j.} - \bar{Y}_{..} \right)^2, \qquad (4.47)$$

where $\hat{\mu}_j = \bar{Y}_{j.} = \sum_{i=1}^{n} Y_{ji}/n$ estimates the jth level mean μ_j and $\hat{\mu} = \bar{Y}_{..} = \sum_{j=1}^{a} \bar{Y}_{j.}/a = \sum_{j=1}^{a} \sum_{i=1}^{n} Y_{ji}/(na)$ estimates the overall mean μ. The treatment sum of squares $SS_{\text{Treatment}}$ measures variability among treatments, or factor levels, and its large values indicate significant differences among treatment effects. Of course, "large" is relative, so we need to compare that variability to the within-treatment variability measured by the *error, or residual, sum of squares*

$$SS_E = \sum_{j=1}^{a} \sum_{i=1}^{n} \left(Y_{ji} - \bar{Y}_{j.} \right)^2. \qquad (4.48)$$

A direct comparison is performed in the ANOVA table shown in Table 4.5, which is analogous to the ANOVA table shown in Table 4.1 for a regression model. The total sum of squares SS_{Total} is defined as

$$SS_{\text{Total}} = \sum_{j=1}^{a} \sum_{i=1}^{n} \left(Y_{ji} - \bar{Y}_{..} \right)^2, \qquad (4.49)$$

Table 4.5 The ANOVA Table for a One-Factor Fixed-Effect Experiment

Source of Variability	Sum of Squares	Degrees of Freedom	Mean Square	F-Statistic
Factor (treatments)	$SS_{\text{Treatment}}$	$a-1$	$MS_{\text{Treatment}} = \dfrac{SS_{\text{Treatment}}}{a-1}$	$F_0 = \dfrac{MS_{\text{Treatment}}}{MS_E}$
Residual (error)	SS_E	$na-a$	$MS_E = \dfrac{SS_E}{na-a}$	
Total	F	$na-1$		

and we have

$$SS_{Total} = SS_{Treatment} + SS_E, \qquad (4.50)$$

which is analogous to equation (4.10) for a regression model.

In a way analogous to the regression case, the *mean square error* MS_E, defined in Table 4.5, is an unbiased estimator of σ^2, that is, $E(MS_E) = \sigma^2$. On the other hand, the *treatment mean square* $MS_{Treatment}$ is an unbiased estimator of σ^2, that is, $E(MS_{Treatment}) = \sigma^2$, only under the null hypothesis H_0 defined in (4.46). In general, we have

$$E(MS_{Treatment}) = \sigma^2 + \frac{n\sum_{j=1}^{a}\tau_j^2}{a-1}. \qquad (4.51)$$

This means that the large values of the F-statistic defined as

$$F_0 = \frac{MS_{Treatment}}{MS_E} \qquad (4.52)$$

will suggest that the null hypothesis H_0 is not true. Under the null hypothesis H_0, the F-statistic F_0 follows the F-distribution with $(a-1)$ and $(na-a)$ degrees of freedom. The α-level F-test of significance of Factor A rejects the null hypothesis H_0 when $F_0 \geq F_{a-1,na-a}(\alpha)$, where $F_{a-1,na-a}(\alpha)$ is the upper (100α)th percentile from the F-distribution with $(a-1)$ and $(na-a)$ degrees of freedom.

4.3.1.2 Two-Factor Experiments

Let us now add a second Factor B with levels coded into $1, 2, \ldots, b$, where b is the number of levels. The treatment effects for Factor B will be denoted as β_k, $k = 1, \ldots, b$, where $\sum_{k=1}^{b}\beta_k = 0$. An *additive fixed-effect model* can now be written as

$$Y_{jki} = \mu + \tau_j + \beta_k + \varepsilon_{jki}, \quad \text{where } j = 1, \ldots, a, \; k = 1, \ldots, b, \text{ and } i = 1, \ldots, n,$$
$$(4.53)$$

and ε_{jki} are independent random variables from the normal distribution $N(0, \sigma^2)$. The model is called additive because the impact of each treatment is expressed by simply adding the treatment effect τ_j or β_k to the overall mean μ in the model. These treatment effects are often called *main effects*. In order to explain this concept, let's consider a 2^2 factorial design shown in Figure 4.20, where the two levels of each factor are coded as 1 and 2. With only two levels of each factor, we could also define the *factor effect* as an impact on Y of switching from level 1 to 2, which, for Factor A, is equal to $\tau_2 - \tau_1 = 2\tau_2$ since $\tau_1 = -\tau_2$. Based on the additive model (4.53), the impact of Factor A is always the same, no matter what the level of Factor B is. The same applies to the impact $\beta_2 - \beta_1 = 2\beta_2$ of Factor B, which adds the same amount to the response Y independently of the Factor A level.

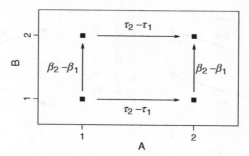

Figure 4.20 Additivity of the factor effects impacting Y independently of each other, no matter what the level of the other factor is.

The additive model (4.53) does not take into account possible interactions between the factors in the way that they impact the response Y. In order to make the model more general, we can introduce interaction terms and write the model as

$$Y_{jki} = \mu + \tau_j + \beta_k + (\tau\beta)_{jk} + \varepsilon_{jki}, \quad \text{where } j = 1, \ldots, a, \ k = 1, \ldots, b,$$

$$\text{and } i = 1, \ldots, n, \tag{4.54}$$

and $(\tau\beta)_{jk}$, $j = 1, \ldots, a$ and $k = 1, \ldots, b$, are the interaction terms. There are $a \cdot b$ interaction terms. The interaction terms are needed in the presence of synergy between factors when the combined effect of two factors is above and beyond the sum of the individual main effects. With positive synergy, when the combined effect is larger than the sum of the main effects, the interaction term is positive. A negative synergy is described by a negative interaction term. Denote by μ_{jk} the population mean $E(Y_{jki})$, which is the mean response when Factor A is at level j and Factor B is at level k. We can write $\mu_{jk} = \mu + \tau_j + \beta_k + (\tau\beta)_{jk}$. The interaction term $(\tau\beta)_{jk}$ is that part of μ_{jk} that is not already explained by the main effects τ_j and β_k. The model (4.54) can also be written as

$$Y_{jki} = \mu_{jk} + \varepsilon_{jki}, \quad \text{where } j = 1, \ldots, a, \ k = 1, \ldots, b, \text{ and } i = 1, \ldots, n. \tag{4.55}$$

The population mean μ_{jk} can be estimated independently in each *cell*, or combination of levels of both factors, by a sample average of observations in that cell, that is, by $\widehat{\mu}_{jk} = \bar{Y}_{jk\cdot} = \sum_{i=1}^{n} Y_{jki}/n$. We assume here that there are at least two runs in each cell, that is, $n \geq 2$.

Let us explain these concepts using an experiment with two factors, each at two levels. In Figure 4.21a, the two levels of Factor A are shown on the horizontal axis, and the two levels of Factor B are marked as two lines connecting the estimated responses $\widehat{\mu}_{jk}$ shown on the vertical axis. The mean responses were estimated as $\widehat{\mu}_{11} = 1$, $\widehat{\mu}_{21} = 3$, $\widehat{\mu}_{12} = 4$, and $\widehat{\mu}_{22} = 6$. Consequently, the estimated overall mean is $\widehat{\mu} = 3.5$. It is useful here to estimate the mean response $\widehat{\mu}_{1\cdot} = (\widehat{\mu}_{11} + \widehat{\mu}_{12})/2$ at level 1 of Factor A, averaged over the two levels of the other factor. This gives us a sense of how level 1 of Factor A impacts the response, all other things being equal. We obtain $\widehat{\mu}_{1\cdot} = 2.5$,

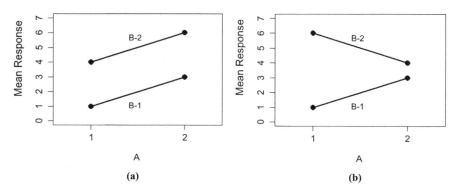

Figure 4.21 Interaction plots for two different experiments. In panel (a), there are no interactions between the factors, which is evident from the parallel lines. In panel (b), the lines are not parallel, which indicates presence of interactions between the factors.

and the treatment effect is estimated as $\tau_1 = \widehat{\mu}_{1.} - \widehat{\mu} = -1$. In a similar fashion, we obtain $\widehat{\mu}_{2.} = 4.5$ and $\widehat{\tau}_2 = 1$. For Factor B, we calculate $\widehat{\mu}_{.1} = (\widehat{\mu}_{11} + \widehat{\mu}_{21})/2 = 2$, which leads to $\widehat{\beta}_1 = -1.5$. For the second level of Factor B, we obtain $\widehat{\mu}_{.2} = 5$ and $\widehat{\beta}_2 = 1.5$.

In order to estimate interactions, we recall that $\mu_{jk} = \mu + \tau_j + \beta_k + (\tau\beta)_{jk}$, which can be written as $(\tau\beta)_{jk} = \mu_{jk} - \mu - \tau_j - \beta_k$. Again, we can see that $(\tau\beta)_{jk}$ is what is left from μ_{jk} once the effect of the overall mean μ and the main effects τ_j and β_k are subtracted. We can now estimate $\widehat{(\tau\beta)}_{11} = \widehat{\mu}_{11} - \widehat{\mu} - \widehat{\tau}_1 - \widehat{\beta}_1 = 1 - 3.5 - (-1) - (-1.5) = 0$. In a similar way, we obtain $\widehat{(\tau\beta)}_{12} = \widehat{(\tau\beta)}_{21} = \widehat{(\tau\beta)}_{22} = 0$. We say that there are no interactions between the two factors in this experiment. Hence, the additive model of the form (4.53) is suitable for these data. For real data, we would not usually get exact zero values and the significance of interactions would be tested by an appropriate F-test.

The plot in Figure 4.21a is called an *interaction plot* because it allows a visual detection of interactions. If the two lines are parallel, the additive model holds, and we say that there are no interactions. Figure 4.21b shows an interaction plot based on a different data set. Here the lines are not parallel, showing substantial interactions. Let us check this algebraically by performing calculations analogous to those shown previously. Here $\widehat{\mu}_{11} = 1, \widehat{\mu}_{21} = 3, \widehat{\mu}_{12} = 6$, and $\widehat{\mu}_{22} = 4$. Consequently, $\widehat{\mu} = 3.5$. For Factor A, we calculate $\widehat{\mu}_{1.} = 3.5, \tau_1 = 0, \widehat{\mu}_{2.} = 3.5$, and $\widehat{\tau}_2 = 0$, and for Factor B, we obtain $\widehat{\mu}_{.1} = 2, \widehat{\beta}_1 = -1.5, \widehat{\mu}_{.2} = 5$, and $\widehat{\beta}_2 = 1.5$. An interaction term is now estimated as $\widehat{(\tau\beta)}_{11} = \widehat{\mu}_{11} - \widehat{\mu} - \widehat{\tau}_1 - \widehat{\beta}_1 = 1 - 3.5 - 0 - (-1.5) = -1$. In a similar way, we obtain $\widehat{(\tau\beta)}_{12} = 1, \widehat{(\tau\beta)}_{21} = 1$, and $\widehat{(\tau\beta)}_{22} = -1$. The statistical significance of these estimated interactions would depend on the sample size and the variability in the error term, as expressed by σ^2.

Both the model (4.55) and the equivalent model (4.54) are called *fully saturated models* because they do not impose any additional structure on the cell means. The additive model (4.53) is not fully saturated because it makes the additional assumption about the additive property of the mean response. In order to estimate all parameters in the fully saturated model, we would usually want at least two runs in each cell, that is, $n \geq 2$. For $n = 1$, one strategy is to assume an additive model, so that some degrees of freedom are available for the mean square error, and the statistical significance of the main effects of the factors can be tested.

4.3.1.3 Multifactor Experiments

The concepts of the previous subsection extend to designs with more than two factors. Such models may have two-way interactions of two factors, three-way interactions of three factors, and so on. We often use an abbreviated notation for interactions, where AB denotes a two-way interaction between Factors A and B. All models discussed here can be written as regression models with the use of the so-called dummy variables, but the details are beyond the scope of this brief section. This means we can also calculate and analyze residuals to check the model assumptions in the same way as discussed for regression models in Section 4.2. For many models for data from a balanced full factorial design, including the saturated model and the additive model, the diagonal elements of the hat matrix **H** defined in (4.23) are all equal to each other. Hence, all residuals have the same variance, and there is no need to use the externally studentized residuals defined in (4.43). We can simply use the classic residuals defined as the observation minus the fitted value.

Although the ANOVA table becomes more complex in multifactor designs, each factor and each interaction can still be tested for significance with an appropriate F-test. So, the general concepts are similar to those explained earlier, and we can evaluate significance of the main factors and interactions based on the p-values produced by a suitable computer routine. Here we want to show some results based on the experiment described earlier in Example 4.5.

Example 4.5 (cont.). In the spectrometer experiment described earlier, we have three factors. Factor A (the tiles) is at three levels ($a = 3$), while Factor B (the spectrometers) and Factor C (the operators) are both at two levels, that is, $b = 2$ and $c = 2$. There are two runs for each combination of factors, so $n = 2$. The response variable Y is the reflectance at the wavelength of 550 nm. We first fit a fully saturated model, that is, with the three main effects A, B, and C, the three two-way interactions AB, BC, and AC, and the three-way interaction ABC. Before performing any statistical inference on the model, we need to verify the model assumptions by inspecting various residual plots.

One of those plots is shown in Figure 4.22, where the residuals are plotted against the fitted values. It turns out that the dot plotted on the left-hand side (with a very small fitted value of almost zero) represents eight overlapping residuals with very small values. We could elicit this information by using jitter as discussed in Section 2.4.4. We conclude that the variability in residuals is highly dependent on the magnitude of the observation. This contradicts the assumption of constant variance σ^2. A physical interpretation of this situation is that the reflectances of the white tile are larger than those for the gray tile and much larger than those for the black tile. For a different view, Figure 4.23 shows the residuals plotted against the tile number (1 = white, 2 = gray, 3 = black). Each tile was measured eight times, so there are eight residuals for each tile.

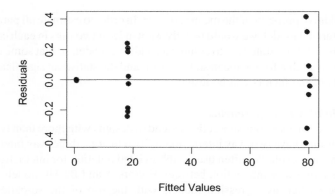

Figure 4.22 The residuals for the model of reflectance as the response (see Example 4.5) are plotted against the fitted values.

However, for the black tile, the residuals are so small that they all show as one dot. On the other hand, the residuals for the white tile are much larger, and those for the gray tile are in between. This is the same grouping of residuals seen earlier in Figure 4.22.

We may suspect that the measurement error is proportional to the magnitudes of observations, which is typically the case. This means that we need to modify our model. In such cases, a transformation of the response variable often helps. In this case, it turns out that a logarithm of reflectance should be used as the response variable. While any base of the logarithm could be used here, the most convenient choice is that of the natural logarithm as we will see later on.

We used the natural logarithm $\ln(Y)$ as the response, and we fitted a fully saturated model with the same three factors with interactions. The resulting residuals are plotted in Figure 4.24. This time, there is only a small difference in variability at the three levels, which is acceptable. Other plots of residuals also suggest that the model is acceptable. Note that the absolute values of residuals are not larger than 0.014. Due to the natural logarithmic scale used on the response, this translates to the approximate

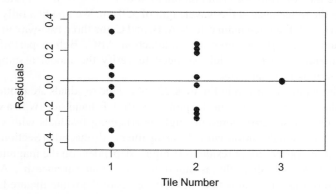

Figure 4.23 The residuals for the model of reflectance as the response (see Example 4.5) are plotted against the tile number (1 = white, 2 = gray, 3 = black).

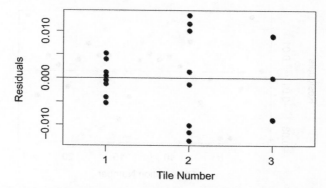

Figure 4.24 The residuals for the model of the natural logarithm of reflectance (see Example 4.5) are plotted against the tile number (1 = white, 2 = gray, 3 = black).

error of up to 1.4% with respect to the magnitude of the original response Y. This calculation is based on the fact that $\exp(\varepsilon) \approx 1 + \varepsilon$ for small ε. Here, $\exp(0.014) \approx 1.014$. If we used logarithm to the base 10 as a transformation on Y, the residuals would have magnitudes up to 0.006, and the calculation $10^{0.006} \approx 1.014$ would lead to the same conclusion.

We can now inspect the ANOVA table shown in Table 4.6. The first column shows the sources of variability in $Ln(Y)$, that is, the three factors, three two-way interactions, and one three-way interaction (the star is notation for an interaction, not multiplication). "Error" relates to the variability in the ε error term, that is, the remaining variability that is not explained by the model factors. This is analogous to the Residual Error line in the ANOVA table for a regression model, such as the one shown in Table 4.1. The last column shows the p-values for testing the significance of the terms in the model. (See formula (3.33) for the definition of the p-values.) A threshold value of $\alpha = 0.05$ is often used for interpretation of the p-values. Based on

Table 4.6 The ANOVA Table for a Fully Saturated Model of Ln(Y) with the Three Factors, as Discussed in Example 4.5

Source	DF	SS	MS	F	P
Operator	1	0.0007	0.0007	6.08	0.030
Spectrometer	1	0.0007	0.0007	5.66	0.035
Tile	2	103.9224	51.9612	445623.81	0.000
Operator* Spectrometer	1	0.0002	0.0002	1.83	0.201
Operator*Tile	2	0.0000	0.0000	0.19	0.831
Spectrometer* Tile	2	0.0014	0.0007	6.00	0.016
Operator* Spectrometer* Tile	2	0.0002	0.0001	0.89	0.438
Error	12	0.0014	0.0001		
Total	23	103.9271			

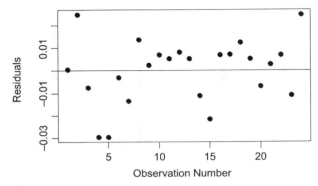

Figure 4.25   A plot of the residuals versus the time order of observations reveals no patterns, indicating that no drift had been observed during the course of the experiment.

that threshold, we would decide that all three main factors and the interaction between Spectrometer and Tile are significant.

However, with seven instances of testing significance, one could argue that a more correct value of $\alpha = 0.05/7 = 0.007$ should be used here (based on the Bonferroni arguments discussed in Section 6.2.2). In that case, only Tile is the significant factor. (One could instead learn that even if the other factors were assumed statistically significant, their impact here is very small and practically insignificant.) We then assume only Tile as the model factor, and fit the model again, so that a final model check can be performed.

One goal of this study was to investigate the spectrometer drift over time. This is why we want to calculate the model residuals, which can be regarded as the log reflectance values adjusted for the differences in the tiles. A plot of the residuals versus time order of observations is shown in Figure 4.25. We do not observe any patterns in that plot that would suggest a drift. Note that the absolute values of residuals are not larger than 0.03. As discussed earlier, this translates to approximately 3% error with respect to the magnitude of the original response Y. This increase from the previous value of 1.4% is caused by dropping from the original model terms whose average mean square, while not statistically significant, was still larger than the original error mean square.   □

4.3.2   Other Topics in Experimental Design

There are many other experimental designs and related models beyond those discussed here. If one deals with many factors, but the number of feasible runs is limited, one can consider *fractional factorial designs* that eliminate some of the runs from the standard grid of the full factorial designs. This reduction in the number of runs is done carefully so that the main effects and some interactions can still be estimated.

If we are not interested in the impact of specific factor levels, but instead we would like to understand a general impact of those levels when they vary within a certain population, we would use *random-effect models*. A typical example would be an operator as a random factor because we may want to know how the response Y changes when various operators perform the experimental runs, but we may not be interested in the performance of specific people. Another important consideration is the limitation on the desired randomization in the experiment. These issues are dealt with by designs such as *randomized block designs* and *split-plot designs.*

Another situation arises when levels of one factor are nested within levels of another factor. Consider an experiment on lightweight eye tracking headgear, where several headgear models are being compared. We would also want to know how the eye tracking headgear performs on various people. An ideal solution would be to test each model on the same set of people, with each person testing all models. However, the headgear models are owned and operated by different universities, which prefer to test the equipment on their own students. We still want to design a consistent experiment across the universities so that the study will be done in the same way at each university. However, the subjects wearing the headgear will have to be different at each university. The subjects are nested here within the headgear models, and the statistical analysis of such data needs to recognize that fact.

These and many other topics on the design and analysis of experiments are covered thoroughly in many specialized book on this topic, including Box et al. (2005), Kutner et al. (2005), and Montgomery (2008). Brief and less mathematical treatments can be found in Antony (2003) and Barrentine (1999).

SUPPLEMENT 4A. VECTOR AND MATRIX ALGEBRA

This supplement gives some basic facts about vectors and matrices. More details can be found in Hadi (1996) and Harville (1997).

Vectors

Within the scope of this book, we can think of a vector as being represented by a line segment with a defined direction or as an arrow connecting an initial point A with a terminal point B. We will mostly consider vectors with an initial point at the origin. In this case, the coordinates of the point B can be regarded as the coordinates of the vector. A vector with n coordinates (or elements) $x_1, x_2, \ldots, x_n$ is written as a column vector

$$\mathbf{x} = \begin{bmatrix} x_1 \\ x_2 \\ \vdots \\ x_n \end{bmatrix}. \tag{4.56}$$

We will often think about $\mathbf{x}$ as a vector (with the initial point at the origin) or a point in the n-dimensional space $\mathbb{R}^n$, and we will use the two interpretations interchangeably. In both cases, we will write $\mathbf{x} \in \mathbb{R}^n$. In order to write a vector as a row vector, we use a transpose operation, $\mathbf{x}^T$, that changes a column vector to a row vector, that is,

$\mathbf{x}^T = [x_1, x_2, \ldots, x_n]$. Multiplication of a vector $\mathbf{x}$ by a number (or scalar) $c \in \mathbb{R}$ and addition of two vectors of the same dimension are done element by element as follows:

$$c\mathbf{x} = \begin{bmatrix} cx_1 \\ cx_2 \\ \vdots \\ cx_n \end{bmatrix}, \qquad \mathbf{x} + \mathbf{y} = \begin{bmatrix} x_1 + y_1 \\ x_2 + y_2 \\ \vdots \\ x_n + y_n \end{bmatrix}. \tag{4.57}$$

The *length* (or *norm*) $\|\mathbf{x}\|$ of a vector $\mathbf{x}$ is the Euclidean distance of its terminal point (also denoted by $\mathbf{x}$) to the origin, that is, it can be calculated as

$$\|\mathbf{x}\| = \sqrt{\sum_{i=1}^{n} x_i^2}. \tag{4.58}$$

An *angle* between vectors $\mathbf{x}$ and $\mathbf{y}$ is defined geometrically in the usual way, and its cosine can be calculated as

$$\cos(\theta) = \frac{\mathbf{x}^T\mathbf{y}}{\|\mathbf{x}\|\|\mathbf{y}\|}, \tag{4.59}$$

where $\mathbf{x}^T\mathbf{y}$ is the *inner, or scalar, product* of two vectors defined as $\mathbf{x}^T\mathbf{y} = \sum_{i=1}^{n} x_iy_i$. The vector length $\|\mathbf{x}\|$ can also be expressed using the inner product as follows:

$$\|\mathbf{x}\| = \sqrt{\sum_{i=1}^{n} x_i^2} = \sqrt{\mathbf{x}^T\mathbf{x}}. \tag{4.60}$$

A distance between two points $\mathbf{x}$ and $\mathbf{y}$ can be expressed as $\|\mathbf{x} - \mathbf{y}\|$ since the vector $(\mathbf{x} - \mathbf{y})$ connects the two points.

Definition 4A.1 A set of vectors $\mathbf{x}_1, \mathbf{x}_2, \ldots, \mathbf{x}_k$ is said to be *linearly dependent* if and only if there exist constants $c_1, c_2, \ldots, c_k$ not all zero, such that

$$\sum_{j=1}^{k} c_j\mathbf{x}_j = 0, \tag{4.61}$$

which means that one of the vectors can be represented as a linear combination of the other vectors.

Definition 4A.2 A set of vectors $\mathbf{x}_1, \mathbf{x}_2, \ldots, \mathbf{x}_k$ is said to be *linearly independent* if and only if the condition

$$\sum_{j=1}^{k} c_j \mathbf{x}_j = 0 \tag{4.62}$$

implies that the constants $c_1, c_2, \ldots, c_k$ are all equal to zero. This means that none of the vectors can be expressed as a linear combination of the other vectors.

Definition 4A.3 Two vectors $\mathbf{x}$ and $\mathbf{y}$ are collinear if there exists $c \neq 0$ such that $\mathbf{y} = c\mathbf{x}$.

Definition 4A.4 A *projection of a point* $\mathbf{x} \in \mathbb{R}^n$ *on a closed set* $A \subset \mathbb{R}^n$ is a point $\mathbf{a}^* \in A$ that is closest to $\mathbf{x}$ among all points in A, that is,

$$\|\mathbf{x} - \mathbf{a}^*\| = \min\{\|\mathbf{x} - \mathbf{a}\| : \mathbf{a} \in A\}. \tag{4.63}$$

Definition 4A.5 A *projection of a vector* $\mathbf{x} \in \mathbb{R}^n$ *on a vector* $\mathbf{y} \in \mathbb{R}^n$ is a vector $c\mathbf{y}$, where $c \in R$, such that

$$\|\mathbf{x} - c\mathbf{y}\| = \min\{\|\mathbf{x} - a\mathbf{y}\| : a \in \mathbb{R}\}. \tag{4.64}$$

A projection on $\mathbf{y}$ is a projection on the line defining the direction of $\mathbf{y}$, that is, $\{c\mathbf{y} : c \in \mathbb{R}\}$. The operation of the projection on $\mathbf{y}$ is denoted by $P_\mathbf{y}$, that is, $P_\mathbf{y}(\mathbf{x})$ is the projection of $\mathbf{x}$ on $\mathbf{y}$.

Property 4A.1 The projection of $\mathbf{x}$ on $\mathbf{y}$ is given by the formula $P_\mathbf{y}(\mathbf{x}) = (\mathbf{x}^T\mathbf{y}/\mathbf{y}^T\mathbf{y})\mathbf{y}$.

Property 4A.2 For a unit length vector $\mathbf{v}$, the projection of $\mathbf{x}$ on $\mathbf{v}$ is given by the formula $P_\mathbf{v}(\mathbf{x}) = (\mathbf{x}^T\mathbf{v})\mathbf{v}$.

Matrices

A matrix is a rectangular array of numbers such as $\mathbf{X} = \begin{bmatrix} -2 & 1 & 5 & -1 \\ 4 & 3 & 0 & 8 \end{bmatrix}$. The matrix $\mathbf{X}$ consists of two rows and four columns. It has a total of eight elements or entries. The transpose operation, $\mathbf{X}^T$, of a matrix $\mathbf{X}$ changes the columns into rows, so that the first column of $\mathbf{X}$ becomes the first row of $\mathbf{X}^T$, the second column becomes the second row, and so forth. For the above example of the matrix $\mathbf{X}$, we have

$$\mathbf{X}^T = \begin{bmatrix} -2 & 4 \\ 1 & 3 \\ 5 & 0 \\ -1 & 8 \end{bmatrix}. \tag{4.65}$$

A general n by p matrix has n rows and p columns and can be written as

$$
\mathbf{X} = \left[x_{ij} \right] = \begin{bmatrix} x_{11} & x_{12} & \cdots & x_{1p} \\ x_{21} & x_{22} & \cdots & x_{2p} \\ \vdots & \vdots & \ddots & \vdots \\ x_{n1} & x_{n2} & \cdots & x_{np} \end{bmatrix},
\tag{4.66}
$$

where the x_{ij} element is in the ith row and the jth column. Matrices are convenient for representing data in an organized format consistent with the format of a statistical database with n observations in rows and p variables in columns. They are also used for many other purposes. It is worth recalling that an n by p matrix represents a linear transformation from a p-dimensional space $\mathbb{R}^p$ to an n-dimensional space $\mathbb{R}^n$ defined as $\mathbf{Xv}$ on a p-dimensional vector $\mathbf{v}$.

Definition 4A.6 A matrix $\mathbf{X}$ is called *symmetric* if and only if $\mathbf{X} = \mathbf{X}^{\mathrm{T}}$.

Definition 4A.7 A matrix is called a square matrix if the number of rows is equal to the number of columns.

When multiplying a matrix $\mathbf{X}$ by a constant c, we multiply each element by c, which results in

$$
c\mathbf{X} = \begin{bmatrix} cx_{11} & cx_{12} & \cdots & cx_{1p} \\ cx_{21} & cx_{22} & \cdots & cx_{2p} \\ \vdots & \vdots & \ddots & \vdots \\ cx_{n1} & cx_{n2} & \cdots & cx_{np} \end{bmatrix}.
\tag{4.67}
$$

When adding two matrices, we add them element by element:

$$
\mathbf{X} + \mathbf{Y} = \begin{bmatrix} x_{11} + y_{11} & x_{12} + y_{12} & \cdots & x_{1p} + y_{1p} \\ x_{21} + y_{21} & x_{22} + y_{22} & \cdots & x_{2p} + y_{2p} \\ \vdots & \vdots & \ddots & \vdots \\ x_{n1} + y_{n1} & x_{n2} + y_{n2} & \cdots & x_{np} + y_{np} \end{bmatrix}.
\tag{4.68}
$$

We define multiplication of two matrices as

$$
\left[x_{ij} \right]_{n \times k} \cdot \left[y_{ij} \right]_{k \times p} = \left[z_{ij} \right]_{n \times p}, \quad \text{where } z_{ij} = \sum_{t=1}^{k} x_{it} y_{tj}.
\tag{4.69}
$$

The element z_{ij} is the scalar product of the ith row of the matrix $\mathbf{X}$ and the jth column of the matrix $\mathbf{Y}$. In order to multiply the two matrices, the number of columns of the first matrix needs to be equal to the number of rows of the second matrix. The matrix multiplication is not commutative, that is, typically $\mathbf{XY} \neq \mathbf{YX}$.

Definition 4A.8 A square matrix $\mathbf{I}$ consisting of values of 1 on the diagonal and zeros elsewhere is called the *identity matrix*. For any matrix $\mathbf{X}$ (of the same dimension as $\mathbf{I}$), we have $\mathbf{I} \cdot \mathbf{X} = \mathbf{X} \cdot \mathbf{I} = \mathbf{X}$.

Definition 4A.9 If there exists a matrix $\mathbf{A}$ such that $\mathbf{A} \cdot \mathbf{X} = \mathbf{X} \cdot \mathbf{A} = \mathbf{I}$, then $\mathbf{A}$ is called the *inverse* of $\mathbf{X}$ and is denoted by $\mathbf{X}^{-1}$.

Definition 4A.10 A matrix $\mathbf{X}$ is called orthogonal if and only if $\mathbf{X} \cdot \mathbf{X}^{\mathrm{T}} = \mathbf{X}^{\mathrm{T}} \cdot \mathbf{X} = \mathbf{I}$, or $\mathbf{X}^{\mathrm{T}} = \mathbf{X}^{-1}$.

Note that columns $\boldsymbol{b}_j$ of an orthogonal matrix $\mathbf{X}$ are orthogonal to each other ($\mathbf{b}_i^{\mathrm{T}} \mathbf{b}_j = 0$ for $i \neq j$) and have unit length ($\mathbf{b}_i^{\mathrm{T}} \mathbf{b}_i = 1$).

Property 4A.3 Any orthogonal matrix $\mathbf{X}$ can be written as a product of a rotation matrix and a permutation of coordinates.

For a matrix $\mathbf{X} = \left[x_{ij} \right]_{n \times k}$, its elements x_{ii} with the row and column index the same are called diagonal elements.

Definition 4A.11 A matrix $\mathbf{X}$ is called diagonal if its off-diagonal elements are all equal to zero.

Definition 4A.12 The determinant of a square $p \times p$ matrix $\mathbf{A} = \{a_{ij}\}$, denoted by $|\mathbf{A}|$, is a scalar defined by the following recursive formula:

$$|\mathbf{A}| = \begin{cases} a_{11} & \text{if } p = 1, \\ \displaystyle\sum_{j=1}^{p} a_{ij} |\mathbf{A}_{ij}| (-1)^{i+j} & \text{if } p > 1, \end{cases} \qquad (4.70)$$

where $\mathbf{A}_{ij}$ is the $(p-1) \times (p-1)$ matrix obtained by deleting the ith row and jth column of $\mathbf{A}$, and i is any integer such that $1 \leq i \leq p$.

Definition 4A.13 When $|\mathbf{A}| = 0$, $\mathbf{A}$ is called *singular*, otherwise it is called *nonsingular*.

Property 4A.4 An inverse of a matrix $\mathbf{A}$ exists, if and only if $\mathbf{A}$ is nonsingular.

For any two square matrices $\mathbf{A}$ and $\mathbf{B}$ of the same dimensions, the determinant of their product is equal to the product of their determinants, that is,

$$|\mathbf{AB}| = |\mathbf{A}||\mathbf{B}|. \qquad (4.71)$$

For a diagonal matrix, its determinant is equal to the product of its diagonal elements.

Definition 4A.14 The maximum number of linearly independent rows (or columns) of a matrix $\mathbf{A}$ is called the *rank* of $\mathbf{A}$ and is denoted by Rank($\mathbf{A}$).

Property 4A.5 For any matrix $\mathbf{A}$, the number of linearly independent rows is equal to the number of linearly independent columns. This is why we can use either rows or columns in the definition of the rank.

Property 4A.6 For any matrices $\mathbf{A}$ and $\mathbf{B}$, we have

  a. $\text{Rank}(\mathbf{AA}^T) = \text{Rank}(\mathbf{A}^T\mathbf{A}) = \text{Rank}(\mathbf{A})$.
  b. $\text{Rank}(\mathbf{AB}) \leq \min\{\text{Rank}(\mathbf{A}), \text{Rank}(\mathbf{B})\}$, if $\mathbf{AB}$ exists.
  c. $\text{Rank}(\mathbf{A} + \mathbf{B}) \leq \text{Rank}(\mathbf{A}) + \text{Rank}(\mathbf{B})$.

Property 4A.7 For any square $p \times p$ matrix $\mathbf{A}$, the matrix $\mathbf{A}$ is singular, that is, $|\mathbf{A}| = 0$, if and only if $\text{Rank}(\mathbf{A}) < p$.

Eigenvalues and Eigenvectors of Matrices

A matrix $\mathbf{A}$ is said to have an eigenvalue λ and a corresponding eigenvector $\mathbf{x} \neq 0$, if $\mathbf{A}\mathbf{x} = \lambda\mathbf{x}$. One can show that a $p \times p$ square symmetric matrix $\mathbf{A}$ has p pairs of real-valued eigenvalues and corresponding normalized eigenvectors

$$\lambda_1, \mathbf{e}_1 \quad \lambda_2, \mathbf{e}_2 \quad \ldots \quad \lambda_p, \mathbf{e}_p, \tag{4.72}$$

that is, $\mathbf{A}\mathbf{e}_i = \lambda_i\mathbf{e}_i$ and $\mathbf{e}_i^T\mathbf{e}_i = 1$ for $i = 1, \ldots, p$. The eigenvectors can be chosen to be mutually orthogonal (perpendicular) and are unique (up to a constant) unless two or more eigenvalues are equal.

Spectral Decomposition of Matrices

The spectral decomposition of a $p \times p$ square symmetric matrix $\mathbf{A}$ is given by

$$\mathbf{A} = \lambda_1\mathbf{e}_1\mathbf{e}_1^T + \lambda_2\mathbf{e}_2\mathbf{e}_2^T + \cdots + \lambda_p\mathbf{e}_p\mathbf{e}_p^T, \tag{4.73}$$

where $\lambda_i, \mathbf{e}_i$, $i = 1, \ldots, p$, are the eigenvalues and normalized eigenvectors of $\mathbf{A}$. Let $\mathbf{P}$ be a matrix consisting of the eigenvectors $\mathbf{e}_i$, $i = 1, \ldots, p$, as columns, that is, $\mathbf{P} = [\mathbf{e}_1, \mathbf{e}_2, \ldots, \mathbf{e}_p]$. The spectral decomposition can then be written as

$$\mathbf{A} = \mathbf{P}\mathbf{\Lambda}\mathbf{P}^T, \tag{4.74}$$

where $\mathbf{\Lambda}$ is a diagonal matrix with λ_i, $i = 1, \ldots, k$, eigenvalues on the diagonal. The matrix $\mathbf{P}^T$ can be regarded as a matrix of transformation from the original system of coordinates to a new system of coordinates in which the transformation $\mathbf{A}$ becomes diagonal. The matrix $\mathbf{P} = \left(\mathbf{P}^T\right)^{-1}$ transforms back to the original system of coordinates.

Property 4A.8 The determinant of a square matrix $\mathbf{A}$ is equal to the product of its eigenvalues, that is,

$$|\mathbf{A}| = \prod_{i=1}^{p} \lambda_i. \tag{4.75}$$

Positive Definite Matrices

The expression $\mathbf{x}^T\mathbf{A}\mathbf{x}$, where $\mathbf{A}$ is a matrix and $\mathbf{x}^T = [x_1, x_2, \dots, x_n]$ is a vector of n variables, is called a *quadratic form* because it has only square terms x_i^2 and product terms $x_i x_j$.

Definition 4A.15 A symmetric matrix $\mathbf{A}$ (and its quadratic form $\mathbf{x}^T\mathbf{A}\mathbf{x}$) is said to be *nonnegative definite* if

$$\mathbf{x}^T\mathbf{A}\mathbf{x} \geq 0 \text{ for all } \mathbf{x} \in \mathbb{R}^n. \tag{4.76}$$

Definition 4A.16 A symmetric matrix $\mathbf{A}$ (and its quadratic form $\mathbf{x}^T\mathbf{A}\mathbf{x}$) is said to be *positive definite* if

$$\mathbf{x}^T\mathbf{A}\mathbf{x} > 0 \text{ for all } \mathbf{x} \neq \mathbf{0}. \tag{4.77}$$

Property 4A.9 A symmetric matrix $\mathbf{A}$ is nonnegative definite if and only if every eigenvalue of $\mathbf{A}$ is nonnegative.

Property 4A.10 A symmetric matrix $\mathbf{A}$ is positive definite if and only if every eigenvalue of $\mathbf{A}$ is positive.

A proof of both properties is given as Problem 4.10.

Property 4A.11 (Maximization Lemma). Let $\mathbf{A}$ be a positive definite matrix and $\mathbf{d}$ a given vector. Then for an arbitrary vector $\mathbf{x} \neq \mathbf{0}$, we have

$$\max_{\mathbf{x} \neq \mathbf{0}} \frac{\left(\mathbf{x}^T\mathbf{d}\right)^2}{\mathbf{x}^T\mathbf{A}\mathbf{x}} = \mathbf{d}^T\mathbf{A}^{-1}\mathbf{d}, \tag{4.78}$$

with the maximum attained when $\mathbf{x} = c\mathbf{A}^{-1}\mathbf{d}$ for any constant $c \neq \mathbf{0}$.

A Square Root Matrix

A square root of a matrix $\mathbf{A}$ is defined as a matrix $\mathbf{D}$ such that

$$\mathbf{D}\mathbf{D} = \mathbf{A}. \tag{4.79}$$

The matrix $\mathbf{D}$ is denoted by $\mathbf{A}^{1/2}$. If a matrix $\mathbf{A}$ is nonnegative definite, with the spectral decomposition

$$\mathbf{A} = \sum_{i=1}^{k} \lambda_i \mathbf{e}_i \mathbf{e}_i^T, \tag{4.80}$$

we can calculate the square root matrix $\mathbf{A}^{1/2}$ as follows:

$$\mathbf{A}^{1/2} = \sum_{i=1}^{k} \sqrt{\lambda_i} \mathbf{e}_i \mathbf{e}_i^T. \tag{4.81}$$

Using the matrix $\mathbf{P} = [\mathbf{e}_1, \mathbf{e}_2, \ldots, \mathbf{e}_k]$, we can also write

$$\mathbf{A}^{1/2} = \mathbf{P}\mathbf{\Lambda}^{1/2}\mathbf{P}^T. \tag{4.82}$$

In a similar fashion, the inverse matrix of a nonsingular matrix $\mathbf{A}$ can be obtained from the formula

$$\mathbf{A}^{-1} = \mathbf{P}\mathbf{\Lambda}^{-1}\mathbf{P}^T. \tag{4.83}$$

Definition 4A.17 The *trace* of a $p \times p$ square matrix $\mathbf{A} = \left[a_{ij} \right]$ is the sum of its diagonal elements, namely,

$$\text{Trace}(\mathbf{A}) = \sum_{j=1}^{p} a_{jj}. \tag{4.84}$$

Trace has the following cyclic property.

Property 4A.12 $\text{Trace}(\mathbf{ABC}) = \text{Trace}(\mathbf{BCA}) = \text{Trace}(\mathbf{CAB})$, provided the products of the matrices exist.

SUPPLEMENT 4B. RANDOM VECTORS AND MATRICES

We provide here some basic notation and properties of random vectors and matrices. When observing values of a single variable, we think of them as realizations of a random variable X representing a population model. If the observations come from a normal distribution, we would assume that X follows a normal distribution. When observing values of p variables, we would use multiple random variables $X_1, X_2, \ldots, X_p$ representing a population model. The variables can be organized into a random vector $\mathbf{X} = \left[X_1, X_2, \ldots, X_p \right]^T$. The vector of the population means of the p variables and the matrix of the population variances and covariances are defined as

$$\mu = E(\mathbf{X}) = \begin{bmatrix} E(X_1) \\ E(X_2) \\ \vdots \\ E(X_p) \end{bmatrix} \text{ and } \Sigma = \text{Var}(\mathbf{X}) = E\Big[(\mathbf{X}-\mu)(\mathbf{X}-\mu)^{\mathrm{T}}\Big]$$

$$= \begin{bmatrix} \sigma_{11} & \sigma_{12} & \cdots & \sigma_{1p} \\ \sigma_{21} & \sigma_{22} & \cdots & \sigma_{2p} \\ \vdots & \vdots & \ddots & \vdots \\ \sigma_{p1} & \sigma_{p2} & \cdots & \sigma_{pp} \end{bmatrix}_{p \times p}, \tag{4.85}$$

where $\sigma_{jk} = E\big[(X_j-\mu_j)(X_k-\mu_k)\big]$. The matrix $\Sigma = \text{Var}(\mathbf{X})$ is called the population variance–covariance (or covariance) matrix. The population mean vector μ and the population variance–covariance matrix Σ are the parameters that give a basic description of the joint multivariate distribution of the random vector $\mathbf{X}$.

Let us consider a nonrandom q by p matrix $\mathbf{A}$ and two p-dimensional random vectors $\mathbf{X}$ and $\mathbf{Y}$. We then have the following properties.

Property 4B.1

a. $E(\mathbf{AX}) = \mathbf{A}E(\mathbf{X})$.
b. $E(\mathbf{X} + \mathbf{Y}) = E(\mathbf{X}) + E(\mathbf{Y})$.
c. $\text{Var}(\mathbf{AX}) = \mathbf{A}\,\text{Var}(\mathbf{X})\mathbf{A}^T$.

The first two properties above are simple consequences of property (2.12), and the third one is a multivariate equivalent of property (2.32). The proof is given as Problem 4.12. Since a q by p matrix $\mathbf{A}$ represents a linear transformation from the p-dimensional space $\mathbb{R}^p$ to the q-dimensional space $\mathbb{R}^q$, $\mathbf{AX}$ is a q-dimensional random vector. Property 4B.1c tells us how the variability in $\mathbf{X}$ propagates through a linear transformation $\mathbf{A}$, resulting in the given variability in the vector $\mathbf{AX}$. This process is often referred to as *error propagation*.

Let $\mathbf{X}_1 = \begin{bmatrix} X_{11}, \ldots, X_{1p} \end{bmatrix}$ and $\mathbf{X}_2 = \begin{bmatrix} X_{21}, \ldots, X_{2q} \end{bmatrix}$ be two random vectors of dimensions p and q, respectively. Then we define the p by q covariance matrix $\text{Cov}(\mathbf{X}_1, \mathbf{X}_2)$ as a matrix of all possible covariances between all components of $\mathbf{X}_1$ and all components of $\mathbf{X}_2$. That is, $\text{Cov}(\mathbf{X}_1, \mathbf{X}_2) = \big[\text{Cov}(X_{1j}, X_{2k})\big]_{j=1,\ldots,p;k=1,\ldots,q}$. We can also write the covariance matrix using vector notation as

$$\text{Cov}(\mathbf{X}_1, \mathbf{X}_2) = E\big[(\mathbf{X}_1-E(\mathbf{X}_1))(\mathbf{X}_2-E(\mathbf{X}_2))^T\big], \tag{4.86}$$

which is analogous to a similar vector formula for the variance–covariance matrix in (4.85). We also have the following property (the proof is given as Problem 4.13).

Property 4B.2 $\text{Cov}(\mathbf{AX}_1, \mathbf{BX}_2) = \mathbf{A}\,\text{Cov}(\mathbf{X}_1, \mathbf{X}_2)\mathbf{B}^{\mathsf{T}}$.

This property can be thought of as propagation of covariance when the vectors $\mathbf{X}_1$ and $\mathbf{X}_2$ are subjected to linear transformations.

Property 4B.3 Any variance–covariance matrix is symmetric and nonnegative definite. It is usually positive definite, except for some degenerate distributions.

Property 4B.4 Any variance–covariance matrix has its square root matrix that can be calculated based on equation (4.82).

Sphering

It is sometimes desirable to decorrelate the distribution of a random vector $\mathbf{X}$. The variance–covariance matrix $\text{Var}(\mathbf{X}) = \mathbf{\Sigma}$ is usually positive definite, except for some degenerate distributions. If $\mathbf{\Sigma}$ is positive definite, we can calculate the inverse $\mathbf{\Sigma}^{-1/2}$ of the square root matrix. We can now decorrelate $\mathbf{X}$ by pre-multiplying it by $\mathbf{\Sigma}^{-1/2}$ in order to obtain

$$\mathbf{Y} = \mathbf{\Sigma}^{-1/2}\mathbf{X}. \tag{4.87}$$

We have $\text{Var}(\mathbf{Y}) = \mathbf{\Sigma}^{-1/2}\mathbf{\Sigma}\,\mathbf{\Sigma}^{-1/2} = \mathbf{I}$, which means that the components of $\mathbf{Y}$ are uncorrelated and standardized. This process is often called whitening or sphering. The term whitening is associated with white noise, which is why we prefer to use that term only in the context of noise. In many imaging applications, the variance–covariance matrix is calculated based on the whole data set, which includes not only noise but also other sources of variability such as the spatial variability. In such cases, we will call this process sphering. The name is justified by the fact that the ellipses of constant density in the normal distribution become circles when the variance–covariance matrix is the identity matrix $\mathbf{I}$ (see Section 5.7.1).

PROBLEMS

4.1. Let's assume that we have n pairs (x_i, y_i), $i = 1, \dots, n$, of values of two variables following the model (4.2) with Assumptions 4.1 and 4.2. Can we conclude that all realizations y_i, $i = 1, \dots, n$, of the response variable exhibit the normal distribution? That is, would a normal probability plot suggest normality of all y responses?

4.2. Show that the least-squares estimator $\mathbf{b}$ given by (4.20) is an unbiased estimator of $\boldsymbol{\beta}$ (Property 4.1a). *Hint*: Use the definition $\mathbf{b} = (\mathbf{X}^{\mathsf{T}}\mathbf{X})^{-1}\mathbf{X}^{\mathsf{T}}\mathbf{Y}$ and Property 4B.1a.

4.3. Show that the least-squares estimator $\mathbf{b}$ given by (4.20) has the variance–covariance matrix $\text{Var}(\mathbf{b}) = \sigma^2(\mathbf{X}^{\mathsf{T}}\mathbf{X})^{-1}$ (Property 4.1b). *Hint*: Use the definition $\mathbf{b} = (\mathbf{X}^{\mathsf{T}}\mathbf{X})^{-1}\mathbf{X}^{\mathsf{T}}\mathbf{Y}$ and Property 4B.1c.

4.4. Show that the variance–covariance matrix of the vector of residuals is given by $\mathrm{Var}(\mathbf{e}) = \sigma^2(\mathbf{I}-\mathbf{H})$ (formula (4.25)). *Hint*: Use Property 4B.1c.

4.5. Show that the mean squared error of prediction $\mathrm{MSE}(c) = E(Y-c)^2$ is minimized by $c = E(Y)$. *Hint*: Calculate the derivative of $\mathrm{MSE}(c)$ with respect to c.

4.6. Derive the formula in Property 4A.1 using geometry and formula (4.59).

4.7. Prove Property 4A.2 from Property 4A.1.

4.8. Calculate (by hand) eigenvalues and eigenvectors for the matrix $A = \begin{bmatrix} 1 & 4 \\ 4 & 5 \end{bmatrix}$. *Hint*: Follow these steps:
   a. The equation $\mathbf{Ax} = \lambda\mathbf{x}$ can be written as $(\mathbf{A}-\lambda\mathbf{I})\mathbf{x} = \mathbf{0}$.
   b. The system of equations $(\mathbf{A}-\lambda\mathbf{I})\mathbf{x} = \mathbf{0}$ has a nonzero solution $\mathbf{x} \neq 0$, only if the determinant of $(\mathbf{A}-\lambda\mathbf{I})$ is equal to zero.
   c. Find λ solutions to $|\mathbf{A}-\lambda\mathbf{I}| = 0$. These are eigenvalues.
   d. For each eigenvalue, find $\mathbf{x}$ such that $(\mathbf{A}-\lambda\mathbf{I})\mathbf{x} = \mathbf{0}$. This is the eigenvector associated with the eigenvalue. Was the value of $\mathbf{x}$ unique?

4.9. Find the spectral decomposition of the matrix $\mathbf{A}$ from Problem **4.8**.

4.10. Prove Properties 4A.9 and 4A.10. *Hint*: Use the spectral decomposition of $\mathbf{A}$ to represent $\mathbf{x}^{\mathrm{T}}\mathbf{Ax}$ as a linear combination (with λ_i coefficients) of squares.

4.11. Check if the matrix $\mathbf{A}$ from Problem 4.8 is positive definite. *Hint*: Use Property 4A.10.

4.12. Prove Property 4B.1.

4.13. Prove Property 4B.2.

4.14. Show that Property 4B.1c is a special case of Property 4B.2.

CHAPTER 5

Fundamentals of Multivariate Statistics

5.1 INTRODUCTION

In earlier chapters, we have seen examples where multiple variables were recorded on the same objects or experimental units. The traditional univariate statistical methods, such as those discussed in Chapter 3, can still be used on such data by analyzing one variable at a time. In Chapter 4, we have seen the use of multiple variables, where one of them, a response, was modeled as a function of other variables (predictors). One common feature of all models discussed in Chapter 4 was that the model uncertainty, expressed by the epsilon term, was described by a univariate distribution in the one-dimensional space ℝ of real numbers. Once a model is fitted to the data, we usually calculate the fitted values and then calculate the residuals as the differences between the observations and the fitted values. For the models discussed in Chapter 4, the residuals are scalars from the one-dimensional space ℝ of real numbers. These are the reasons why such models are not regarded as part of multivariate analysis.

Starting with this chapter, we are going to discuss multivariate methods that describe and visualize the simultaneous relationships among many variables. They are based on underlying probability models using multivariate probability distributions in order to deal with multiple response variables. When the residuals are calculated, they are multivariate vectors rather than scalars. The tools discussed here will be used in subsequent chapters for modeling of multivariate data. In order to understand multivariate methods, let us first look at a multivariate data set in the following example.

Example 5.1 Consider spectral reflectance Tiles Data as measured using an X-Rite Series 500 Spectrodensitometer. The data were collected by measuring each of 12 tiles in the BCRA II Series Calibration tiles. Each observation consists of 31 values of reflectance measured in 31 spectral bands over the spectral range from 400 to 700 nm at 10 nm increments. Each of the 12 tiles was measured four times for a total of

---

Statistics for Imaging, Optics, and Photonics, Peter Bajorski.
© 2012 John Wiley & Sons, Inc. Published 2012 by John Wiley & Sons, Inc.

Table 5.1   A Subset of the Whole Data Set Discussed in Example 5.1

Observation No.	W400	W410	W420	W430	W440	W450	W460	W470	W480	W490
1	3.15	3.59	4.27	5.15	6.27	7.79	10.04	13.44	18.21	25.31
2	3.15	3.59	4.28	5.15	6.25	7.79	10.03	13.43	18.18	25.27
3	3.14	3.56	4.24	5.13	6.25	7.80	10.04	13.47	18.26	25.44
4	3.13	3.55	4.24	5.13	6.25	7.79	10.03	13.46	18.26	25.44
5	2.37	2.15	1.94	1.72	1.52	1.34	1.20	1.08	0.99	0.92
6	2.32	2.08	1.88	1.67	1.47	1.29	1.15	1.04	0.95	0.89
7	2.34	2.09	1.89	1.68	1.48	1.31	1.16	1.05	0.96	0.89
8	2.34	2.12	1.90	1.69	1.48	1.31	1.17	1.05	0.96	0.89
9	4.94	5.90	4.75	3.27	2.38	1.96	1.85	1.92	2.23	2.93
10	4.97	5.91	4.77	3.27	2.38	1.97	1.85	1.93	2.24	2.94
11	4.94	5.94	4.77	3.27	2.38	1.97	1.84	1.93	2.24	2.94
12	5.00	5.96	4.77	3.28	2.39	1.97	1.84	1.93	2.24	2.94

[a]Twelve observations are shown with values on the first 10 variables from W400 to W490 representing the wavelengths from 400 to 490 nm.

48 multivariate observations. Table 5.1 shows a subset of the whole data set. Each row represents an observation, with the first four rows representing the four repeated measurements of the first tile, followed by four measurements of the second tile, and so on. Each column represents one spectral band, which is treated as a variable. So, the whole data set consists of 48 observations on 31 variables. All 48 observations are shown graphically in Figure 5.1 as 48 curves plotted as functions of the wavelength. The four repeated observations for a given tile are very close to each other, and consequently the curves overlap almost perfectly. Thus, in Figure 5.1, one can see only 12 distinct curves for the 12 different tiles.   □

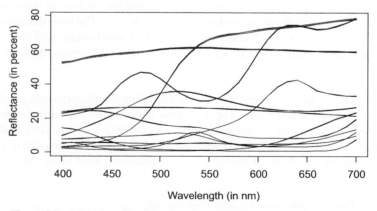

Figure 5.1   A function plot of 48 spectral curves as functions of the wavelength.

5.2 THE MULTIVARIATE RANDOM SAMPLE

In Chapters 2 and 3, we were explaining the duality, where the observations need to be treated sometimes as numbers and sometimes as random variables. The same type of statistical thinking applies to multivariate observations. When observing p different characteristics of a process or a population, we would describe them as p random variables or *components* $X_1, X_2, \ldots, X_p$. These components may correspond to the $p = 10$ spectral bands shown in Table 5.1. The components can be organized into a random vector $\mathbf{X} = [X_1, X_2, \ldots, X_p]^T$. (Supplements 4A and 4B provide information about the vector and matrix operations that will be used throughout this chapter.) When specific assumptions are made about the multivariate distribution of $\mathbf{X}$, we call it a population model of the observed characteristics. In Supplement 4B, we discussed some properties of such random vectors and their distributions. The population model will often be described by the population mean, or expected value, $\boldsymbol{\mu} = E(\mathbf{X})$ and the population variance–covariance matrix $\boldsymbol{\Sigma} = \mathrm{Var}(\mathbf{X})$.

In Chapter 3, we treated a sample as a collection of n independent random variables $X_1, X_2, \ldots, X_n$. In the multivariate case, we have n independent random vectors $\mathbf{X}_1, \mathbf{X}_2, \ldots, \mathbf{X}_n$, each following the same distribution of the random vector $\mathbf{X}$ described in the previous paragraph. These random vectors can be placed as rows into the following random matrix:

$$
\mathcal{X}_{(n \times p)} =
\begin{matrix}
X_1 & X_2 & \cdots & X_p \\
\hline
\end{matrix}
\begin{bmatrix}
X_{11} & X_{12} & \cdots & X_{1p} \\
X_{21} & X_{22} & \cdots & X_{2p} \\
\vdots & \vdots & \ddots & \vdots \\
X_{n1} & X_{n2} & \cdots & X_{np}
\end{bmatrix}
=
\begin{bmatrix}
\mathbf{X}_1^T \\
\mathbf{X}_2^T \\
\vdots \\
\mathbf{X}_n^T
\end{bmatrix},
\tag{5.1}
$$

where the vectors $\mathbf{X}_i^T$, $i = 1, \ldots, n$, are the row vectors representing the p-dimensional observations. The row of components $X_1, X_2, \ldots, X_p$ written above the horizontal line symbolizes the fact that the first column corresponds to values of the first component X_1, the second column corresponds to X_2, and so on. Table 5.1 is an example of a realization matrix of the random matrix $\mathcal{X}$ with $n = 12$ observations on $p = 10$ components.

We will use the descriptive statistics defined earlier in Chapter 2, but they will now be organized into appropriate vectors and matrices. To this end, we want to introduce the following multivariate notation. For the jth component (or column), we denote the sample mean and sample variance as

$$
\overline{X}_j = \frac{1}{n} \sum_{i=1}^n X_{ij} \quad \text{and} \quad s_j^2 = s_{jj} = \frac{1}{n-1} \sum_{i=1}^n (X_{ij} - \overline{X}_j)^2, \quad j = 1, 2, \ldots, p, \tag{5.2}
$$

respectively. The sample covariance between the jth and kth variables will be denoted as

$$s_{jk} = \frac{1}{n-1} \sum_{i=1}^{n} (X_{ij} - \overline{X}_j)(X_{ik} - \overline{X}_k), \quad j = 1, 2, \ldots, p \text{ and } k = 1, 2, \ldots, p. \quad (5.3)$$

Note that we use the lowercase notation for the covariances s_{jk} even though they are random variables in this context. The double notation for variance as $s_j^2 = s_{jj}$ reflects the tradition of denoting the variance as the square of the standard deviation, as well as the fact that the covariance of a variable with itself is equal to the variance. In order to have a scale-independent version of covariance, we define the sample correlation coefficient between two variables as

$$r_{jk} = \frac{s_{jk}}{\sqrt{s_{jj}}\sqrt{s_{kk}}} = \frac{\sum_{i=1}^{n} (X_{ij} - \overline{X}_j)(X_{ik} - \overline{X}_k)}{\sqrt{\sum_{i=1}^{n} (X_{ij} - \overline{X}_j)^2}\sqrt{\sum_{i=1}^{n} (X_{ik} - \overline{X}_k)^2}}, \quad j = 1, 2, \ldots, p$$

and $k = 1, 2, \ldots, p$. $\qquad\qquad\qquad\qquad\qquad\qquad\qquad\qquad\qquad (5.4)$

This is the sampling version of the population correlation coefficient defined in formula (2.34). We assume here that the sample variances s_{jj}, $j = 1, 2, \ldots, p$, are positive. The descriptive statistics introduced here can be arranged into vectors and matrices as follows:

$$\overline{\mathbf{X}} = \begin{bmatrix} \overline{X}_1 \\ \overline{X}_2 \\ \vdots \\ \overline{X}_p \end{bmatrix}, \quad \underset{(p \times p)}{\mathbf{S}} = \begin{bmatrix} s_{11} & s_{12} & \cdots & s_{1p} \\ s_{21} & s_{22} & \cdots & s_{2p} \\ \vdots & \vdots & \ddots & \vdots \\ s_{p1} & s_{p2} & \cdots & s_{pp} \end{bmatrix}, \quad \underset{(p \times p)}{\mathbf{R}} = \begin{bmatrix} 1 & r_{12} & \cdots & r_{1p} \\ r_{21} & 1 & \cdots & r_{2p} \\ \vdots & \vdots & \ddots & \vdots \\ r_{p1} & r_{p2} & \cdots & 1 \end{bmatrix},$$

$$(5.5)$$

where $\overline{\mathbf{X}}$ is the p-dimensional vector of sample means, $\mathbf{S}$ is the p by p sample variance–covariance matrix (also called the covariance matrix), and $\mathbf{R}$ is the p by p sample correlation matrix. Instead of the previous element-by-element formulas 5.2 and 5.3, we can also use the following vector calculations:

$$\overline{\mathbf{X}} = \frac{1}{n} \sum_{i=1}^{n} \mathbf{X}_i \quad \text{and} \quad \mathbf{S} = \frac{1}{n-1} \sum_{i=1}^{n} (\mathbf{X}_i - \overline{\mathbf{X}})(\mathbf{X}_i - \overline{\mathbf{X}})^{\mathrm{T}}. \quad (5.6)$$

It is often convenient to replace formulas (5.6) with direct calculations on the matrix $\mathcal{X}$. Let $\mathbf{1}_n$ be an n-dimensional vector with all coordinates equal to 1. With this notation, we have the following useful formulas:

$$\overline{\mathbf{X}} = \frac{1}{n}\mathscr{X}^{\mathrm{T}}\mathbf{1}_n, \qquad \mathscr{X}_c = \mathscr{X} - \mathbf{1}_n \cdot \overline{\mathbf{X}}^{\mathrm{T}}, \qquad \mathbf{S} = \frac{1}{n-1}\mathscr{X}_c^{\mathrm{T}}\mathscr{X}_c, \qquad (5.7)$$

where $\mathscr{X}_c$ is the matrix of the mean-centered values of all variables obtained by subtracting the sample mean of a given variable from all of its observations.

Another useful matrix is a p by p diagonal matrix $\mathbf{D}$ of variances s_{ii}, $i = 1, \ldots, p$, on the diagonal. Its square root $\mathbf{D}^{1/2}$ is the *sample standard deviation matrix*

$$\mathbf{D}^{1/2}_{(p \times p)} = \begin{bmatrix} \sqrt{s_{11}} & 0 & . & . & . & 0 \\ 0 & \sqrt{s_{22}} & 0 & . & . & 0 \\ . & & 0 & . & . & . \\ . & & & . & . & . \\ . & & & . & . & 0 \\ 0 & 0 & . & . & 0 & \sqrt{s_{pp}} \end{bmatrix}. \qquad (5.8)$$

We can now establish the following relationship between the sample variance–covariance matrix and the correlation matrix:

$$\mathbf{S} = \mathbf{D}^{1/2}\mathbf{R}\mathbf{D}^{1/2} \qquad (5.9)$$

(see Problem 5.7 for a hint on derivation). Since we assume here that the variances s_{jj}, $j = 1, 2, \ldots, p$, are positive, we can also calculate the inverse matrix $\mathbf{D}^{-1/2}$. By pre-multiplying and post-multiplying both sides of (5.9) by $\mathbf{D}^{-1/2}$, we obtain the following matrix formula for the correlation matrix:

$$\mathbf{R} = \mathbf{D}^{-1/2}\mathbf{S}\mathbf{D}^{-1/2}. \qquad (5.10)$$

Since each element of the sample $\mathbf{X}_1, \mathbf{X}_2, \ldots, \mathbf{X}_n$ follows the same distribution, we have $E(\mathbf{X}_i) = \boldsymbol{\mu}$ for all $i = 1, \ldots, n$. It is then easy to see that $E(\overline{\mathbf{X}}) = \boldsymbol{\mu}$, which means that $\overline{\mathbf{X}}$ is an unbiased estimator of $\boldsymbol{\mu}$. One can show that the variance–covariance matrix of $\overline{\mathbf{X}}$ is equal to $(1/n)\boldsymbol{\Sigma}$, which is a multivariate equivalent of formula (2.42). One can also formulate multivariate equivalents of the laws of large numbers discussed in Section 2.7. For example, $\overline{\mathbf{X}}$ approaches $\boldsymbol{\mu}$ as n tends to infinity, and its distribution can be approximated by a multivariate normal distribution that we will define in Section 5.7.

When treating samples as sets of numbers rather than random variables, we are mostly going to use lowercase letters. However, for matrices, we still want to use capital letters. The data matrix like the one described in Table 5.1 will be denoted by $\mathscr{X}$.

Example 5.2 This is a continuation of Example 2.4, and we again use a subset of Printing Data explained in Appendix B. There are eight cyan patches (at maximum gradation) on each printed page. Those patches are treated as observations here. This time, we use spectral reflectances of the cyan patches measured in 31 spectral bands, treated as variables, from 400 to 700 nm. For each patch, we have one 31-dimensional observation vector (or spectral curve) that we can denote by $\mathbf{x}_i$, $i = 1, \ldots, 8$. The eight observation vectors $\mathbf{x}_1, \mathbf{x}_2, \ldots, \mathbf{x}_8$ can be placed into an 8×31 matrix $\mathscr{X}$. This is a data set representing just one page. However, there are 21 different pages—three pages printed immediately after calibration, and then 18 pages printed after 14 hours of idle time. We will need to perform our multivariate calculations 21 times.

For each page, we can calculate the mean vector of the eight spectra. We denote it as $\bar{\mathbf{x}}_k$, where $k = 1, \ldots, 21$ is the index for the kth page. In order to investigate the variability of patches within each page, we can calculate the sample variance–covariance matrices $\mathbf{S}_k$, $k = 1, \ldots, 21$, each based on the eight observations (patches) from a given page. Each page is now described by the mean vector $\bar{\mathbf{x}}_k$ and the sample variance–covariance matrix $\mathbf{S}_k$. In order to represent this information graphically, the resulting 21 mean vector spectra are shown in Figure 5.2. All curves overlap considerably, which does not give us much information about page-to-page variability. All we can see in the plot is the shape of the spectral characteristic of cyan.

In order to create a more interesting graph, we can calculate the overall mean $\bar{\bar{\mathbf{x}}} = \sum_{k=1}^{21} \bar{\mathbf{x}}_k$ of the page mean vectors $\bar{\mathbf{x}}_k$ shown in Figure 5.2. We can then subtract the overall mean from all groups in order to calculate the mean deviations $\mathbf{d}_k = \bar{\mathbf{x}}_k - \bar{\bar{\mathbf{x}}}$, $k = 1, \ldots, 21$, shown in Figure 5.3, where the three solid lines represent the three pages printed immediately after the calibration. The dashed line represents the first page printed after the idle period, and the dotted lines represent the remaining pages. We can see that the first four printed pages are quite different from the remaining pages.

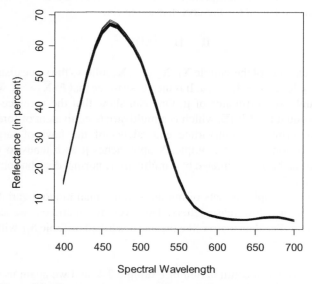

Figure 5.2 Twenty-one mean spectral curves discussed in Example 5.2.

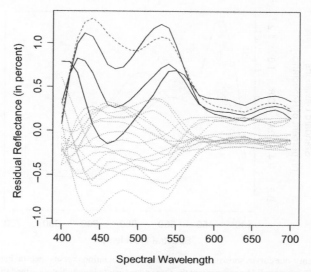

Figure 5.3 Twenty-one curves showing the spectral mean residual reflectances (in percent) as discussed in Example 5.2. The three solid lines represent the three pages printed immediately after calibration, and the dashed line represents the first page printed right after 14 h of idle time.

Some readers might be tempted to subtract the first-page mean instead of the overall mean. Such an approach is not consistent with statistical thinking and results in more noisy data.

The 21 matrices S_k, $k = 1, \ldots, 21$, of dimensions 31 by 31 are not easy to display, but we can concentrate on the variability within each band and consider the sample standard deviations, that is, the square roots of the diagonal elements of S_k, $k = 1, \ldots, 21$. Those values are plotted in Figure 5.4 as 21 curves—each curve being a function of 31 spectral bands. The general magnitudes of variability in the first three pages (plotted as solid lines in Figure 5.4) are not much different from those in the other pages. However, the pattern as a function of the spectral bands is quite different from the consistent pattern for the other pages.

Investigation of all patterns discussed in this example is somewhat difficult due to the high dimensionality of data. If each curve could be characterized with one or two numbers, we could more easily see patterns in many of such curves. The tools for the dimensionality reduction will be discussed in Chapter 7. Some insight can also be obtained from Problem 5.1. □

5.3 MULTIVARIATE DATA VISUALIZATION

In Section 2.3, we discussed scatter plots for representing two-dimensional data. When dealing with multiple variables, we can create a *scatter plot matrix*, which is a matrix of scatter plots for all possible pairs of variables as shown in Figure 5.5. This graphical tool is explained in the following example.

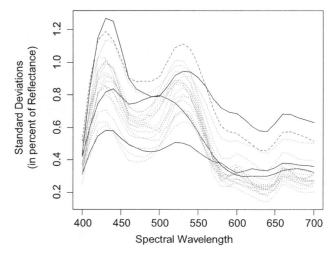

Figure 5.4 Twenty-one curves showing the sample standard deviations versus spectral bands as discussed in Example 5.2. The three solid lines represent the three pages printed immediately after calibration, and the dashed line represents the first page printed right after 14 h of idle time.

Example 5.3 This is a continuation of Example 2.5, where we considered one RGB image (shown in Figure 2.14) from the Eye Tracking data set explained in Appendix B. The scatter plot matrix in Figure 5.5 represents three variables—the intensities in the three channels: Red, Green, and Blue. The middle panel in the first row shows the scatter plot (previously shown in Figure 2.15) of Red versus Green values. The variable Red is shown on the vertical axis in all plots in the first row and on the horizontal axis in all plots

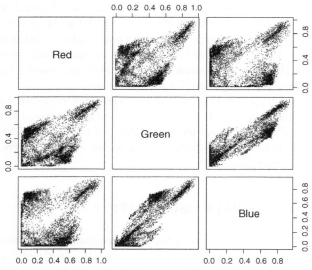

Figure 5.5 A scatter plot matrix of Red, Green, and Blue intensities in 16,384 pixels of an RGB image used in Example 5.3.

in the first column. In the same fashion, Green is on the vertical axis in all plots in the second row and on the horizontal axis in all plots in the second column. Similarly, Blue is associated with the last row and the last column. The bottom left panel is a scatter plot of Blue versus Red, and the top right panel is a scatter plot of Red versus Blue. The two plots are mirror images of each other, and they both might be helpful, even though they contain the same information. In this case, they both look almost identical because they are symmetric with respect to the diagonal. The two plots (mirror images) look different when plotting Green versus Blue. Due to some discreteness of data, we used the jitter, as discussed in Example 2.5. □

Another way to represent multivariate data is a *color matrix*, where small color patches are placed in a pattern of rows and columns representing observations and variables (or vice versa) as shown in Figure 5.6 and discussed in Example 5.4 below. The patch color represents the value based on several possible scales. Figure 5.6 uses a simple gray scale, but better results can be achieved with other scales shown in Figure 5.7.

Example 5.4 This is a follow-up on Example 5.2. In Figure 5.3, we saw four curves with values larger than those of the remaining 17 curves. Since the curves represent pages that were printed in a given order, it would be interesting to check for potential time-order related patterns. However, it is difficult to represent the time order of so many curves in that plot. To verify such trends, we can use the color matrix shown in Figure 5.6, where each row represents one curve (or one printed page), and the columns represent the spectral bands. A darker color indicates a larger value. There is an increasing trend in the first four pages (in Bands 3–15), where we see lighter color in the first two pages and darker colors in the next two. For Pages 5–21, we cannot see any specific trends over time. Instead, the values oscillate up and down (lighter and darker shades oscillate). □

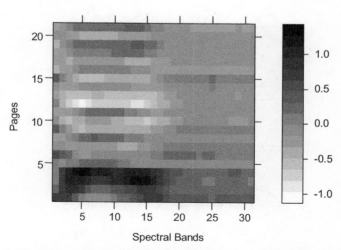

Figure 5.6 A color matrix for spectral mean residual reflectances (in percent) shown in Figure 5.3 and discussed in Example 5.4. A darker color indicates a larger value.

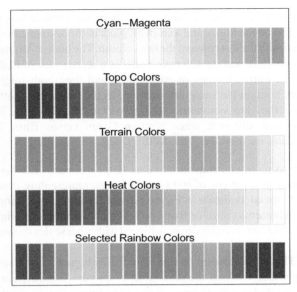

Figure 5.7   Some potential color scales that can be used for representation of quantitative data.

Another method for visualization of multivariate data is to create a symbol for each observation that would express the values of multiple variables through various features of the symbol. One example is a *star plot*, where each ray of the star is plotted with the length proportional to the value of a given variable as shown in Figure 5.8 and discussed in the following example.

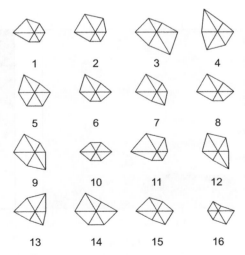

Figure 5.8   A star plot of ratings of six print-on-demand books based on 16 observers as discussed in Example 5.5.

Example 5.5 Consider Print-on-Demand Data explained in Appendix B. Sixteen observers rated overall image quality of six print-on-demand books on a scale from 1 to 5 (with ratings being 1 = very low satisfaction, 5 = very high satisfaction). Ratings for a given book are treated as values of one variable, for a total of six variables for six books. The data set consists of 16 six-dimensional observation vectors. Figure 5.8 shows a star plot of the data. Each star represents one observer. The horizontal ray to the right shows the value of the first variable, or the first book rated, and then in the counterclockwise direction, the rays show the subsequent variables. The shape of a given star tells us which books were favored by a given person. For example, Observer 1 favored Book 4, and Observer 3 favored Book 6. Stars 5 and 6 have similar shapes, which means that those observers have similar preferences.

The observers were also asked how much they would be willing to pay for this quality of book as a memento of the observer's vacation. For each observer, an average of the six prices for six books was calculated and recorded as the Vacation Price variable. We also know the age of each observer. Based on two quantitative variables, Vacation Price and Age, we can create a scatter plot, where each observation is represented as a star from Figure 5.8. Such a scatter plot is shown in Figure 5.9, where eight variables are represented—two variables represented by the system of coordinates and six variables by the star rays. Using that plot, we can determine if the observers of similar age have similar book preferences, and how much they are willing to pay for the books. For example, Observers 2 and 11, marked in Figure 5.9, are similar in their Ratings (similar stars) as well as in Vacation Price and Age (because the stars are close to each other). □

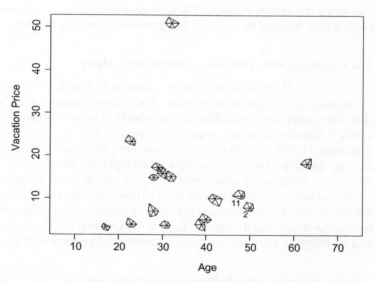

Figure 5.9 A scatter plot based on Example 5.5 data showing eight variables—two variables represented by the system of coordinates and six variables by the star rays. The same stars are also shown in Figure 5.8.

When dealing with three-dimensional data, we could use a three-dimensional scatter plot, but such plots are difficult to represent on a two-dimensional display. With dynamic graphics, where we can rotate the data, some feel for the third dimension can be achieved.

5.4 THE GEOMETRY OF THE SAMPLE

Geometric intuition plays an important role in understanding data and their statistical modeling. This is why visualization of data is an important part of statistics. In p-dimensional data, we often think of each observation as a point in the p-dimensional space $\mathbb{R}^p$. The whole data set is thought of as a cloud of points in that space. For $p = 2$, a scatter plot shows the whole data set as a cloud of points on a plane. The geometric interpretation in the p-dimensional space $\mathbb{R}^p$ will be important throughout this book. However, in this section, we are going to discuss a different type of geometric interpretation. Specifically, instead of an observation being represented as a point, we will represent each variable as a point or vector.

In Section 5.2, we organized the whole data set into an n by p nonrandom matrix $\mathscr{X}$, where the p columns represented p variables. We will now denote those columns as n-dimensional vectors $\mathbf{y}_1, \mathbf{y}_2, \ldots, \mathbf{y}_p$. Hence, the data matrix can be written as $\mathscr{X} = \begin{bmatrix} \mathbf{y}_1 & \mathbf{y}_2 & \cdots & \mathbf{y}_p \end{bmatrix}$. The vectors $\mathbf{y}_i$, $i = 1, \ldots, p$, can be represented as points in the n-dimensional space $\mathbb{R}^n$. The number of observations n is usually fairly large, so it may seem challenging to try to find interpretations in such high-dimensional space. One fact that will help here is that any p vectors are embedded in a p-dimensional subspace of $\mathbb{R}^n$. For example, when analyzing the first two variables, we can concentrate on the two vectors $\mathbf{y}_1$ and $\mathbf{y}_2$. The two vectors generate a two-dimensional plane, which makes it easier to visualize the relationship between the two vectors.

5.4.1 The Geometric Interpretation of the Sample Mean

Let $\mathbf{v} = [x_1, x_2, \ldots, x_n]^T$ be one of the column vectors in $\mathbf{X}$. That is, we can think of $x_1, x_2, \ldots, x_n$ as a single sample on one variable. If we want to characterize the sample by using just one number, say c, what should this number be? One way to think about this question in the n-dimensional space is to consider a vector $c \cdot \mathbf{1}_n = [c, \ldots, c]^T$, where $\mathbf{1}_n$ is an n-dimensional vector with all coordinates equal to 1 and c is an arbitrary scalar. Characterizing the sample by a single scalar c is like replacing the set $x_1, x_2, \ldots, x_n$ with a set $c, \ldots, c$ of repeated values of c. Hence, it would make sense to make $x_1, x_2, \ldots, x_n$ and $c, \ldots, c$ as close to each other as possible. This can be done by an orthogonal projection of $\mathbf{v}$ on $\mathbf{1}_n$ because the result of the projection is defined as the point from the one-dimensional space $\{c \cdot \mathbf{1}_n : c \in \mathbb{R}\}$ such that it is the closest point to $\mathbf{v}$. Based on Property 4A.1, we can write an orthogonal projection of $\mathbf{v}$ on $\mathbf{1}_n$ as

$$\text{Proj}_{\mathbf{1}_n}(\mathbf{v}) = \frac{\mathbf{v}^T \mathbf{1}_n}{\mathbf{1}_n^T \mathbf{1}_n} \mathbf{1}_n = \frac{\sum_{i=1}^n x_i}{n} \mathbf{1}_n = \overline{x} \mathbf{1}_n. \tag{5.11}$$

We can say informally that the sample mean $\bar{x}$ is a projection of the sample vector on $\mathbf{1}_n$ (more precisely, $\bar{x}\mathbf{1}_n$ is the projection). Clearly, $c = \bar{x}$ is the best single number that characterizes the sample in the sense of the Euclidean distance closeness in the $\mathbb{R}^n$ space. That is, $c = \bar{x}$ is a number that minimizes the mean square error of prediction (MSE)

$$\text{MSE}(c) = \frac{1}{n} \sum_{i=1}^{n} (x_i - c)^2 \tag{5.12}$$

when we try to predict a sample value by using a single number c (see also Problem 4.5).

Example 5.6 Consider a small data set of three observations on two variables given by the following matrix:

$$\mathscr{X} = \begin{bmatrix} 1 & 5 \\ 2 & -3 \\ 6 & -2 \end{bmatrix}. \tag{5.13}$$

When concentrating on the first variable, we have $\mathbf{v} = \begin{bmatrix} 1 & 2 & 6 \end{bmatrix}^{\text{T}}$, $\bar{x} = (1 + 2 + 6)/3 = 3$, and

$$\text{Proj}_{1_n}(\mathbf{v}) = 3 \cdot \mathbf{1}_n = \begin{bmatrix} 3 \\ 3 \\ 3 \end{bmatrix}. \tag{5.14}$$

$\square$

5.4.2 The Geometric Interpretation of the Sample Standard Deviation

Once we calculate the sample mean $\bar{x}$ from the sample defined by the vector $\mathbf{v}$ (see the previous section for notation), we can remove its impact by calculating a vector of deviations (or residuals) $\mathbf{d} = \mathbf{v} - \bar{x} \cdot \mathbf{1}_n$. The deviation vector $\mathbf{d}$ contains the remaining information (about variability) in the sample after the location parameter ($\bar{x}$) information is taken into account. Since $\bar{x} \cdot \mathbf{1}_n$ is an orthogonal projection of $\mathbf{v}$, it is clear that $\mathbf{d}$ is orthogonal to $\bar{x} \cdot \mathbf{1}_n$ (see Figure 5.10 and Problem 5.2). This is related to the fact that the coordinates of $\mathbf{d}$ sum up to 0 (see Problem 2.2).

We can say that $\mathbf{v} = \bar{x} \cdot \mathbf{1}_n + \mathbf{d}$ is a partitioning of $\mathbf{y}$ into two orthogonal components $\bar{x} \cdot \mathbf{1}_n$ and $\mathbf{d}$. This geometric interpretation is also helpful in explaining the concept of degrees of freedom. We say that there are n degrees of freedom in the original sample vector $\mathbf{v}$ because the sample elements can be anywhere in the n-dimensional space $\mathbb{R}^n$. On the other hand, the deviation vector $\mathbf{d}$ is constrained to an $(n-1)$-dimensional subspace orthogonal to $\mathbf{1}_n$. This is why we say that $\mathbf{d}$ has $(n-1)$ degrees of freedom and $\bar{x} \cdot \mathbf{1}_n$ has one degree of freedom. We can also say that we lose one degree of freedom in $\mathbf{d}$ for the constraint that the total of all its coordinates equals zero.

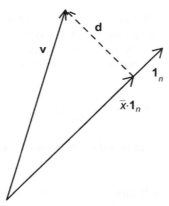

Figure 5.10 The vector $\bar{x} \cdot \mathbf{1}_n$ shown as an orthogonal projection of $\mathbf{v}$ on $\mathbf{1}_n$. The deviation vector $\mathbf{d}$ is orthogonal to $\mathbf{1}_n$.

Example 5.6 (cont.) For the previously used sample $\mathbf{v} = \begin{bmatrix} 1 & 2 & 6 \end{bmatrix}^{\mathrm{T}}$, we obtain $\mathbf{d} = \begin{bmatrix} -2 & -1 & 3 \end{bmatrix}^{\mathrm{T}}$, which is perpendicular (orthogonal) to $3 \cdot \mathbf{1}_n$ (or $\mathbf{1}_n$) because $\mathbf{d} \cdot (\bar{x} \cdot \mathbf{1}_n) = 3 \cdot [(-2) + (-1) + 3] = 0$. The vector $\mathbf{v}$ has three degrees of freedom, while $\mathbf{d}$ has two degrees of freedom. □

We can now calculate the sample variance as

$$s^2 = \frac{1}{n-1} \sum_{i=1}^{n} (x_i - \bar{x})^2 = \frac{1}{n-1} \mathbf{d}^{\mathrm{T}} \mathbf{d} = \frac{1}{n-1} \|\mathbf{d}\|^2, \qquad (5.15)$$

where $\|\mathbf{d}\|$ stands for the length of $\mathbf{d}$. This means that the sample variance is proportional to the squared length of $\mathbf{d}$. Consequently, the sample standard deviation is proportional to the length of $\mathbf{d}$, that is,

$$s = \frac{1}{\sqrt{n-1}} \|\mathbf{d}\|. \qquad (5.16)$$

5.4.3 The Geometric Interpretation of the Sample Correlation Coefficient

Let us now consider two variables with their respective samples represented by the jth and kth columns $\mathbf{v}_j$ and $\mathbf{v}_k$ of $\mathcal{X}$ defined in 5.1. The sample correlation coefficient of the two variables can then be written as

$$r_{jk} = \frac{s_{jk}}{\sqrt{s_{jj}}\sqrt{s_{kk}}} = \frac{\sum_{i=1}^{n}(x_{ij} - \bar{x}_j)(x_{ik} - \bar{x}_k)}{\sqrt{\sum_{i=1}^{n}(x_{ij} - \bar{x}_j)^2}\sqrt{\sum_{i=1}^{n}(x_{ik} - \bar{x}_k)^2}} = \frac{\mathbf{d}_j^{\mathrm{T}}\mathbf{d}_k}{\sqrt{\mathbf{d}_j^{\mathrm{T}}\mathbf{d}_j}\sqrt{\mathbf{d}_k^{\mathrm{T}}\mathbf{d}_k}}$$

$$= \frac{\|\mathbf{d}_j\| \cdot \|\mathbf{d}_k\|\cos(\theta_{jk})}{\|\mathbf{d}_j\| \cdot \|\mathbf{d}_k\|} = \cos(\theta_{jk}), \qquad (5.17)$$

and

$$s_{jk} = \sqrt{s_{jj}}\sqrt{s_{kk}}\cos(\theta_{jk}), \tag{5.18}$$

where θ_{jk} is the angle between the vectors $\mathbf{d}_j$ and $\mathbf{d}_k$. This means that for perfectly correlated variables, the respective deviation vectors are collinear. On the other hand, for uncorrelated variables ($r_{jk} = 0$), the deviation vectors are orthogonal. This geometric interpretation allows for a better understanding of mutual correlations among variables such as those discussed in Problem 5.3.

Example 5.6 (cont.)   For the previously used data matrix $\mathscr{X}$, we obtain the deviations of the two variables as $\mathbf{d}_1 = \begin{bmatrix} -2 & -1 & 3 \end{bmatrix}^T$ and $\mathbf{d}_2 = \begin{bmatrix} 5 & -3 & -2 \end{bmatrix}^T$. The correlation between the two variables is calculated as

$$r_{12} = \frac{\mathbf{d}_1^T\mathbf{d}_2}{\sqrt{\mathbf{d}_1^T\mathbf{d}_1}\sqrt{\mathbf{d}_2^T\mathbf{d}_2}} = \frac{-13}{\sqrt{14}\sqrt{38}} = -0.5636. \tag{5.19}$$

The angle between the two deviation vectors is equal to $\theta_{12} = 124.3° = 2.1695$ rad because $\cos(\theta_{12}) = -0.5636$. $\square$

5.5   THE GENERALIZED VARIANCE

The sample variance–covariance matrix $\mathbf{S}$ defined in equations (5.5) and (5.6) describes variability in multivariate data. However, for large dimensionalities p, especially prevalent in imaging applications, the matrix $\mathbf{S}$ is very large, and it is difficult to interpret it as a whole matrix. In statistics, we always try to summarize the information in data with a small number of summary or descriptive statistics. Here, we will try to summarize the information contained in $\mathbf{S}$. We will discuss two ways to do this—by defining the generalized sample variance and the total variability.

Definition 5.1   The *generalized sample variance* (GSV) is defined as the determinant of the sample variance–covariance matrix $\mathbf{S}$ denoted as $|\mathbf{S}|$.

Based on Property 4A.8, we can calculate the GSV as a product of the eigenvalues of $\mathbf{S}$, that is,

$$|\mathbf{S}| = \prod_{i=1}^{p} \lambda_i. \tag{5.20}$$

The following example demonstrates how the GSV can be interpreted.

Example 5.7   Consider a simple case of two-dimensional data, where the sample variance of both variables is 1, and their sample correlation coefficient is equal to r, where $-1 \le r \le 1$. This means that $\mathbf{S} = \begin{bmatrix} 1 & r \\ r & 1 \end{bmatrix}$ in our multivariate notation.

The GSV is then equal to $|\mathbf{S}| = 1 - r^2$. When both variables are perfectly correlated, we have $r = \pm 1$, and the matrix $\mathbf{S}$ is singular, that is, $|\mathbf{S}| = 0$. For two perfectly correlated variables, the data are effectively one dimensional because all information is contained in one variable in the sense that the values of the second variable can be precisely calculated from the first variable. In a scatter plot, perfectly correlated data would line up along a straight line. We can say that there is no two-dimensional variability, which is expressed by the GSV equal to zero. Clearly, we get the maximum GSV for an uncorrelated case with $r = 0$. Figure 5.11 shows four scatter plots in order to demonstrate examples of data with various values of r and the associated values of $|\mathbf{S}|$. Each scatter plot is based on $n = 1000$ data points. We would say that the GSV measures the amount of two-dimensional variability. □

We will now discuss a case more general than the one discussed in Example 5.7. Consider two-dimensional data with an arbitrary sample variance–covariance matrix $\mathbf{S}$, which can always be represented as

$$\mathbf{S} = \begin{bmatrix} s_1^2 & rs_1s_2 \\ rs_1s_2 & s_2^2 \end{bmatrix}, \tag{5.21}$$

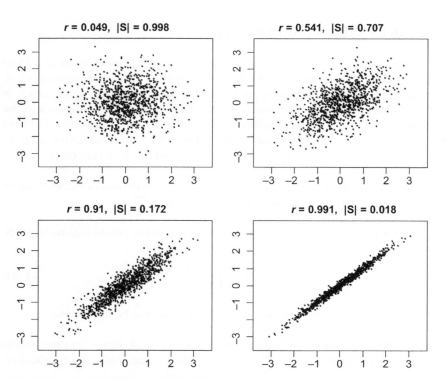

Figure 5.11 Scatter plots of four sets of data with various values of the sample correlation coefficient r and the associated values of the generalized sample variance $|\mathbf{S}|$ as discussed in Example 5.7.

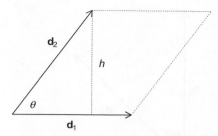

Figure 5.12 Two deviation vectors $\mathbf{d}_1$ and $\mathbf{d}_2$ defining a parallelogram.

where s_i, $i = 1, 2$, are the sample standard deviations of the two variables, and r is the sample correlation coefficient of the two variables. From the n-dimensional geometry, we know that $s_i^2 = \|\mathbf{d}_i\|^2/(n-1)$ and $r = \cos(\theta)$, where $\mathbf{d}_i$ is the vector of deviations for the ith variable and θ is the angle between $\mathbf{d}_1$ and $\mathbf{d}_2$ (see Figure 5.12). Hence, the GSV can be calculated as $|\mathbf{S}| = s_1^2 s_2^2 (1 - r^2) = \|\mathbf{d}_1\|^2 \|\mathbf{d}_2\|^2 (1 - \cos^2 \theta)/(n-1)^2$. Since the height of the parallelogram shown in Figure 5.12 is equal to $h = \|\mathbf{d}_2\|\sin(\theta)$, we obtain $|\mathbf{S}| = \|\mathbf{d}_1\|^2 h^2/(n-1)^2 = (\text{Area})^2/(n-1)^2$, where Area denotes the area of the parallelogram generated by the deviation vectors $\mathbf{d}_1$ and $\mathbf{d}_2$. For highly correlated variables, the angle θ is small, which means a small Area for given lengths of the deviation vectors, and consequently small GSV $|\mathbf{S}|$.

These interpretations can be generalized to higher dimensions. For p variables, the p deviation vectors generate a p-dimensional parallelotope. The parallelogram in Figure 5.12 is an example of a two-dimensional parallelotope defined by the deviation vectors $\mathbf{d}_1$ and $\mathbf{d}_2$. For $p = 3$, a parallelotope is called a parallelepiped. The formula for the GSV of p-dimensional data takes the form

$$|\mathbf{S}| = \frac{(\text{Volume})^2}{(n-1)^p}, \tag{5.22}$$

where Volume is the volume of a p-dimensional parallelotope defined by the p deviation vectors as its edges.

We conclude that the GSV depends not only on the lengths of deviation vectors (i.e., standard deviations of variables), but also on the angles among them (i.e., correlations). The generalized variance gets larger for uncorrelated variables and smaller for correlated variables. However, for p-dimensional data, we need to take into account more complex relationships than those that can be described by pairwise correlations among variables. We can define p deviation vectors, which in general span a p-dimensional subspace in the $(n-1)$-dimensional subspace of $\mathbb{R}^n$ orthogonal to $\mathbf{1}_n$ (we assume here that $p \leq n - 1$). However, if the p-dimensional data are, in fact, confined to a $(p-1)$-dimensional subspace (we call this condition *perfect multi-collinearity*), then the p deviation vectors span a subspace of only up to $(p-1)$ dimensions. Consequently, the p-dimensional volume used in formula 5.22 is zero and the GSV is also zero. Such multicollinearity may not be obvious from the pairwise correlations among variables as demonstrated by the following example.

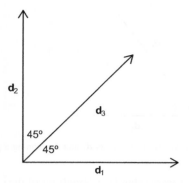

Figure 5.13   Three deviation vectors $\mathbf{d}_1$, $\mathbf{d}_2$, and $\mathbf{d}_3$ used in Example 5.8.

Example 5.8   Consider $n \geq 3$ observations on three variables—each with the same variance, let's say, equal to 1. Assume further that the three deviation vectors are positioned on a two-dimensional plane in the way shown in Figure 5.13, that is, the angle between $\mathbf{d}_3$ and any of the first two deviation vectors is $45°$ and the deviation vectors $\mathbf{d}_1$ and $\mathbf{d}_2$ are orthogonal. Clearly, the vector $\mathbf{d}_3$ is a linear combination of $\mathbf{d}_1$ and $\mathbf{d}_2$, which is an example of perfect multicollinearity. Since the sample variances are 1 and $\cos(45°) = 2^{-1/2}$, we obtain $s_{13} = s_{23} = 2^{-1/2}$ from equation 5.18. The sample variance–covariance matrix is then equal to

$$
\mathbf{S} = \begin{bmatrix} 1 & 0 & 2^{-1/2} \\ 0 & 1 & 2^{-1/2} \\ 2^{-1/2} & 2^{-1/2} & 1 \end{bmatrix}.
\tag{5.23}
$$

The eigenvalues of $\mathbf{S}$ can be calculated as $\lambda_1 = 2$, $\lambda_2 = 1$, and $\lambda_3 = 0$ (see Problem 5.5). None of the pairs of variables are perfectly correlated. However, the GSV is equal to $|\mathbf{S}| = 0$ (see equation 5.20), which confirms the perfect multicollinearity that can be seen in Figure 5.13.   □

Multicollinearity is also present when the number of observations n is not larger than the dimensionality p, that is, $n \leq p$. To investigate this case, we use the geometric interpretation in p dimensions, where the n observations are represented as points in $\mathbb{R}^p$. Any n points lie in an $(n - 1)$ affine subspace, or hyperplane, L. Hence, the mean of all points also lies in the same subspace, and the variability of the points around the mean is also confined to L. We can say that the variability in the direction orthogonal to L is zero. This is expressed more formally by the following property.

Property 5.1   Let $\mathbf{S}$ be the sample variance–covariance matrix calculated from n p-dimensional observations. Then

$$
\text{Rank}(\mathbf{S}) \leq \min(n - 1, p).
\tag{5.24}
$$

When $n \leq p$, $\text{Rank}(\mathbf{S}) \leq n - 1 < p$ and $|\mathbf{S}| = 0$ (see Property 4A.7).

We now consider a general case of the perfect multicollinearity, either due to a small number of observations or due to the collinearity of variables. In that case, the GSV is equal to zero, indicating that there is zero p-dimensional variability. However, there is still some amount of lower dimensional variability. In order to measure that variability, we define *k-dimensional generalized sample variance* (kGSV) as

$$\text{kGSV} = \prod_{i=1}^{k} \lambda_i = \lambda_1 \cdots \lambda_k, \tag{5.25}$$

that is, the product of the first k eigenvalues of the p by p variance–covariance matrix $\mathbf{S}$, where $k \leq p$. When $k = p$, kGSV is the same as GSV (see equation 5.20). The kGSV is most useful in cases when $\lambda_{k+1} = \cdots = \lambda_p = 0$. For the variance–covariance matrix $\mathbf{S}$ used in Example 5.8, $\lambda_1 = 2$, $\lambda_2 = 1$, and $\lambda_3 = 0$, and consequently the GSV is equal to $|\mathbf{S}| = 0$, but a two-dimensional GSV is equal to 2.

Another measure of variability in multivariate data is the total variability defined as

$$\text{Total Variability} = \text{Trace}(\mathbf{S}) = \sum_{i=1}^{p} s_{ii}, \tag{5.26}$$

where s_{ii} is the variance of the ith variable, which measures the amount of variability in the direction of the ith axis when the observations are viewed as points in the p-dimensional space $\mathbb{R}^p$. The total variability is best interpreted in the p-dimensional geometry, unlike the generalized sample variance, which was mostly interpreted in the n-dimensional geometry.

We can say that the Total Variability measures the total amount of "linear" variability in all p orthogonal directions of the p axes. This also turns out to be equal to the total amount of "linear" variability in any set of p orthonormal directions (or basis vectors). In order to see why, note that a matrix of the transformation between any two sets of orthonormal basis vectors is an orthogonal matrix. If the data are realizations of a random vector $\mathbf{X}$ in one orthonormal basis, then they will be expressed as $\mathbf{BX}$ after the transformation, where $\mathbf{B}$ is an orthogonal matrix. From Property 4B.1c, the variance–covariance matrix of the transformed data is equal to $\mathbf{BSB}^T$. From Property 4A.12, we have $\text{Trace}(\mathbf{BSB}^T) = \text{Trace}(\mathbf{SB}^T\mathbf{B}) = \text{Trace}(\mathbf{S})$, which means that the Total Variability is the same when calculated with respect to both systems of coordinates.

In particular, we can take $\mathbf{B} = \mathbf{P}^T$, where $\mathbf{P}$ is the matrix of eigenvectors of $\mathbf{S}$. Based on the spectral decomposition (equation (4.74)), we obtain $\mathbf{BSB}^T = \mathbf{\Lambda}$, where $\mathbf{\Lambda}$ is a diagonal matrix with λ_i, $i = 1, \ldots, p$, eigenvalues on the diagonal. We conclude that the Total Variability can also be calculated as the sum of the eigenvalues, that is,

$$\text{Total Variability} = \text{Trace}(\mathbf{\Lambda}) = \sum_{i=1}^{p} \lambda_i. \tag{5.27}$$

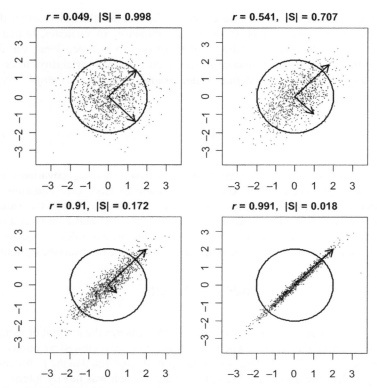

Figure 5.14   Scatter plots of data used in Figure 5.11 with lengths of arrows equal to two times standard deviation of the variability in given directions. Circles of radius 2 are plotted for comparison.

For the variance–covariance matrix $\mathbf{S} = \begin{bmatrix} 1 & r \\ r & 1 \end{bmatrix}$ used in Example 5.7, the Total Variability is equal to 2, so it does not depend on r. This means that for all four cases shown in Figure 5.11, the Total Variability is the same. Figure 5.14 shows the same data with the arrows plotted along the eigenvectors of $\mathbf{S}$, which are rotated clockwise by 45° with respect to the axes of the standard system of coordinates. As explained in the two paragraphs above equation 5.27, the variability in those directions is described by the eigenvalues. The length of each arrow is equal to two times standard deviation (to reflect the two-sigma rule) of the variability in a given direction. The circles have radius 2, so that we can see graphically which standard deviations are larger than 1 (i.e., when the arrow goes beyond the circle). The shorter arrow in the fourth case (the bottom right corner) is too short to be visible. The sum of the variances in the two directions is equal to the Total Variability (equal to 2) in all cases. We can say that the reduced variability in one rotated direction is compensated by increased variability in the other (orthogonal) direction.

When investigating variability, we sometimes want to eliminate the impact of variability of individual variables. We can then standardize each variable by subtracting its mean and then dividing by its sample standard deviation ($\sqrt{s_{ii}}$). One situation when this should be done is when the variables in the data set are in different units, and consequently their direct comparison is not meaningful. The sample variance–covariance matrix calculated on the standardized data is equal to the sample correlation matrix $\mathbf{R}$. The GSV of the standardized data can then be calculated as the *generalized sample correlation* $|\mathbf{R}|$. Let us use the n-dimensional geometry and denote the deviation vector of the ith standardized variable by $\mathbf{f}_i$. We can now use a formula analogous to (5.22), that is,

$$|\mathbf{R}| = \frac{(\text{Standardized Volume})^2}{(n-1)^p}, \qquad (5.28)$$

where Standardized Volume is the volume of a p-dimensional parallelotope defined by the p deviation vectors $\mathbf{f}_i$, $i = 1, \ldots, p$. Since $\mathbf{f}_i$ is a standardized version of the deviation vector $\mathbf{d}_i$ of the ith original variable, the two vectors are collinear, and we have $\mathbf{d}_i = \sqrt{s_{ii}}\mathbf{f}_i$. Consequently, $\|\mathbf{d}_i\|^2 = s_{ii}\|\mathbf{f}_i\|^2$ and

$$(\text{Volume})^2 = (\text{Standardized Volume})^2 \prod_{i=1}^{p} s_{ii} \qquad (5.29)$$

because the two parallelotopes have exactly the same shape (the same angles between the edges), and they differ only by the scale factors equal to the standard deviations of the variables. This shows the following relationship between the generalized sample variance and the generalized sample correlation:

$$|\mathbf{S}| = |\mathbf{R}| \cdot \prod_{i=1}^{p} s_{ii}, \qquad (5.30)$$

which could also be verified algebraically (see Problem 5.6). In order to study further properties of $|\mathbf{R}|$, define $\mathbf{g}_i = \mathbf{f}_i/\sqrt{n-1}$. Since the variance of the standardized variables is 1, we have $1 = \|\mathbf{f}_i\|^2/(n-1)$ and consequently $\|\mathbf{g}_i\| = 1$. Let G be a parallelotope generated by the unit length vectors $\mathbf{g}_i$, $i = 1, \ldots, p$. It is clear that G has exactly the same shape as the parallelotope defined by $\mathbf{f}_i$'s and

$$\text{Volume of } G = \left(\sqrt{n-1}\right)^{-p} (\text{Standardized Volume}). \qquad (5.31)$$

Hence,

$$|\mathbf{R}| = (\text{Volume of } G)^2. \qquad (5.32)$$

This means that the generalized sample correlation $|\mathbf{R}|$ depends only on the angles between the unit length vectors $\mathbf{g}_i$, which are the same as the angles between the deviation vectors $\mathbf{d}_i$. Cosines of those angles are the correlations between variables, which means that $|\mathbf{R}|$ depends only on the correlations between the variables. This, of course, can also be concluded directly from the definition of the matrix $\mathbf{R}$, which consists of correlations. Nevertheless, formula (5.32) is helpful for a better understanding of $|\mathbf{R}|$. For example, the parallelotope G with unit length edges has the largest volume when the vectors $\mathbf{g}_i$ are orthogonal, that is, when $\mathbf{R}$ is an identity matrix. Hence, the largest possible value for $|\mathbf{R}|$ is 1, and we obtain

$$0 \leq |\mathbf{R}| \leq 1. \tag{5.33}$$

The generalized sample correlation $|\mathbf{R}|$ measures multicollinearity in data without the impact of variability of individual variables. Values close to 0 indicate high multicollinearity—with perfect multicollinearity when $|\mathbf{R}| = 0$. Values close to 1 indicate lack of multicollinearity. The concept of kGSV can also be used for standardized data when $\mathbf{R}$ is singular. Since the matrix $\mathbf{R}$ has values of 1 on the diagonal, the Total Variability of the standardized variables is always equal to the dimensionality p of the data. The following example demonstrates practical use of the Total Variability and kGSV on the original and standardized data.

Example 5.9 In Example 5.1, we used a 31-dimensional data set consisting of spectral curves as observations. For each calibration tile, four measurements were taken. The resulting data on each tile can be stored in a 4 by 31 matrix $\mathbf{X}$. The variability in those measurements is due to a measurement error, which can be described by the variance–covariance matrix $\mathbf{S}$. We usually desire $(n - 1) \geq p$, so that $\mathbf{S}$ is nonsingular. Since here we have $(n - 1) = 3 < 31 = p$, the rank of $\mathbf{S}$ is not larger than 3 (see Property 5.1) and $\mathbf{S}$ is singular. This means that the GSV is 0 for all tiles. However, we can use three-dimensional kGSV in order to characterize and compare the measurement errors in various tiles. Figure 5.15 shows a scatter plot of the logarithm of the kGSV values ($k = 3$) versus the logarithm of the Total Variability for the 12 tiles numbered in their order in the original data set as shown in Table 5.2. We can clearly see strong correlation in the scatter plot, which is not surprising because the kGSV depends on the eigenvalues of $\mathbf{S}$, which in turn depend on the overall amount of variability measured by the Total Variability. In order to assess

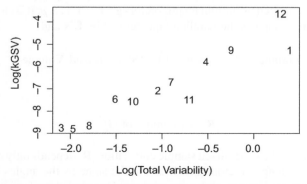

Figure 5.15   A scatter plot of the logarithm (to base 10) of the kGSV values ($k = 3$) versus the logarithm of the Total Variability for the 12 tiles numbered in Table 5.2.

Table 5.2 Colors of the 12 Tiles Used in Example 5.9 in Their Original Order in the Data Set

Tile Number	Color Name
1	Yellow
2	Maroon
3	Dark green
4	Light green
5	Green blue
6	Dark blue
7	Medium blue
8	Brown
9	Pink
10	Dark gray
11	Medium gray
12	Light gray

multicollinearity in the data without the impact of variability of individual variables, one can calculate the kGSV based on **R** as the product of the three positive eigenvalues of **R**. The logarithms of those values are plotted in Figure 5.16 versus the logarithms of the Total Variability for the 12 tiles. We observe the highest level of multicollinearity in spectral bands for Tile 11 and the lowest multicollinearity for Tiles 3 and 8. The tiles' colors are provided in Table 5.2 as a reference.   ☐

5.6 DISTANCES IN THE P-DIMENSIONAL SPACE

In statistics, we often study the question of statistical significance. The answer to that question depends on whether the observations are close to what we expect them to be (according to a null hypothesis). A crucial element is to define what "close" means.

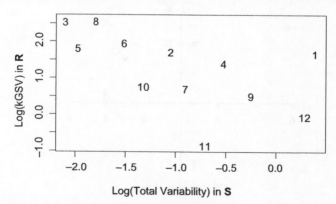

Figure 5.16 A scatter plot of the logarithm (to base 10) of the kGSV values ($k = 3$) calculated for the standardized data based on **R** and plotted versus the logarithm of the Total Variability for the 12 tiles numbered in Table 5.2.

This is why we are interested here in the concept of distance. Since p-dimensional observations can be represented as points in a p-dimensional space, we need to define distances in that space.

Let $\mathbf{x}$, $\mathbf{y}$, and $\mathbf{z}$ be three points in $\mathbb{R}^p$. A distance in $\mathbb{R}^p$ is any real-valued function $d(\mathbf{x}, \mathbf{y})$ such that

1. $d(\mathbf{x}, \mathbf{y}) = d(\mathbf{y}, \mathbf{x})$.
2. $d(\mathbf{x}, \mathbf{y}) > 0$ if $\mathbf{x} \neq \mathbf{y}$.
3. $d(\mathbf{x}, \mathbf{y}) = 0$ if $\mathbf{x} = \mathbf{y}$.
4. $d(\mathbf{x}, \mathbf{y}) \leq d(\mathbf{x}, \mathbf{z}) + d(\mathbf{z}, \mathbf{y})$ (triangle inequality).

The best known distance is the Euclidean distance defined as

$$d(\mathbf{x}, \mathbf{y}) = \sqrt{(\mathbf{x} - \mathbf{y})^\mathrm{T}(\mathbf{x} - \mathbf{y})} = \sqrt{(x_1 - y_1)^2 + (x_2 - y_2)^2 + \cdots + (x_p - y_p)^2},$$

(5.34)

where $\mathbf{x} = [x_1, x_2, \ldots, x_p]^\mathrm{T}$ and $\mathbf{y} = [y_1, y_2, \ldots, y_p]^\mathrm{T}$. According to the Euclidean distance, all directions in $\mathbb{R}^p$ are equally important. In practice, this is not always the case. In order to see why, consider a bivariate data set shown in Figure 5.17. The data are realizations of a random vector $\mathbf{X} = [X_1, X_2]$ with the mean $\boldsymbol{\mu} = [0, 0]$ and the variance–covariance matrix $\boldsymbol{\Sigma} = \begin{bmatrix} 1 & 0 \\ 0 & 9 \end{bmatrix}$. We can see less variability in the horizontal (x_1) direction with the standard deviation of 1 and more variability in the vertical (x_2) direction with the standard deviation of 3. Hence, in assessing whether points A and B are outliers (i.e., not coming from the underlying distribution of the majority of the points), we should take into account the variability being different in the x_1 and x_2

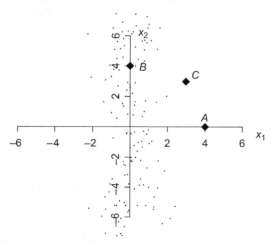

Figure 5.17   An example of uncorrelated data with the standard deviations of 1 and 3 in the horizontal (x_1) and vertical (x_2) directions, respectively.

directions. In order to do that, we can calculate the standardized distance by dividing the horizontal distance by the horizontal standard deviation of 1 and the vertical distance by the vertical standard deviation of 3. The resulting standardized distances of points A and B from the origin are $\sqrt{(4/1)^2 + (0/3)^2} = 4$ and $\sqrt{(0/1)^2 + (4/3)^2} = 4/3$, respectively. We will discuss the exact interpretation of those distances later on, but at this point we can see that A is much farther away from the origin in the sense of the standardized distance. Hence, A is more likely than B to be an outlier. This is consistent with the fact that point B is well within the cloud of observations, and A is an extreme observation. For point $C = [3, 3]$, both the vertical and horizontal components need to be calculated resulting in a standardized distance from the origin equal to $\sqrt{(3/1)^2 + (3/3)^2} = \sqrt{10} \approx 3.16$. This value reflects the less extreme position of C relative to A. In a general setting of a population model with a diagonal variance–covariance matrix Σ, we would standardize by dividing by the standard deviations $\sqrt{\sigma_{ii}}$. The standardized distance between the points $\mathbf{x} = [x_1, x_2, \ldots, x_p]^{\mathrm{T}}$ and $\mathbf{y} = [y_1, y_2, \ldots, y_p]^{\mathrm{T}}$ would then be defined as

$$d(\mathbf{x}, \mathbf{y}) = \sqrt{\frac{(x_1 - y_1)^2}{\sigma_{11}} + \frac{(x_2 - y_2)^2}{\sigma_{22}} + \cdots + \frac{(x_p - y_p)^2}{\sigma_{pp}}}. \tag{5.35}$$

So far, we have assumed that the coordinates of $\mathbf{X}$ in the population model were uncorrelated (diagonal Σ). In order to account for correlations, we want to use a general case of the variance–covariance matrix Σ and define the *statistical*, or *Mahalanobis*, distance as

$$d_M(\mathbf{x}, \mathbf{y}) = \sqrt{(\mathbf{x} - \mathbf{y})^T \Sigma^{-1} (\mathbf{x} - \mathbf{y})}. \tag{5.36}$$

One can show that (5.35) is a special case of (5.36) when the matrix Σ is diagonal (this is given as Problem 5.4). If the components of $\mathbf{X}$ are uncorrelated and have unit variances, we have $\Sigma = \mathbf{I}$, and the Mahalanobis distance becomes the Euclidean distance. For a fixed point $\mathbf{x}$, we may want to find the set of $\mathbf{y}$ points equidistant from $\mathbf{x}$. For the Euclidean distance, this leads to a circle. For the Mahalanobis distance, the set of equidistant points can be written as $\{\mathbf{y} \in \mathbb{R}^p : (\mathbf{x} - \mathbf{y})^T \Sigma^{-1} (\mathbf{x} - \mathbf{y}) = r^2\}$, which is a p-dimensional ellipsoid centered at $\mathbf{x}$ and having axes $\pm c \sqrt{\lambda_i} \cdot e_i$, where $\Sigma e_i = \lambda_i \cdot e_i$ for $i = 1, \ldots, p$ (i.e., λ_i and e_i are eigenvalues and eigenvectors of Σ, respectively).

In practice, the variance–covariance matrix Σ is usually unknown, but it can be estimated from data using $\mathbf{S}$. The resulting Mahalanobis distance is then defined as

$$d_M(\mathbf{x}, \mathbf{y}) = \sqrt{(\mathbf{x} - \mathbf{y})^T \mathbf{S}^{-1} (\mathbf{x} - \mathbf{y})}. \tag{5.37}$$

In some sources, the expression $(\mathbf{x} - \mathbf{y})^T \mathbf{S}^{-1}(\mathbf{x} - \mathbf{y})$ is called the Mahalanobis distance. This is incorrect because the expression is not a distance in the sense of satisfying the triangle inequality. The expression $(\mathbf{x} - \mathbf{y})^T \mathbf{S}^{-1}(\mathbf{x} - \mathbf{y})$ should be called a squared Mahalanobis distance.

Example 5.10 In Example 3.1, we considered $n = 104$ pixels, or observations, from an image of a monochromatic tile. For each pixel, we had reflectance values in three wide spectral bands. We can now represent that data as a 104 by 3 matrix $\mathscr{X}$. As a first step in exploring the data, we inspect a scatter plot matrix shown in Figure 5.18. We can see that Band 3 is correlated with the other two bands, which do not seem to be correlated with each other. This is confirmed by the calculated sample variance–covariance and correlation matrices

$$\mathbf{S} = \begin{bmatrix} 0.067 & 0.007 & 0.068 \\ 0.007 & 0.057 & 0.056 \\ 0.068 & 0.056 & 0.135 \end{bmatrix} \quad \text{and} \quad \mathbf{R} = \begin{bmatrix} 1 & 0.11 & 0.71 \\ 0.11 & 1 & 0.64 \\ 0.71 & 0.64 & 1 \end{bmatrix}. \quad (5.38)$$

Based on Figure 5.18, there are no outliers in the data set. For further investigation of the data, we can calculate the sample mean vector $\bar{\mathbf{x}} = [25.02, 37.51, 75.02]^T$ as a center point and then the Euclidean distance of each point to the center $\bar{\mathbf{x}}$. The resulting 104 Euclidean distances are plotted in Figure 5.19, and again none of the observations seems unusual. In order to take into account the covariance structure of the data, we can now calculate the Mahalanobis distances, which are shown in Figure 5.20. We can see that four pixels have distinctly larger Mahalanobis distances, which suggests that

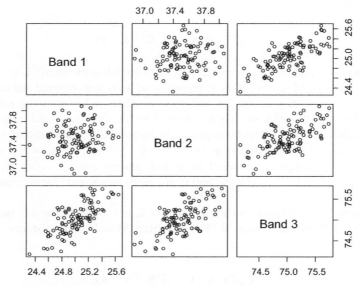

Figure 5.18 A scatter plot matrix of the three-dimensional data used in Example 5.10.

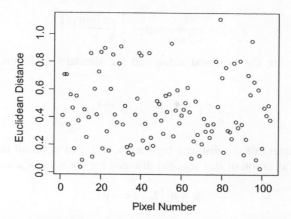

Figure 5.19 Euclidean distances of 104 pixels from the center point $\bar{x}$ plotted versus pixel number. See Example 5.10.

they might be outliers. These potential outliers were discovered only after we used the information about the variance–covariance matrix and calculated the Mahalanobis distances. This example will be further discussed in the next section. ☐

5.7 THE MULTIVARIATE NORMAL (GAUSSIAN) DISTRIBUTION

5.7.1 The Definition and Properties of the Multivariate Normal Distribution

The univariate normal (or Gaussian) distribution is defined by the probability density function

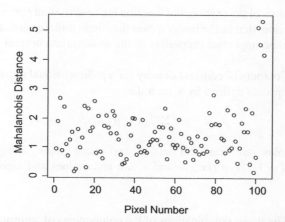

Figure 5.20 Mahalanobis distances of 104 pixels from the center point $\bar{x}$ plotted versus pixel number. See Example 5.10.

$$f(x) = \frac{1}{\sqrt{2\pi\sigma^2}} \exp\left[-\frac{1}{2}\left(\frac{x-\mu}{\sigma}\right)^2\right],$$

where μ and σ are the expected value and the standard deviation, respectively. Note that

$$\left(\frac{x-\mu}{\sigma}\right)^2 = (x-\mu)\left(\sigma^2\right)^{-1}(x-\mu) \tag{5.39}$$

measures the square of the distance between x and μ in standard deviation units. A natural generalization of that squared distance for the vectors $\mathbf{x}$ and $\boldsymbol{\mu}$ is

$$(\mathbf{x}-\boldsymbol{\mu})^{\mathrm{T}}\boldsymbol{\Sigma}^{-1}(\mathbf{x}-\boldsymbol{\mu}), \tag{5.40}$$

which is the squared Mahalanobis distance between $\mathbf{x}$ and $\boldsymbol{\mu}$ as defined in equation (5.37). This leads to a natural generalization of the normal distribution to the multivariate normal distribution.

Definition 5.2 The *multivariate normal (or Gaussian) distribution* with the mean vector $\boldsymbol{\mu}$ and the variance–covariance matrix $\boldsymbol{\Sigma}$ is defined by the probability density function

$$f(\mathbf{x}) = \frac{1}{(2\pi)^{p/2}\sqrt{|\boldsymbol{\Sigma}|}} \exp\left[-\frac{1}{2}(\mathbf{x}-\boldsymbol{\mu})^{\mathrm{T}}\boldsymbol{\Sigma}^{-1}(\mathbf{x}-\boldsymbol{\mu})\right] \tag{5.41}$$

and is denoted by $N_p(\boldsymbol{\mu}, \boldsymbol{\Sigma})$.

Definition 5.3 A random vector $\mathbf{X}$ following the normal distribution $N(\mathbf{0}, \sigma^2\mathbf{I})$ is called the white Gaussian noise.

The constant in front of the exponential function is chosen so that $f(\mathbf{x})$ has the property of a density function, that is, the integral over the whole p-dimensional space is equal to 1. Here are some important properties of the multivariate normal distribution.

Property 5.2 Contours of constant density for a p-dimensional normal distribution $N_p(\boldsymbol{\mu}, \boldsymbol{\Sigma})$ are ellipsoids defined by $\mathbf{x}$ such that

$$(\mathbf{x}-\boldsymbol{\mu})^T\boldsymbol{\Sigma}^{-1}(\mathbf{x}-\boldsymbol{\mu}) = c^2. \tag{5.42}$$

These ellipsoids are centered at $\boldsymbol{\mu}$ and have axes $\pm c\sqrt{\lambda_i} \cdot e_i$, where $\boldsymbol{\Sigma}e_i = \lambda_i \cdot e_i$ for $i = 1, \ldots, p$ (i.e., λ_i and e_i are eigenvalues and eigenvectors of $\boldsymbol{\Sigma}$, respectively).

Property 5.3 The linear combinations of the components of a normally distributed $\mathbf{X}$, that is, $\sum_{i=1}^{p} c_i X_i$, are normally distributed. For example, both X_1 and $X_3 + 2X_4$ are normal. This means that a projection on a straight line also follows a normal distribution.

Property 5.4 All subsets of the components of the normally distributed random vector **X** have normal distribution. For example, the vector $[X_2, X_5, X_6]$ is normal.

Properties 5.3 and 5.4 can be generalized to the following property about a linear transformation of the normal distribution.

Property 5.5 If **A** is a q by p nonrandom matrix, **d** is a q-dimensional nonrandom vector, and **X** is a p-dimensional random vector following a normal distribution, then $\mathbf{Y} = \mathbf{AX} + \mathbf{d}$ also follows a normal distribution. From Property 4B.1, we also conclude more specifically that if **X** follows $N_p(\boldsymbol{\mu}, \boldsymbol{\Sigma})$, then **Y** follows $N_q(\mathbf{A}\boldsymbol{\mu} + \mathbf{d}, \mathbf{A}\boldsymbol{\Sigma}\mathbf{A}^\mathsf{T})$. This means that a projection of a normal distribution on any affine subspace also follows a normal distribution.

Property 5.6 Let $\mathbf{X}_1$ and $\mathbf{X}_2$ be two independent random vectors (not necessarily normal) of dimensions p and q, respectively. Then the p by q covariance matrix $\mathrm{Cov}(\mathbf{X}_1, \mathbf{X}_2)$ (defined in Supplement 4B) is equal to the matrix **0** consisting of all zeros. (We say that independence always implies zero covariance and consequently zero correlation.)

Property 5.7 Let $\mathbf{X} = \begin{bmatrix} \mathbf{X}_1 \\ \mathbf{X}_2 \end{bmatrix}$ be a $(p + q)$-dimensional random vector consisting of p- and q-dimensional subvectors $\mathbf{X}_1$ and $\mathbf{X}_2$. Assume that **X** follows a multivariate normal distribution with the variance–covariance matrix $\boldsymbol{\Sigma} = \begin{bmatrix} \boldsymbol{\Sigma}_{11} & \boldsymbol{\Sigma}_{12} \\ \boldsymbol{\Sigma}_{21} & \boldsymbol{\Sigma}_{22} \end{bmatrix}$. The matrix $\boldsymbol{\Sigma}_{12}$ is the covariance matrix $\mathrm{Cov}(\mathbf{X}_1, \mathbf{X}_2)$. Then $\mathbf{X}_1$ and $\mathbf{X}_2$ are independent if and only if $\boldsymbol{\Sigma}_{12} = \mathbf{0}$. (We say that zero covariance (or correlation) implies independence under normality assumption.)

Property 5.8 Let $\mathbf{X} = \begin{bmatrix} \mathbf{X}_1 \\ \mathbf{X}_2 \end{bmatrix}$ be a $(p + q)$-dimensional random vector consisting of p- and q-dimensional subvectors $\mathbf{X}_1$ and $\mathbf{X}_2$. Assume that **X** follows a multivariate normal distribution with the mean $\boldsymbol{\mu} = \begin{bmatrix} \boldsymbol{\mu}_1 \\ \boldsymbol{\mu}_2 \end{bmatrix}$ and the variance–covariance matrix $\boldsymbol{\Sigma} = \begin{bmatrix} \boldsymbol{\Sigma}_{11} & \boldsymbol{\Sigma}_{12} \\ \boldsymbol{\Sigma}_{21} & \boldsymbol{\Sigma}_{22} \end{bmatrix}$ such that $|\boldsymbol{\Sigma}_{22}| > 0$. Then the conditional distribution of $\mathbf{X}_1$ given that $\mathbf{X}_2 = \mathbf{x}_2$ is normal with

$$\text{Mean} = \boldsymbol{\mu}_1 + \boldsymbol{\Sigma}_{12}\boldsymbol{\Sigma}_{22}^{-1}(\mathbf{x}_2 - \boldsymbol{\mu}_2) \tag{5.43}$$

and

$$\text{Covariance Matrix} = \boldsymbol{\Sigma}_{11} - \boldsymbol{\Sigma}_{12}\boldsymbol{\Sigma}_{22}^{-1}\boldsymbol{\Sigma}_{12}^\mathsf{T}. \tag{5.44}$$

Proofs of the above properties can be found in Johnson and Wichern (2007) and Anderson (2003).

Property 5.9 Let $\mathbf{X}$ be distributed as $N_p(\boldsymbol{\mu}, \boldsymbol{\Sigma})$ with $|\boldsymbol{\Sigma}| > 0$. The random variable $(\mathbf{X} - \boldsymbol{\mu})^T \boldsymbol{\Sigma}^{-1} (\mathbf{X} - \boldsymbol{\mu})$ has the chi-squared distribution with p degrees of freedom.

Proof. Based on Property 5.5, $\mathbf{Z} = \boldsymbol{\Sigma}^{-1/2} (\mathbf{X} - \boldsymbol{\mu})$ is standard normal $N_p(\mathbf{0}, \mathbf{I})$. Note that $(\mathbf{X} - \boldsymbol{\mu})^T \boldsymbol{\Sigma}^{-1} (\mathbf{X} - \boldsymbol{\mu}) = \mathbf{Z}^T \mathbf{Z} = \sum_{i=1}^p Z_i^2$, where Z_i are the components of $\mathbf{Z}$. Since Z_i follow $N(0, 1)$, $\sum_{i=1}^p Z_i^2$ follows the chi-squared distribution with p degrees of freedom (see Appendix A). $\qquad\Box$

5.7.2 Properties of the Mahalanobis Distance

Property 5.9 allows an interpretation of the Mahalanobis distance $d_M(\mathbf{x}, \boldsymbol{\mu}) = \sqrt{(\mathbf{x} - \boldsymbol{\mu})^T \boldsymbol{\Sigma}^{-1} (\mathbf{x} - \boldsymbol{\mu})}$. In Section 2.6, we discussed the rules of two and three sigma under the assumption of normality. In a one-dimensional space ($p = 1$), the Mahalanobis distance can be written as $d_M(x, \mu) = |x - \mu| / \sigma$, which can be interpreted as a distance in standard deviation units. Consequently, for a random variable X following the normal distribution $N(\mu, \sigma)$, we have $P\{d_M(X, \mu) \leq 2\} \approx 0.95$ and $P\{d_M(X, \mu) \leq 3\} \approx 0.997$, which are the two- and three-sigma rules expressed with the help of the Mahalanobis distance.

In a p-dimensional space, we have a similar interpretation of the Mahalanobis distance as a distance in "standard deviation units," but we need more "standard deviations" to cover the same probability. We conclude from Property 5.9 that for $\mathbf{X}$ distributed as $N(\boldsymbol{\mu}, \boldsymbol{\Sigma})$, we have $P\{d_M(\mathbf{X}, \boldsymbol{\mu}) \leq k\} = G_p(k^2)$, where G_p is the CDF of the chi-squared distribution with p degrees of freedom. By taking the values of k equal to 2 and 3, we obtain the p-dimensional equivalents of the two- and three-sigma rules.

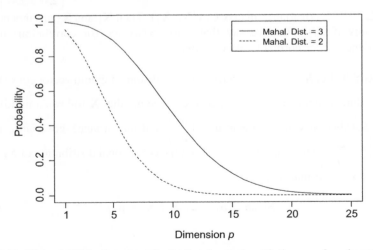

Figure 5.21 The probabilities that a normally distributed vector is not further away from the mean than 2 (or 3) based on the Mahalanobis distance as a function of the dimension p.

Figure 5.21 shows the probability $G_p(k^2)$ as a function of the dimensionality p for the two values of k as the Mahalanobis distance threshold.

We can see that the probabilities decrease rather rapidly to very small values. This is a symptom of the so-called "curse of dimensionality," which in our context means that the central part of the distribution around μ seems increasingly empty. We used the word "seems" in the previous sentence because the real reason for this effect is the increased volume of the high-dimensional space. According to the multivariate normal distribution, the region around μ has a higher density function than any other region in the same space. If we think about a large set of data points in $\mathbb{R}^p$ as realizations of a multivariate normal distribution, then the density of points (the number of points per unit volume) will be highest around μ. However, an increasing dimensionality will result in higher volume and reduced density of points (assuming the fixed total number of points), not only around μ but also in all regions of the $\mathbb{R}^p$ space. In other words, the probability $P\{d_M(\mathbf{X}, \mathbf{y}) \leq k\}$ for any fixed point $\mathbf{y} \in \mathbb{R}^p$, not only $\mathbf{y} = \mu$, decreases to zero as p increases to infinity. (Our previous argument about the decreasing density of points is not sufficient here, but it turns out that the volume enclosed by the ellipsoid $d_M(\mathbf{X}, \mathbf{y}) = k$ increases much slower than the volume of the space.)

For further interpretation of the Mahalanobis distance, we can take a different approach and ask how large an ellipsoid we need in order to cover a certain probability. Again from Property 5.9, we can see that in order to cover $(1 - \alpha)$ probability with an ellipsoid of "radius" k (given by $(\mathbf{x} - \mu)^{\mathrm{T}}\Sigma^{-1}(\mathbf{x} - \mu) = k^2$), we need $k = \sqrt{\chi_p^2(\alpha)}$, where $\chi_p^2(\alpha)$ is the upper (100α) percentile from the chi-squared distribution with p degrees of freedom. The values of k (understood as the Mahalanobis distance threshold) are shown in Figure 5.22.

The rules of two and three sigma and their generalization as shown in Figure 5.22 can be used for identification of outliers in normal samples. However, they should be

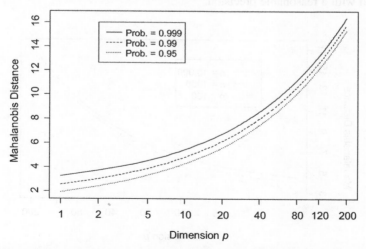

Figure 5.22 The Mahalanobis distance threshold needed in order to enclose a given probability (0.95, 0.99, and 0.999 for the three curves) is plotted as a function of the dimension p.

used only as general guidelines and only after considering the sample size, as discussed in Section 3.7. A more direct approach is to calculate the probability of a given extreme observation. If the probability is very small, we can call the extreme observation an outlier. Assume that we observe a sample of size n from the normal distribution $N_p(\mathbf{\mu}, \mathbf{\Sigma})$, where $\mathbf{\mu}$ and $\mathbf{\Sigma}$ are known. For each observation, we can then calculate its Mahalanobis distance from $\mathbf{\mu}$. If the largest of those Mahalanobis distances is larger than a certain threshold L, we can call that extreme observation an outlier. In order to calculate L, we want to check how likely it is that the largest Mahalanobis distance is greater than L. Let $\mathbf{X}_1, \mathbf{X}_2, \ldots, \mathbf{X}_n$ be i.i.d. random vectors from $N_p(\mathbf{\mu}, \mathbf{\Sigma})$. Those random vectors describe our random sample. Let $Y_i = (\mathbf{X}_i - \mathbf{\mu})^T \mathbf{\Sigma}^{-1} (\mathbf{X}_i - \mathbf{\mu})$, $i = 1, \ldots, n$, be the squared Mahalanobis distances as random variables. The variables Y_i are i.i.d. from the chi-squared distribution with p degrees of freedom. We can now calculate the probability that the largest squared Mahalanobis distance is larger than L^2 as follows:

$$P\left\{ \max_{1 \leq i \leq n} Y_i > L^2 \right\} = 1 - P\left\{ \max_{1 \leq i \leq n} Y_i \leq L^2 \right\} = 1 - \prod_{i=1}^{n} P\{Y_i \leq L^2\} = 1 - \left[G_p(L^2) \right]^n,$$

(5.45)

where G_p is the CDF of the chi-squared distribution with p degrees of freedom as defined earlier. For this probability to be equal to a (usually small) value of α, we need $L = \sqrt{\chi_p^2\left((1-\alpha)^{1/n}\right)}$. The values of the threshold L are plotted in Figure 5.23 as a function of the dimension p for various sample sizes and $\alpha = 0.05$.

Note that these calculations assume that the $\mathbf{\Sigma}$ matrix is known precisely. In practice, we should use this approximate rule only for large sample sizes n when $\mathbf{\Sigma}$ is estimated with a reasonable precision.

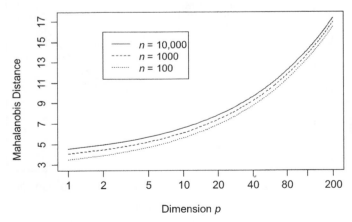

Figure 5.23 The Mahalanobis distance threshold needed in order to call an observation an outlier is plotted as a function of the dimension p for various sample sizes and $\alpha = 0.05$.

In Example 5.10, we calculated $n = 104$ Mahalanobis distances and plotted them in Figure 5.20. Four of those values were distinctly larger than the remaining values. We can now compare those values to $L = \sqrt{\chi_3^2\left((1 - 0.05)^{1/104}\right)} = 4.21$. Three out of the four suspected outliers had Mahalanobis distances larger than 4.21, and the fourth one was slightly lower at 4.17. For more precise calculations, one should take into account that the fourth value is a fourth order statistic rather than the maximum, but those calculations are much more complex (the resulting threshold would be lower than 4.21, of course). An alternative approach is to use formula (5.45) with $L = 4.17$, which gives the probability of 0.058. The four points also turn out to be next to each other spatially as pixels. Considering the above facts, we would decide that all four observations are outliers.

PROBLEMS

5.1. Use the data set from Example 5.2 and investigate the variability of patches within pages, which were characterized by the sample variance–covariance matrices $\mathbf{S}_k$, $k = 1, \ldots, 21$, each based on the eight observations (patches) from a given page. Use the generalized sample variance (or kGSV defined in (5.25), if needed) of the variance–covariance and correlation matrices and the Total Variability defined in (5.26). This analysis is similar to the analysis performed in Example 5.9.

5.2. Using the notation of Section 5.4.2, show algebraically that $\mathbf{d}$ is orthogonal to $\mathbf{1}_n$ (and $\overline{x} \cdot \mathbf{1}_n$).

5.3. Consider three random variables X, Y, and Z. Assume that X is not correlated with Y (Corr$(X, Y) = 0$), and Y is not correlated with Z. What can you tell about the correlation between X and Z? Find all possible values for Corr(X, Z). Use the n-dimensional geometric interpretation to answer this question.

5.4. Show that formula (5.35) is a special case of (5.36) when the matrix Σ is diagonal.

5.5. Consider the sample variance–covariance matrix

$$\mathbf{S} = \begin{bmatrix} 1 & 0 & 2^{-1/2} \\ 0 & 1 & 2^{-1/2} \\ 2^{-1/2} & 2^{-1/2} & 1 \end{bmatrix} \tag{5.46}$$

used in Example 5.8. Show by hand calculations that the eigenvalues of $\mathbf{S}$ are $\lambda_1 = 2$, $\lambda_2 = 1$, and $\lambda_3 = 0$, and the normalized eigenvectors are $\mathbf{e}_1 = [0.5, 0.5, 2^{-1/2}]$, $\mathbf{e}_2 = [2^{-1/2}, -2^{-1/2}, 0]$, and $\mathbf{e}_3 = [0.5, 0.5, -2^{-1/2}]$.

5.6. Verify algebraically formula (5.30) saying that $|\mathbf{S}| = |\mathbf{R}| \cdot \prod_{i=1}^{p} s_{ii}$. *Hint*: Use equations (5.9) and (4.71).

5.7. Show that $\mathbf{S} = \mathbf{D}^{1/2}\mathbf{R}\mathbf{D}^{1/2}$ (see equation (5.9)). *Hint*: Use $s_{jk} = r_{jk}\sqrt{s_{jj}}\sqrt{s_{kk}}$ based on the definition of the correlation coefficient. Also note that post-multiplying by a diagonal matrix is equivalent to multiplying the jth column vector by the jth diagonal element. Pre-multiplying by a diagonal matrix is equivalent to multiplying the jth row vector by the jth diagonal element.

5.8. Use the infrared astronomy data introduced in Example 3.3.

   a. Create a scatter plot for both types of objects (stars) based on the H and K band magnitudes. Mark the two groups with different symbols.

   b. Do you think it would be challenging to distinguish the two groups based on only H band magnitudes? What if only K band magnitudes were used? Is it easier to distinguish the two groups based on the scatter plot from point a?

   c. Repeat points a and b for the pair of J and H band magnitudes.

   d. Repeat points a and b for the pair of J and K band magnitudes.

5.9. Use the infrared astronomy data introduced in Example 3.3. This is a three-dimensional data set ($p = 3$) with bands J, H, and K treated as three variables. Calculate the mean vector $\bar{\mathbf{x}}$ and the matrices $\mathbf{S}$, $\mathbf{R}$, and $\mathbf{D}^{1/2}$ given in equations (5.5) and (5.8) for

   a. The sample of C AGB stars.

   b. The sample of H II regions.

5.10. Use the infrared astronomy data from Problem 5.9. Calculate the generalized sample variance $|\mathbf{S}|$ for the two samples—C AGB stars and H II regions. What do the numbers tell us about the two samples?

5.11. Use the infrared astronomy data from Problem 5.9. Calculate the generalized sample correlation $|\mathbf{R}|$ for the two samples—C AGB stars and H II regions. What do the numbers tell us about the two samples?

5.12. Use the infrared astronomy data from Problem 5.9.

   a. Calculate the mean vector $\bar{\mathbf{x}}$ for the sample of C AGB stars. Find the Euclidean distances of all observations (both C AGB stars and H II regions) from $\bar{\mathbf{x}}$. Create a dot plot of the distances that would show the two samples separately (like in Figure 2.12a). Draw conclusions.

   b. Calculate the variance–covariance matrix $\mathbf{S}$ for the sample of C AGB stars. Find the Mahalanobis distances of all observations (both C AGB stars and H II regions) from $\bar{\mathbf{x}}$. Create a dot plot analogous to the one in point a. Draw conclusions.

    c. Create a scatter plot of the Euclidean distances from point a (on the vertical axis) versus the Mahalanobis distances from point b. Use different colors and/or symbols for the two samples. Draw conclusions.

    d. Should you consider some of the observations in the whole data set as outliers based on the Euclidean and/or Mahalanobis distances calculated in points a and b?

5.13. Repeat the tasks described in Problem 5.12, but this time calculate the mean vector $\bar{\mathbf{x}}$ and the variance–covariance matrix $\mathbf{S}$ for the sample of H II regions. Are your conclusions different this time?

CHAPTER 6

Multivariate Statistical Inference

6.1 INTRODUCTION

In Chapter 3, we discussed statistical inference about a univariate distribution that served as a model for observations on one single characteristic. In Examples 3.2 and 3.3, we were dealing with three characteristics (spectral bands) defined as three different random variables with potentially different distributions. In that case, we simply repeated our univariate analysis three times, and the conclusions were drawn separately for each variable or spectral band. An important question is whether such an analysis is sufficient or perhaps we are losing some important information by not taking into account mutual relationships among the three spectral bands. We will find the answer in this chapter. As one might expect, we will need to model the relationships among variables by using multivariate distributions, especially the multivariate normal distribution discussed earlier in Section 5.7.

6.2 INFERENCES ABOUT A MEAN VECTOR

Throughout this chapter we are going to deal with a random sample $\mathbf{X}_1, \mathbf{X}_2, \ldots, \mathbf{X}_n$ of p-dimensional random vectors following a p-dimensional normal distribution $N_p(\boldsymbol{\mu}, \boldsymbol{\Sigma})$ with the mean vector $\boldsymbol{\mu}$ and the variance–covariance matrix $\boldsymbol{\Sigma}$. We will start with the hypothesis testing inference to check if the data are consistent with a predetermined value $\boldsymbol{\mu}_0$, such as one coming from a specification for a physical quantity being measured.

6.2.1 Testing the Multivariate Population Mean

Let us start with recalling a univariate scenario of a random sample $X_1, X_2, \ldots, X_n$ from a normal distribution $N(\mu, \sigma^2)$. In order to test the null hypothesis

Statistics for Imaging, Optics, and Photonics, Peter Bajorski.
© 2012 John Wiley & Sons, Inc. Published 2012 by John Wiley & Sons, Inc.

$H_0 : \mu = \mu_0$ versus $H_1 : \mu \neq \mu_0$, we would use an absolute value of the following t-statistic:

$$t = \frac{(\overline{X} - \mu_0)}{s/\sqrt{n}} = \sqrt{n}\frac{(\overline{X} - \mu_0)}{s}, \tag{6.1}$$

which has a Student's t-distribution with $(n - 1)$ degrees of freedom under the null hypothesis $H_0 : \mu = \mu_0$. Note that the absolute value $|t|$ is proportional to $|\overline{X} - \mu_0|/s$, which can be interpreted as the standardized distance between $\overline{X}$ and μ_0. Using an absolute value of the t-statistic is equivalent to using its square, which could be written in the following form:

$$t^2 = n\left[\frac{(\overline{X} - \mu_0)}{s}\right]^2 = n(\overline{X} - \mu_0)\left(s^2\right)^{-1}(\overline{X} - \mu_0). \tag{6.2}$$

It turns out that t^2 follows an F-distribution with 1 and $(n - 1)$ degrees of freedom (see Appendix A), so it could be used as a statistic equivalent to $|t|$. We are now going to generalize this statistic in order to deal with a random sample $\mathbf{X}_1, \mathbf{X}_2, \ldots, \mathbf{X}_n$ of p-dimensional vectors following a p-dimensional normal distribution $N_p(\mu, \Sigma)$. For testing the null hypothesis

$$H_0 : \mu = \mu_0 \text{ versus } H_1 : \mu \neq \mu_0, \tag{6.3}$$

we can use a natural generalization of t^2 to the multivariate case, that is,

$$T^2 = n(\overline{\mathbf{X}} - \mu_0)^{\mathrm{T}}\mathbf{S}^{-1}(\overline{\mathbf{X}} - \mu_0), \tag{6.4}$$

where $\overline{\mathbf{X}}$ is the vector of the sample means of all p-variables and $\mathbf{S}$ is the sample variance–covariance matrix. The statistic T^2 is proportional to the squared Mahalanobis distance $(\overline{\mathbf{X}} - \mu_0)^{\mathrm{T}}\mathbf{S}^{-1}(\overline{\mathbf{X}} - \mu_0)$ between $\overline{\mathbf{X}}$ and μ_0, so it measures how far $\overline{\mathbf{X}}$ is from the hypothesized mean value μ_0 in a way that is adjusted for the different variances of the p components and their correlations. Under the null hypothesis $H_0 : \mu = \mu_0$, the statistic T^2 is distributed as

$$\frac{(n - 1)p}{(n - p)}F_{p,n-p}, \tag{6.5}$$

where $F_{p,n-p}$ is a random variable with an F-distribution with p and $(n - p)$ degrees of freedom. This means that a T^2-test will reject H_0 when

$$T^2 \geq c_0 = \frac{(n - 1)p}{(n - p)}F_{p,n-p}(\alpha), \tag{6.6}$$

where $F_{p,n-p}(\alpha)$ is the upper 100α percentile from the F-distribution with p and $(n - p)$ degrees of freedom.

Example 6.1 Let us consider the three-band data from a small image (8 by 13 pixels) of a monochromatic tile. This data set was used earlier in Example 3.3, where we tested whether the mean Band 1 reflectance is consistent with the specification value of $\mu_0 = 25.05$. In Example 5.10, we discovered that 4 out of 104 observations were outliers, so we are now going to use only the remaining 100 observations. We can perform calculations for Band 2 and 3 data in order to test the specifications of 37.53% and 74.99% of reflectance in the two bands, respectively. The resulting t-statistics are equal to -0.646 and 0.756, respectively. In both cases, we are not able to reject the null hypothesis because the absolute values of those statistics are smaller than the critical value $t_{\alpha/2, n-1} = t_{0.025, 99} = 1.98$. Based on our univariate analysis repeated three times for the three spectral bands, we conclude no evidence against conforming to the specification.

In order to perform multivariate analysis, we assume that our measurements are realizations of a random sample $X_1, X_2, \ldots, X_n$ ($n = 100$) from a three-dimensional normal distribution $N_3(\mu, \Sigma)$. The vector μ represents "true" reflectances as measured by the sensor. Let us denote the specification values as the vector $\mu_0 = [25.05, 37.53, 74.99]^T$. We assume here that the sensor is calibrated, so that the sensor measurements can be directly compared against the specification values. The null hypothesis $H_0 : \mu = \mu_0$ means that the tile conforms to the specification, while the alternative $H_a : \mu \neq \mu_0$ means that it does not conform. In order to calculate T^2 defined in formula (6.4), we need to calculate $\bar{x} = [25.0252, 37.5146, 75.0180]^T$,

$$S = \begin{bmatrix} 0.0668 & 0.0049 & 0.0730 \\ 0.0049 & 0.0568 & 0.0610 \\ 0.0730 & 0.0610 & 0.1373 \end{bmatrix}, \quad \text{and} \quad S^{-1} = \begin{bmatrix} 405.4 & 376.4 & -382.6 \\ 376.4 & 383.0 & -370.1 \\ -382.6 & -370.1 & 375.0 \end{bmatrix}.$$

$$(6.7)$$

This gives the value of the statistic $T^2 = 177.35$. In order to calculate the threshold value c_0 defined in (6.6), we obtain $F_{3,100-3}(0.05) = 2.6984$ and $c_0 = (100 - 1) \cdot 3 \cdot F_{3,100-3}(0.05)/(100 - 3) = 8.262$. Since $T^2 = 177.35 \geq 8.262 = c_0$, we reject the null hypothesis $H_0 : \mu = \mu_0$ and conclude that the tile does not conform to the specification. The conclusion is significantly different from the previous conclusion based on the three univariate analyses. The reason for the discrepancy is the strong correlation among the three variables (spectral bands), which has not been taken into account in the univariate analyses. □

6.2.2 Interval Estimation for the Multivariate Population Mean

In Section 3.3, we discussed confidence intervals as a way to assess precision of point estimation and to give an interval that is likely to contain the true value of a parameter. When estimating the whole vector of the multivariate population mean $\mu = [\mu_1, \mu_2, \ldots, \mu_p]^T$, we are in fact dealing with a set of p parameters, and we can construct p confidence interval, one for each parameter. We have certain, say 95%,

Table 6.1   The Confidence Intervals for the Population Mean Values in the Three Spectral Bands Based on the Data Discussed in Example 6.2

Estimated Parameter	Lower Bound	Upper Bound
μ_1	24.974	25.077
μ_2	37.467	37.562
μ_3	74.945	75.092

Each interval is at the confidence level of 0.95.

confidence in each single confidence interval, but it is unclear how certain we can be about all intervals jointly. The issue will be studied in this subsection.

Recall that for each single component μ_i of the vector $\boldsymbol{\mu} = \left[\mu_1, \mu_2, \ldots, \mu_p\right]^T$, one can construct the following confidence interval at the $(1 - \alpha)$ confidence level:

$$\overline{X}_i - t_{n-1}(\alpha/2)\sqrt{\frac{s_{ii}}{n}} < \mu_i < \overline{X}_i + t_{n-1}(\alpha/2)\sqrt{\frac{s_{ii}}{n}}, \qquad (6.8)$$

where $\overline{X}_i$ is the sample mean of the ith variable and s_{ii} is the sample variance, that is, the ith element on the diagonal of the sample variance–covariance matrix $\mathbf{S}$.

Example 6.2   Let us consider the three-band data used in Example 6.1. In Example 3.2, we constructed a confidence interval for the mean Band 1 reflectance based on all 104 observations. Since 4 out of those observations turned out to be outliers (see Example 5.10), we will use only the remaining 100 observations here. Using formula (6.8) with $\alpha = 0.05$, we obtain the values for the lower and upper bounds of the confidence intervals for the population mean values μ_1, μ_2, and μ_3 as shown in Table 6.1.

The confidence interval for the Band 1 mean is almost identical to the one calculated in Example 3.2, which means that the four outliers had little impact on the results. Note that the specification values given in the vector $\boldsymbol{\mu}_0 = [25.05, 37.53, 74.99]^T$ are all within the above confidence intervals. This is consistent with our previous univariate conclusion that all three univariate tests do not reject the hypothesized specification values.   □

In Example 6.2, we constructed only three confidence intervals, but for higher dimensions often seen in spectral data, we may need to use a much larger number of confidence intervals. Consider an example of $m = 20$ independent confidence intervals, each at 95% confidence level. If we denote by W the number of unsuccessful confidence intervals out of $m = 20$, then W follows the binomial distribution with $p = \alpha = 0.05$ and $n = m = 20$. We have $E(W) = np = 1$, which means that we can expect, on average, 1 out of 20 confidence intervals to be unsuccessful in covering the parameter. The probability of all 20 confidence intervals being successful is equal to $P(W = 0) = (1 - 0.05)^{20} = 0.358$, which is unacceptably low. One way to increase this probability is to increase the confidence level of each single confidence interval. We will now show how this can be done.

We have assumed so far that the sample $\mathbf{X}_1, \mathbf{X}_2, \ldots, \mathbf{X}_n$ consists of p-dimensional vectors following a p-dimensional normal distribution $N_p(\boldsymbol{\mu}, \boldsymbol{\Sigma})$. We now additionally assume that the p-dimensional components are stochastically independent, that is, the variance–covariance matrix $\boldsymbol{\Sigma}$ is diagonal with variances σ_{ii} on the diagonal. We can then construct $m \le p$ independent confidence intervals for the means as

$$\overline{X}_i - t_{n-1}(\alpha/2)\sqrt{\frac{s_{ii}}{n}} < \mu_i < \overline{X}_i + t_{n-1}(\alpha/2)\sqrt{\frac{s_{ii}}{n}}, \quad i = 1, \ldots, m. \tag{6.9}$$

We would like to make sure that all those confidence intervals are successful in covering the respective parameters with a high probability. From the definition of independent events, the probability of an intersection of independent events is equal to the product of the events' probabilities. Hence, we obtain the following formula:

$$P\{\text{all } m \text{ confidence intervals at the level } (1 - \alpha)$$

$$\text{are successful, that is, they contain } \mu\text{'s}\} = (1 - \alpha)^m.$$

Since this probability describes the joint probability of success, it is called the *joint confidence level*. In order to have the joint confidence level for all m simultaneous confidence intervals at the level of $(1 - \alpha_0)$, we need to construct each of the single, one-at-a-time confidence intervals at the confidence level $(1 - \alpha)$, where

$$\alpha = 1 - (1 - \alpha_0)^{1/m}. \tag{6.10}$$

For example, in order to construct $m = 20$ independent confidence intervals at the joint confidence level of 0.95 ($\alpha_0 = 0.05$), we need to use $\alpha = 0.002561$, which results in an approximate 99.74% confidence level for each one-at-a-time confidence interval.

In practice, we are rarely able to assume independence of the confidence intervals. This is why we are going to introduce the so-called Bonferroni confidence intervals that can be used without any assumptions about the relationships among the intervals. The Bonferroni approach is based on the following Bonferroni inequality.

Result 6.1 (Bonferroni Inequality). For any set of events A_i, $i = 1, \ldots, m$, we have the following boundary on the probability of the intersection of the events:

$$P\left(\bigcap_{i=1}^{m} A_i\right) \ge 1 - \sum_{i=1}^{m} P(A_i^c), \tag{6.11}$$

where A_i^c is a complement of A_i, which means that $P(A_i^c) = 1 - P(A_i)$.

See Problem 6.4 for a hint on how to obtain a proof of the above result. If we denote by A_i the event that the ith $(1 - \alpha)$ confidence interval is successful in covering the parameter it estimates, then $P(A_i) = 1 - \alpha$ and $P(A_i^c) = \alpha$. The intersection of all A_i events describes the joint success of all confidence intervals, and we obtain the following boundary on the joint confidence level:

P(all m confidence intervals at the level $(1-\alpha)$ are successful, i.e., they contain μ_i's)

$$= P\left(\bigcap_{i=1}^{m} A_i\right) \geq (1 - m\alpha). \tag{6.12}$$

In order to make sure that the joint confidence level is at least $1 - \alpha_0$, we need to find the value of α such that $(1 - m\alpha) = 1 - \alpha_0$, that is, $\alpha = \alpha_0/m$. In other words, the m *simultaneous Bonferroni confidence intervals* at the joint confidence level $1 - \alpha_0$ are defined using the following formula:

$$\overline{X}_i - t_{n-1}\left(\frac{\alpha_0}{2m}\right)\sqrt{\frac{S_{ii}}{n}} \leq \mu_i \leq \overline{X}_i + t_{n-1}\left(\frac{\alpha_0}{2m}\right)\sqrt{\frac{S_{ii}}{n}}, \quad i = 1, \ldots, m. \tag{6.13}$$

We can now apply the Bonferroni methodology to the confidence intervals constructed in Example 6.2.

Example 6.2 (cont.) We again want to find the simultaneous confidence intervals for the true reflectances μ_1, μ_2, and μ_3, but this time we apply the Bonferroni methodology in order to achieve the joint confidence level of $1 - \alpha_0 = 0.95$. We use formula (6.13), which results in each one-at-a-time confidence interval with $\alpha = 0.05/3 = 0.0167$. Table 6.2 shows the numeric results for the three confidence intervals.

These confidence intervals are wider, of course, than one-at-a-time confidence intervals calculated before for $\alpha = 0.05$ and shown in Table 6.1. This means that the specification values are even farther from the confidence bounds, which gives even stronger support for the tile conforming to the specification, at least when using the univariate analysis repeated three times. The stronger support can also be explained with the hypothesis testing approach. If we test the specification values as was done in Example 6.1, we calculate the p-values in the usual way, but when multiple tests are done, the Bonferroni approach is to compare those p-values to $\alpha = 0.05/3 = 0.0167$. We can say that a given result becomes less significant in the presence of multiple testing. $\square$

We may wonder if the Bonferroni approach gives us significantly inflated confidence intervals in relation to the case of independent confidence intervals. In order to check that, we can compare the individual $\alpha_{\text{idep}} = 1 - (1 - \alpha_0)^{1/m}$ value for the independent intervals (see (6.10)) with those of the Bonferroni $\alpha_{\text{Bonf}} = \alpha_0/m$. It turns out that α_{idep} is only slightly larger than α_{Bonf}, which means that the nominal

Table 6.2 The Bonferroni Confidence Intervals at the Joint Confidence Level of 0.95 for the Population Mean Values in the Three Spectral Bands Based on the Data Discussed in Example 6.2

Estimated Parameter	Lower Bound	Upper Bound
μ_1	24.962	25.088
μ_2	37.457	37.573
μ_3	74.928	75.108

confidence levels of the individual confidence intervals will be very similar in both cases. One can show that for any $m \geq 1$, we have the following boundaries for the proportion of the two α's:

$$1 \leq \frac{\alpha_{\text{idep}}}{\alpha_{\text{Bonf}}} \leq \frac{-\ln(1 - \alpha_0)}{\alpha_0} \quad \text{for } 0 < \alpha_0 < 1 \tag{6.14}$$

and

$$1 < \frac{-\ln(1 - \alpha_0)}{\alpha_0} \leq \frac{-\ln(1 - 0.05)}{0.05} = 1.026 \quad \text{for } 0 < \alpha_0 \leq 0.05. \tag{6.15}$$

Consequently, α_{idep} is never larger than α_{Bonf} by more than 2.6% for the range of $0 < \alpha_0 \leq 0.05$ that is usually used. In other words, the Bonferroni confidence intervals are not much wider than the confidence intervals assuming independence. This makes the Bonferroni approach very attractive in any context when the exact relationships among the confidence intervals are difficult to take into account. On the other hand, more efficient confidence regions exist for correlated variables. This will be the topic of the next section.

The considerations of the previous paragraphs apply to a set of independent confidence intervals. Now we want to compare a single Bonferroni confidence interval with a single classic confidence interval. It is worthwhile to check how much wider the Bonferroni interval is in relation to the classic one. Note that the length of a Bonferroni confidence interval is proportional to $t_{n-1}(\alpha_0/2m)$, while the length of a single classic confidence interval is proportional to $t_{n-1}(\alpha_0/2)$. The resulting proportion

$$\frac{t_{n-1}(\alpha_0/2m)}{t_{n-1}(\alpha_0/2)} \tag{6.16}$$

is shown in Figure 6.1 as a function of the number m of the Bonferroni confidence intervals for various sample sizes and $\alpha_0 = 0.05$. We can see that $m = 2$ results in an approximately 15% increase in the length of the confidence interval. For a large number of Bonferroni intervals up to $m = 200$, the length of the confidence interval less than doubles, except for very small sample sizes n. The solid line represents the large-sample approximation based on the proportion of the percentiles from the standard normal distribution, that is,

$$\frac{z(\alpha_0/2m)}{z(\alpha_0/2)}. \tag{6.17}$$

6.2.3 T^2 Confidence Regions

Consider bivariate data generated from a bivariate normal distribution with the population mean $\boldsymbol{\mu} = [\mu_1, \mu_2]^{\text{T}}$. When constructing the simultaneous confidence

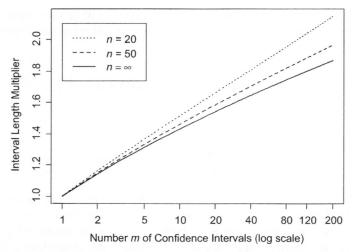

Figure 6.1   The graph shows how many times the Bonferroni confidence intervals are longer than a single confidence interval at the confidence level $\alpha_0 = 0.05$. The values were calculated based on formulas (6.16) and (6.17).

intervals for the two means, we are making a joint statement about plausible values of $\boldsymbol{\mu}$. If a confidence interval $L_1 < \mu_1 < U_1$ for the first mean is considered in isolation, we are not assuming any knowledge about the second mean μ_2. Consequently, in the two-dimensional space of values for $\boldsymbol{\mu}$ plotted in Figure 6.2, the confidence interval corresponds to an infinite vertical strip marked by shaded lines. On the other hand, a confidence interval $L_2 < \mu_2 < U_2$ for the second mean corresponds to an infinite horizontal strip in Figure 6.2 because no knowledge can be assumed about the first mean. A joint statement of both simultaneous confidence intervals is equivalent to an intersection of the two strips in Figure 6.2, resulting in a rectangle as a plausible region

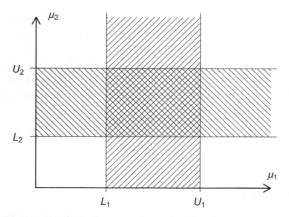

Figure 6.2   Simultaneous confidence intervals for two means resulting in a rectangular confidence region as an intersection.

for μ. The rectangle is called a *confidence region* for μ. The joint confidence level $(1 - \alpha_0)$ of the two simultaneous confidence intervals becomes the confidence level of the rectangular confidence region.

In a general context, a p-dimensional confidence region consists of all p-dimensional vectors that are plausible as values of the mean vector μ. Confidence regions may have various shapes, but some of them might be better than others. We will now investigate this issue.

We again assume that our measurements are realizations of a random sample $X_1, X_2, \ldots, X_n$ from a p-dimensional normal distribution $N_p(\mu, \Sigma)$. In Chapter 3, we constructed a confidence interval for a single parameter as a set of parameter values that would not be rejected if tested in hypothesis testing. We can now use the same concept to define a confidence region related to the T^2-test introduced in Section 6.2.1. Based on the rejection rule defined in (6.6), we can write that μ would not be rejected if

$$n(\overline{\mathbf{X}} - \mu)^T \mathbf{S}^{-1}(\overline{\mathbf{X}} - \mu) < c_0, \qquad (6.18)$$

where $c_0 = [(n-1)p/(n-p)]F_{p,n-p}(\alpha)$. The inequality (6.18) defines a $100(1-\alpha)\%$ T^2confidence region for the mean μ. Based on Property 5.2, the region is ellipsoidal with the center at $\overline{\mathbf{X}}$ and a boundary defined by the p-dimensional ellipsoid $\left\{\mu \in \mathbb{R}^p : n(\overline{\mathbf{X}} - \mu)^T \mathbf{S}^{-1}(\overline{\mathbf{X}} - \mu) = c_0\right\}$.

Example 6.3 Here we use the data set from Example 6.1, where 100 three-dimensional observations were used. Utilizing the calculations of $\overline{x}$ and $\mathbf{S}^{-1}$ done in Example 6.1, we obtain the following 95% ellipsoidal confidence region for μ:

$$100 \cdot \left(\begin{bmatrix} 25.0252 \\ 37.5146 \\ 75.0180 \end{bmatrix} - \mu\right)^T \begin{bmatrix} 405.4 & 376.4 & -382.6 \\ 376.4 & 383.0 & -370.1 \\ -382.6 & -370.1 & 375.0 \end{bmatrix} \left(\begin{bmatrix} 25.0252 \\ 37.5146 \\ 75.0180 \end{bmatrix} - \mu\right) < c_0,$$
$$(6.19)$$

where $c_0 = (100 - 1) \cdot 3 \cdot F_{3,100-3}(0.05)/(100 - 3) = 8.262$.

A disadvantage of confidence regions, as opposed to confidence intervals, is that the inequality (6.19) cannot be written in a simpler form, and consequently, one cannot immediately see which vector values of μ belong to the confidence region. In order to check a specific value μ_0, we need to calculate the left-hand side of (6.19) for $\mu = \mu_0$. If the left-hand side is less than c_0, this means that μ_0 belongs to the confidence region. This procedure is more difficult and less intuitive than simply looking at the simultaneous confidence intervals to check the plausible values for the parameters.

In order to check the specification values $\mu_0 = [25.05, 37.53, 74.99]^T$ considered in Example 6.1, one can calculate the left-hand side of (6.19), which is the same as the value of the T^2 statistic calculated in Example 6.1 as 177.35. That value is larger than $c_0 = 8.262$, and we conclude that the vector $\mu_0 = [25.05, 37.53, 74.99]^T$ is outside of the 95% confidence region for the mean vector μ. In other words, the three specification values are not the plausible true reflectance values as measured by the sensor.   □

We may wonder how the ellipsoidal T^2 confidence regions are related to the simultaneous Bonferroni confidence intervals. This is addressed by the following example.

Example 6.4   Here we considered four data sets, each consisting of $n = 100$ bivariate observations generated from a bivariate normal distribution with the population mean $\mu = [\mu_1, \mu_2]^T$. Each data set represented a different scenario with a different value of the correlation coefficient r between the two variables (the variances were assumed to be the same in all scenarios). The simultaneous Bonferroni confidence intervals for μ_1 and μ_2 were constructed at the joint confidence of 0.95. This resulted in a rectangular confidence region analogous to the one shown in Figure 6.2. The rectangular region is not impacted by the correlation coefficient r as shown by identical rectangles in all four scenarios in Figure 6.3. On the other hand, the elliptical T^2 confidence regions will vary with r because of its impact on the sample variance–covariance matrix $\mathbf{S}$. Figure 6.3 shows the elliptical regions at the same confidence level of 0.95 for four different values of r. For $r \geq 0.9$, there is a significant difference between the rectangular and elliptical confidence regions. We can see large parts of the rectangle that are outside the ellipse. One may wonder which of these two approaches is a more correct representation of the unknown population mean μ. It turns out that the areas outside of the ellipse are less likely to capture the mean μ. This is why the elliptical regions are able to have the same nominal confidence while covering less area, which in turn means being more specific in pinpointing the position of μ. This is analogous to having shorter confidence intervals that give a higher

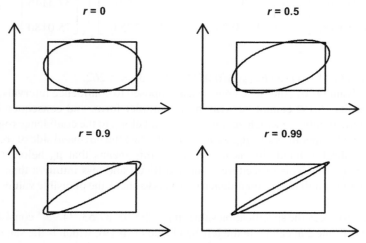

Figure 6.3   A comparison of the rectangular Bonferroni confidence regions with the elliptical T^2 confidence regions at the same nominal joint confidence of 0.95 as discussed in Example 6.4.

precision of estimation in the case of one dimension. Consequently, the elliptical regions are more efficient in capturing μ because they use smaller areas to achieve the same confidence as that of the rectangular region. It should also be mentioned that the actual joint confidence of the Bonferroni confidence intervals would typically be larger than their nominal joint confidence, but there is no easy way to calculate what the exact joint confidence is. This is another reason for the Bonferroni confidence intervals being less efficient than the elliptical confidence regions. □

6.3   COMPARING MEAN VECTORS FROM TWO POPULATIONS

Let us now consider two independent samples from two different populations. We will be interested in understanding differences between the two populations. One way to do this is to see if they have the same mean. This section will deal with statistical inference about the difference in the two means. Here is an example where such statistical inference can be useful.

Example 6.5   We again use the Example 6.1 three-band data from a small image (8 by 13 pixels) of a monochromatic tile. We now want to take into account the spatial structure in the image. The four outliers discussed earlier are all situated in one corner (the white area in the lower right corner in Figure 6.4). After excluding the outliers, we may wonder whether certain areas in the image are significantly different and how different they are from other areas in terms of their reflectance. We define the first six columns of pixels as Sample 1 (this is an 8 by 6 pixel gray area on the left-hand side in Figure 6.4), and the remaining 52 pixels as Sample 2 (marked as black in Figure 6.4). In order to decide if there is a statistically significant difference between the two samples and how large the difference between the true reflectances might be, we need to develop some general methodology, discussed next. □

Figure 6.4   Three areas marked in a small image (8 by 13 pixels). The white area in the lower right corner indicates four outlying pixels. The two areas being compared in Example 6.5 are in gray and black, respectively.

Let us assume that $X_{11}, X_{12}, \ldots, X_{1n_1}$ is a random sample of size n_1 from a p-dimensional normal distribution $N_p(\mu_1, \Sigma_1)$, and $X_{21}, X_{22}, \ldots, X_{2n_2}$ is a random sample of size n_2 from $N_p(\mu_2, \Sigma_2)$. We assume that the two samples are independent of each other. We want to test the null hypothesis

$$H_0 : \mu_1 = \mu_2 \text{ versus } H_1 : \mu_1 \neq \mu_2. \tag{6.20}$$

Hence, we need to estimate $\mu_1 - \mu_2$, the difference between the mean vectors. We can do this by using $\overline{X}_1 - \overline{X}_2$, where $\overline{X}_1$ and $\overline{X}_2$ are sample means from the first and the second samples, respectively. In order to construct the appropriate test, we need to make some assumptions. The following three subsections will consider three versions of those assumptions. More details and proofs of some of the results can be found in Johnson and Wichern (2007).

6.3.1 Equal Covariance Matrices

Here we assume that the variance–covariance matrices in the two populations are the same and equal to $\Sigma = \Sigma_1 = \Sigma_2$. A test to check this assumption will be discussed in Section 6.4. In order to estimate Σ, we use the so-called *pooled variance–covariance matrix* equal to a weighted average of the sample variance–covariance matrices from the two groups, that is,

$$S_{\text{pooled}} = \frac{(n_1 - 1)}{(n_1 + n_2 - 2)} S_1 + \frac{(n_2 - 1)}{(n_1 + n_2 - 2)} S_2. \tag{6.21}$$

We can now define a T^2-type statistic of the form

$$T_2^2 = \left(\overline{X}_1 - \overline{X}_2\right)^T \left[\left(\frac{1}{n_1} + \frac{1}{n_2}\right) S_{\text{pooled}}\right]^{-1} \left(\overline{X}_1 - \overline{X}_2\right). \tag{6.22}$$

When testing the null hypothesis $H_0 : \mu_1 = \mu_2$ versus $H_1 : \mu_1 \neq \mu_2$, we would reject H_0 when $T_2^2 \geq c_2$, where

$$c_2 = \frac{(n_1 + n_2 - 2)p}{(n_1 + n_2 - p - 1)} F_{p, n_1 + n_2 - p - 1}(\alpha). \tag{6.23}$$

The test is not particularly sensitive to slight departures from normality or a few outliers. For more significant departures from the assumptions, one can use a more robust version of the test as developed by Tiku and Singh (1982).

An ellipsoidal $100(1 - \alpha)\%$ T^2 confidence region for $\mu_1 - \mu_2$ is defined by the following inequality:

$$\left(\overline{X}_1 - \overline{X}_2 - (\mu_1 - \mu_2)\right)^T \left[\left(\frac{1}{n_1} + \frac{1}{n_2}\right) S_{\text{pooled}}\right]^{-1} \left(\overline{X}_1 - \overline{X}_2 - (\mu_1 - \mu_2)\right) < c_2. \tag{6.24}$$

Example 6.5 (cont.) We want to test if there is a significant difference and how large the difference is in the mean reflectances in the two areas in the image marked in Figure 6.4. We obtain the following numerical results:

$$\bar{x}_1 - \bar{x}_2 = [-0.0212, -0.0697, -0.0598]^\mathsf{T}, \quad S_{\text{pooled}} = \begin{bmatrix} 0.0674 & 0.0045 & 0.0734 \\ 0.0045 & 0.0561 & 0.0606 \\ 0.0734 & 0.0606 & 0.1378 \end{bmatrix}.$$

$$(6.25)$$

Since the variability within each of the two regions is believed to be due to the measurement error, the difference $\mu_1 - \mu_2$ tells us how different the true reflectances of the two regions are. In order to find the confidence region for $\mu_1 - \mu_2$, we can use formula (6.24), which leads to the following inequality:

$$\left(\begin{bmatrix} -0.0212 \\ -0.0697 \\ -0.0598 \end{bmatrix} - (\mu_1 - \mu_2) \right)^\mathsf{T} \begin{bmatrix} 11018.5 & 10349.6 & -10415.7 \\ 10349.6 & 10566.8 & -10155.0 \\ -10415.7 & -10155.0 & 10190.3 \end{bmatrix}$$

$$\left(\begin{bmatrix} -0.0212 \\ -0.0697 \\ -0.0598 \end{bmatrix} - (\mu_1 - \mu_2) \right) < c_2, \qquad (6.26)$$

where $c_2 = 8.267$, since $F_{3,100-3-1}(0.05) = 2.6994$. In order to check whether $\mu_1 - \mu_2 = 0$ belongs to the confidence region, we calculate the left-hand side of the above inequality for $\mu_1 - \mu_2 = 0$, getting the value 12.256, which is larger than $c_2 = 8.267$. This means that $\mu_1 - \mu_2 = 0$ does not belong to the confidence region, that is, there is a statistically significant difference between the mean reflectance values in the two areas in the image. The inequality (6.26) describes how large the difference $\mu_1 - \mu_2$ is, but it is difficult to see it graphically. If the data were two dimensional, we could plot the ellipsoidal confidence region to get an idea about the possible range of values for $\mu_1 - \mu_2$.

We could also use the equivalent hypothesis testing approach and calculate the statistic T_2^2 defined in (6.22), which is the same as the left-hand side of the inequality (6.26) for $\mu_1 - \mu_2 = 0$, that is, $T_2^2 = 12.256$. Since $T_2^2 = 12.256 > 8.267 = c_2$, we reject the null hypothesis $H_0 : \mu_1 = \mu_2$ and draw the same conclusion that the two areas are different. $\qquad \square$

6.3.2 Unequal Covariance Matrices and Large Samples

Here we no longer assume that the variance–covariance matrices in the two populations are the same, but we need to assume large sample sizes, since the approximation used here is based on the central limit theorem. The assumption of normality is not needed here, although the approximation will usually be more precise under normality. We define a T^2-type statistic of the form

$$T_3^2 = (\overline{\mathbf{X}}_1 - \overline{\mathbf{X}}_2)^T \left[\frac{1}{n_1} \mathbf{S}_1 + \frac{1}{n_2} \mathbf{S}_2 \right]^{-1} (\overline{\mathbf{X}}_1 - \overline{\mathbf{X}}_2). \tag{6.27}$$

When testing the null hypothesis $H_0 : \boldsymbol{\mu}_1 = \boldsymbol{\mu}_2$ versus $H_1 : \boldsymbol{\mu}_1 \neq \boldsymbol{\mu}_2$, we would reject H_0 when $T_3^2 \geq \chi_p^2(\alpha)$, where $\chi_p^2(\alpha)$ is the upper (100α)th percentile from a chi-square distribution with p degrees of freedom.

An ellipsoidal $100(1 - \alpha)\%$ T^2 approximate confidence region for $\boldsymbol{\mu}_1 - \boldsymbol{\mu}_2$ is defined by the following inequality:

$$(\overline{\mathbf{X}}_1 - \overline{\mathbf{X}}_2 - (\boldsymbol{\mu}_1 - \boldsymbol{\mu}_2))^T \left[\frac{1}{n_1} \mathbf{S}_1 + \frac{1}{n_2} \mathbf{S}_2 \right]^{-1} (\overline{\mathbf{X}}_1 - \overline{\mathbf{X}}_2 - (\boldsymbol{\mu}_1 - \boldsymbol{\mu}_2)) < c_3, \tag{6.28}$$

where $c_3 = \chi_p^2(\alpha)$.

Example 6.5 (cont.) We again want to see how different the two areas are. In order to find the confidence region for $\boldsymbol{\mu}_1 - \boldsymbol{\mu}_2$, we can use inequality (6.28), which leads to the following inequality:

$$\left(\begin{bmatrix} -0.0212 \\ -0.0697 \\ -0.0598 \end{bmatrix} - (\boldsymbol{\mu}_1 - \boldsymbol{\mu}_2) \right)^T \begin{bmatrix} 10962.5 & 10274.1 & -10378.8 \\ 10274.1 & 10468.0 & -10099.7 \\ -10378.8 & -10099.7 & 10172.4 \end{bmatrix}$$

$$\left(\begin{bmatrix} -0.0212 \\ -0.0697 \\ -0.0598 \end{bmatrix} - (\boldsymbol{\mu}_1 - \boldsymbol{\mu}_2) \right) < c_3, \tag{6.29}$$

where $c_3 = \chi_3^2(0.05) = 7.8147$. In order to check whether $\boldsymbol{\mu}_1 - \boldsymbol{\mu}_2 = \mathbf{0}$ belongs to the confidence region, we calculate the left-hand side of the above inequality for $\boldsymbol{\mu}_1 - \boldsymbol{\mu}_2 = \mathbf{0}$, getting the value 12.018, which is larger than $\chi_3^2(0.05) = 7.8147$. This means that $\boldsymbol{\mu}_1 - \boldsymbol{\mu}_2 = \mathbf{0}$ does not belong to the confidence region, that is, there is a statistically significant difference between the mean reflectance values in the two areas in the image. As discussed in Section 6.3.1, we could obtain an equivalent result by utilizing the hypothesis testing approach. We can see that the numerical results shown here are not much different from those in Section 6.3.1.   □

6.3.3 Unequal Covariance Matrices and Samples Sizes Not So Large

Here again we do not assume that the variance–covariance matrices in the two populations are the same, and the sample sizes do not need to be as large as those in Section 6.3.2, but we do assume normality again. We use the T_3^2 statistic defined in (6.27). When testing the null hypothesis $H_0 : \boldsymbol{\mu}_1 = \boldsymbol{\mu}_2$ versus $H_1 : \boldsymbol{\mu}_1 \neq \boldsymbol{\mu}_2$, we would reject H_0 when $T_3^2 \geq c_4$, where

$$c_4 = \frac{vp}{(v - p + 1)} F_{p,v-p+1}(\alpha), \tag{6.30}$$

and v is the adjusted number of degrees of freedom calculated as

$$v = \frac{p + p^2}{\sum_{j=1}^{2} (1/n_j) \left\{ \text{Trace}\left(\mathbf{U}_j^2\right) + \left[\text{Trace}\left(\mathbf{U}_j\right)\right]^2 \right\}}, \tag{6.31}$$

where

$$\mathbf{U}_j = \frac{1}{n_j} \mathbf{S}_j \mathbf{W}^{-1} \quad \text{and} \quad \mathbf{W} = \frac{1}{n_1} \mathbf{S}_1 + \frac{1}{n_2} \mathbf{S}_2. \tag{6.32}$$

An ellipsoidal $100(1 - \alpha)\%$ T^2 approximate confidence region for $\boldsymbol{\mu}_1 - \boldsymbol{\mu}_2$ is again defined by the inequality (6.28), except that the threshold c_3 is replaced with c_4 defined by (6.30). More details on this approach can be found in Krishnamoorthy and Yu (2004) and Nel and Van der Merwe (1986).

Example 6.5 (cont.) We again want to test if there is a difference between the two areas. The confidence region for $\boldsymbol{\mu}_1 - \boldsymbol{\mu}_2$ is given by formula (6.29), except that the threshold c_3 needs to be replaced with c_4 calculated from formula (6.30). We obtain $v = 96.95$ and $c_4 = 8.272$. Note that the value of c_4 is very close to c_2 obtained in Section 6.3.1. Since $T_3^2 = 12.018 > 8.272 = c_0$, we again reject $H_0 : \boldsymbol{\mu}_1 = \boldsymbol{\mu}_2$. $\square$

6.4 INFERENCES ABOUT A VARIANCE–COVARIANCE MATRIX

We sometimes deal with two or more independent samples from different populations, and we may need to know if the variance–covariance matrices can be assumed to be the same in those populations. That is, we would like to test the null hypothesis $H_0 : \boldsymbol{\Sigma}_1 = \boldsymbol{\Sigma}_2 = \cdots = \boldsymbol{\Sigma}_g$, where g is the number of populations or groups. When testing the equality of means in two populations in Section 6.3, the methodology is dependent on whether the two variance–covariance matrices are the same.

If the variance–covariance matrices from several populations are the same, we can estimate the joint matrix $\boldsymbol{\Sigma}$ with the pooled estimate analogous to formula (6.21). This time, the estimate is based on the sample variance–covariance matrices $\mathbf{S}_j, j = 1, \ldots, g$, from the g samples with their sizes denoted by $n_j, j = 1, \ldots, g$. The pooled sample variance–covariance matrix is then defined as

$$\mathbf{S}_{\text{pooled}} = \frac{1}{\sum_{j=1}^{g} (n_j - 1)} \sum_{j=1}^{g} (n_j - 1)\mathbf{S}_j. \tag{6.33}$$

If the population variance–covariance matrices are different, then their estimates should be substantially different from the pooled estimate S_{pooled}. The difference is measured by the Box's M statistic (see Box (1950) and Box and Draper (1969)) defined as

$$M = \left[\sum_{j=1}^{g}(n_j - 1)\right] \ln\left|S_{\text{pooled}}\right| - \sum_{j=1}^{g}(n_j - 1)\ln\left|S_j\right|. \tag{6.34}$$

The M statistic is closely related to a likelihood ratio statistic for this testing problem (see Section 10.2 in Anderson (2003)). The approximate α-level Box's M-test rejects $H_0 : \Sigma_1 = \Sigma_2 = \cdots = \Sigma_g$ when $M \geq c$, where

$$c = \frac{1}{1-u}\chi_v^2(\alpha), \qquad v = p(p+1)(g-1)/2, \tag{6.35}$$

and

$$u = \left[\sum_{j=1}^{g}\frac{1}{(n_j - 1)} - \frac{1}{\sum_{j=1}^{g}(n_j - 1)}\right]\left[\frac{(2p^2 + 3p - 1)p}{12v}\right], \tag{6.36}$$

where p is the dimensionality of the data. The approximation works well when each sample size n_j exceeds 20 and p and g do not exceed 5. When these conditions do not hold, other approximations can be used (see Box (1949,1950)).

Example 6.6   In Example 6.5, two areas in an image (see Figure 6.4) were compared by considering the two sets of pixels as independent samples. In order to decide which of the methodologies introduced in Section 6.3 would be most appropriate, we would like to test the equality of the population variance–covariance matrices. The pooled sample variance–covariance matrix was calculated earlier and is shown in (6.25). We also obtain $v = 6$, $u = 0.0332$, $\chi_6^2(0.05) = 12.59$, $c = 13.02$, and $M = 8.67$. Since $M = 8.67 < 13.02 = c$, we do not reject the $H_0 : \Sigma_1 = \Sigma_2$ and accept the possibility that the population variance–covariance matrices are the same. Hence, the methodology introduced in Section 6.2.1 is the most appropriate in this context.   □

6.5   HOW TO CHECK MULTIVARIATE NORMALITY

The assumption of normality is often used in various methods of statistical inference in order to utilize distributional properties of some statistics. It is then of interest to check if a given p-dimensional data set comes from a normal distribution. Assume that our data come from a population model described by a random vector $X = [X_1, \ldots, X_p]$. We want to test the hypothesis that X follows a multivariate normal distribution. One way to do this is to check for some conditions necessary for $\mathbf{X}$ to follow a multivariate normal distribution. Based on Property 5.4, one such condition is that each component X_j, $j = 1, \ldots, p$, follows a univariate normal distribution. In Section 3.5, we described some ways of checking the univariate normality. This procedure should be performed on all p variables. A different procedure is suggested by Property 5.3 saying that any linear combination should also be normal. Assuming we have n multivariate observations, we could generate values of a linear combination by multiplying the n by p data matrix $\mathscr{X}$ by a p-dimensional vector of coefficients $\mathbf{a}$, which is equivalent to projecting the data points on the direction of the vector $\mathbf{a}$. The resulting data given by the n-dimensional vector $\mathbf{y} = \mathscr{X}\mathbf{a}$ should be checked for normality. Various choices for the vector of coefficients $\mathbf{a}$ could be obtained by using random numbers. A procedure proposed by Srivastava (1984) (see also Section 3.5.2 in Srivastava (2002)) amounts to projecting on the vector $\mathbf{a}$ equal to an eigenvector $\mathbf{e}_j$, $j = 1, \ldots, p$, of the sample variance–covariance matrix $\mathbf{S}$. The appeal of this method is that the resulting p variables described by $\mathbf{y}_j = \mathscr{X}\mathbf{e}_j$, $j = 1, \ldots, p$, are uncorrelated (which will become clear in Chapter 7). When the original components X_j, $j = 1, \ldots, p$, are in different physical units, the variables could be first standardized prior to further calculations.

If any of the univariate projections exhibit highly significant non-normality, there is no need for further investigation. We stress the need for highly significant results because when many projection directions are used, one needs to ensure a large joint confidence level as discussed in Section 6.2.2.

If a given data set passes the test of univariate projections, further methods can be used that more explicitly address the multivariate structure of the data distribution. One approach is to consider the squared Mahalanobis distances of observation vectors from the mean vector, that is,

$$D_i^2 = \left(\mathbf{X}_i - \overline{\mathbf{X}}\right)^{\mathrm{T}} \mathbf{S}^{-1} \left(\mathbf{X}_i - \overline{\mathbf{X}}\right), \quad i = 1, \ldots, n. \tag{6.37}$$

As n tends to infinity, the distribution of D_i^2 approaches the chi-square distribution with p degrees of freedom. We could then create a probability plot (see Section 3.6) of the sorted d_i^2 values of D_i^2's versus percentiles from the chi-square distribution. Some authors suggest this procedure for n and $(n - p)$ greater than 25. However, other authors point out that the use of the chi-square distribution could be misleading, and a much larger sample size n is needed for larger p.

A preferred approximation is with the use of a beta distribution, which leads to the following Small's (1978) graphical method. Based on the observed d_i^2 values, calculate

$$u_i = \frac{n d_i^2}{(n - 1)^2}, \tag{6.38}$$

and then plot the sorted u_i, $i = 1, \ldots, n$, values against the percentiles from the beta distribution given by

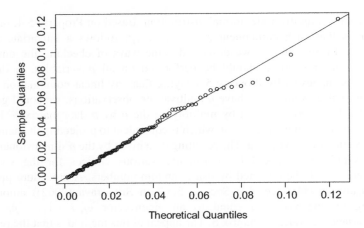

Figure 6.5 A probability plot using the Small's method for Example 6.7 data.

$$v_i = F_{a,b}^{-1}\left(\frac{i - \gamma}{n - \gamma - \beta + 1}\right), \tag{6.39}$$

where

$$a = p/2, \qquad b = (n - p - 1)/2, \qquad \gamma = \frac{a - 1}{2a}, \qquad \beta = \frac{b - 1}{2b}. \tag{6.40}$$

The plotted points should line up along a straight line having slope 1 and going through the origin. Departures from this pattern suggest non-normality of all data or some single observations.

Example 6.7 We again use the Example 6.1 three-band data from a small image (8 by 13 pixels) of a monochromatic tile. The univariate normality was confirmed in Example 3.4 for all 104 pixels in the image, even when including the four outliers identified later on in Example 5.10. Here we are going to investigate only the remaining 100 observations. We can implement the Small's graphical method for checking three-dimensional normality. We obtain the values $a = 1.5$, $b = 48$, $\gamma = 0.1667$, and $\beta = 0.4896$, and the plot of the sorted u_i, $i = 1, \ldots, n$, values against the percentiles given by (6.39) is shown in Figure 6.5. The points line up reasonably close to the line, which is consistent with the normality assumption.

Figure 6.6 shows the probability plot of the squared Mahalanobis distances d_i^2 versus percentiles of the chi-square distribution with $p = 3$ degrees of freedom. The points line up along the plotted line almost as well as those in Figure 6.5. It seems that the chi-square approximation works quite well in this case, probably due to a reasonably sized sample with $n = 100$ for p as small as 3. $\square$

More topics on multivariate statistical inference can be found in Johnson and Wichern (2007), Anderson (2003), Srivastava (2002), and Rencher (2002). A more applied approach can be found in Hardle and Simar (2007).

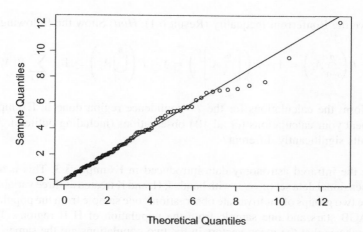

Figure 6.6 A probability plot using the chi-square approximation for Example 6.7 data.

PROBLEMS

6.1. Perform the calculations for the T^2 statistic done in Example 6.1. Repeat your calculations for all 104 observations (including outliers). Are the results significantly different?

6.2. Use the infrared astronomy data introduced in Example 3.3. This is a three-dimensional data set ($p = 3$) with bands J, H, and K treated as three variables.

    a. Your colleague calculated the mean vector $\bar{x}$ for the sample of C AGB stars and sent the results to you in an email. Do the calculations yourself, so that you can find out what the vector coordinates are.

    b. Your colleague was in rush, so he made a mistake and wrote to you that these are typical, average values for H II regions. You look at the vector $\bar{x}$ and wonder if this can indeed be typical for H II regions. So, you consider that vector as your vector μ_0. You define μ as the theoretical mean of a population of H II regions. Of course, you do not have information about the whole population. You have only a sample of 59 H II regions. You want to test the null hypothesis that $H_0 : \mu = \mu_0$. Use the T^2-test to test that hypothesis. Draw conclusions. Can you be sure that your colleague has made a mistake?

    c. Use the data on H II regions in order to construct a T^2 confidence region for the theoretical mean μ of a population of H II regions. Check if the vector μ_0 from point b belongs to that region.

6.3. Do the opposite of calculations in Problem 6.2, that is, calculate μ_0 based on H II regions and test if this is a plausible mean vector for the population of C AGB stars. Draw conclusions.

6.4. Prove the Bonferroni inequality (Result 6.1). *Hint*: Show the following:

$$P\left(\bigcap_{i=1}^{n} A_i\right) = 1 - P\left(\left[\bigcap_{i=1}^{n} A_i\right]^c\right) = 1 - P\left(\bigcup_{i=1}^{n} A_i^c\right) \geq 1 - \sum_{i=1}^{n} P(A_i^c).$$

6.5. Perform the calculations for the T^2 confidence region done in Example 6.3. Repeat your calculations for all 104 observations (including outliers). Are the results significantly different?

6.6. Use the infrared astronomy data introduced in Example 3.3. This is a three-dimensional data set ($p = 3$) with bands J, H, and K treated as three variables. We have two groups of multivariate observations, one sample from the population of C AGB stars and one sample from the population of H II regions. Test the hypothesis that the mean vectors in the two populations are the same.

    a. Here assume that the population variance–covariance matrices are the same in the two populations.

    b. Now assume that the population variance–covariance matrices are not the same in the two populations. Consider the samples large, but not necessarily normal.

    c. Now assume that the population variance–covariance matrices are not the same in the two populations. Consider the samples not so large, but assume they are normal.

6.7. Perform the calculations leading to the implementation of the Small's graphical method as done in Example 6.7. Repeat your calculations for all 104 observations (including outliers). Are the results significantly different?

6.8. Use the infrared astronomy data introduced in Example 3.3. This is a three-dimensional data set ($p = 3$) with bands J, H, and K treated as three variables. We have two groups of multivariate observations, one sample from the population of C AGB stars and one sample from the population of H II regions.

    a. Test the hypothesis that the variance–covariance matrices in the two populations are the same.

    b. Use the Small's graphical method to check if the sample describing C AGB stars comes from a three-dimensional normal distribution in terms of the values in bands J, H, and K.

    c. Repeat point b for the sample of H II regions.

CHAPTER 7

Principal Component Analysis

7.1 INTRODUCTION

In Section 5.5, we considered various ways of measuring overall variability in multivariate data. If p-dimensional observations are represented as points in a p-dimensional space $\mathbb{R}^p$, the overall variability measures the variability of all those points. It is clearly a difficult task to summarize such variability with only one number. In this chapter, we will further investigate the p-dimensional variability by exploring the variability in various directions. This will help in understanding the multidimensional structure of the variability. The simplest way to start is to think about variability in the direction of the axes in $\mathbb{R}^p$. We could start with the first coordinate x_1 and project all points onto the first axis. This is the same as considering a univariate sample of data on the first variable X_1, so it does not reveal anything new. However, investigating other directions will be more interesting as demonstrated by the following example.

Example 7.1 We want to consider a simplified scenario of an implanted radiopaque marker, which is monitored by orthogonal projections on two X-ray screens. Figure 7.1 shows a simplified two-dimensional scenario, where the screens are shown in panel (a) along the normalized vectors v_1 and v_2, and the 150 dots represent measurements collected over a 5-minute interval (once every 2 seconds). Panel (a) shows the physical locations with respect to standard coordinates f_1 and f_2. When a given point is projected on v_1, the distance of that projection from the origin is denoted as x_1. In the same fashion, the projection on v_2 is denoted as x_2. The variables x_1 and x_2 are the mathematical coordinates describing the locations of points based on the measured values on the X-ray screens. Those coordinates are shown in panel (b). The calculations will be done in the mathematical coordinates x_1 and x_2. In order to relate back to the physical coordinates, one needs to use a linear transformation determined by the vectors v_1 and v_2.

Statistics for Imaging, Optics, and Photonics, Peter Bajorski.
© 2012 John Wiley & Sons, Inc. Published 2012 by John Wiley & Sons, Inc.

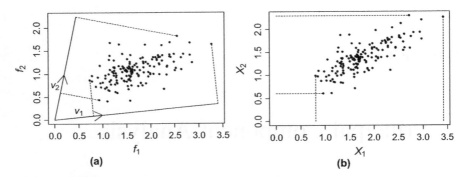

Figure 7.1 Locations of a radiopaque marker as measured by two X-ray screens positioned along the vectors v_1 and v_2. Panel (a) shows the physical locations with respect to standard coordinates f_1 and f_2. Panel (b) shows the locations in the mathematical coordinates x_1 and x_2 as measured on the X-ray screens.

It is known that the marker can oscillate only along a fixed axis. We are assuming here that the measurement errors in the two screens are independent of each other and have the same normal distribution with the mean zero (i.e., the measurements are unbiased). This means that the measurement error variability is the same in all directions in the x_1, x_2 coordinates shown in Figure 7.1b. The oscillation of the marker along a fixed axis will cause additional variability in that direction, which means that we can find the oscillation axis by finding the direction of maximum variability. The remaining variability, causing the points to be outside of that axis, must be due to the measurement error.

The intuitive concept of the variability in a certain direction can be studied by projecting the observation points on a straight line. Let us describe a straight line L in a parametric form as a set of all vectors $t \cdot \mathbf{b} + \mathbf{d}$ parameterized by $t \in R$, where $\mathbf{d}$ is a vector orthogonal to a normalized vector $\mathbf{b}$. Then the orthogonal projection of a point $\mathbf{x} = [x_1, x_2]^T$ on L is given by $(\mathbf{b}^T \cdot \mathbf{x})\mathbf{b} + \mathbf{d}$, which means that the position of the projection on the line L can be described by the scalar product $\mathbf{b}^T \cdot \mathbf{x}$. If $\mathbf{X} = [X_1, X_2]^T$ is a random vector, the random variable $Y = \mathbf{b}^T \cdot \mathbf{X}$ describes the variability in the direction $\mathbf{b}$. We can also think of Y as a linear combination of the components of $\mathbf{X}$.

Figure 7.2 shows an example of points projected on two straight lines. The lines L_1 and L_2 are defined by the normalized vectors $\mathbf{b}_1 = [-1, 4]^T/\sqrt{17}$ and $\mathbf{b}_2 = [7, -1]^T/\sqrt{50}$, respectively. The variability in those two directions can be described by variances of the variables $Y_i = \mathbf{b}_i^T \cdot \mathbf{X}$, $i = 1, 2$ (here, $\mathbf{X}$ has the sampling distribution described by the points in Figure 7.2). For the data shown in Figure 7.2, $\mathrm{Var}(Y_1) = 0.0542$ and $\mathrm{Var}(Y_2) = 0.2199$. It is intuitively clear that we can get even larger variance by projecting on the line P_1 shown in Figure 7.3. In the next section, we will discuss how the direction of maximum variability can be found.

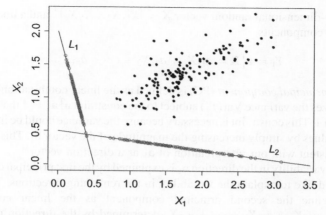

Figure 7.2 Variability of data in two directions as given by projections on the direction vectors.

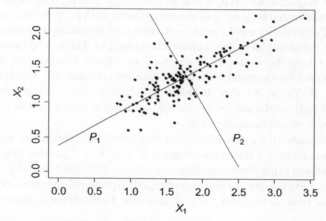

Figure 7.3 The direction of maximum variability (P_1) and the orthogonal direction of the minimum variability (P_2).

7.2 DEFINITION AND PROPERTIES OF PRINCIPAL COMPONENTS

Principal component analysis (PCA) identifies linear combinations of the original variables that contain most of the information, in the sense of variability, contained in the data. The general assumption is that useful information is proportional to the variability. PCA is used for data dimensionality reduction and for interpretation of data.

7.2.1 Definition of Principal Components

In Figure 7.3, we drew an intuitively appealing direction P_1 of maximum variability. The question is: how do we find the direction of the line P_1 mathematically? In order to answer this question, let us introduce more general notation and some definitions.

Consider a p-dimensional random vector $\mathbf{X} = \begin{bmatrix} X_1, X_2, \ldots, X_p \end{bmatrix}^\mathsf{T}$ and a linear combination of its components

$$Y_1 = \mathbf{a}_1^\mathsf{T}\mathbf{X} = a_{11}X_1 + a_{21}X_2 + \cdots + a_{p1}X_p. \tag{7.1}$$

The *first principal component* (PC) is defined as the linear combination $Y_1 = \mathbf{a}_1^\mathsf{T}\mathbf{X}$ that maximizes the variance $\mathrm{Var}(Y_1)$ subject to the constraint $\mathbf{a}_1^\mathsf{T}\mathbf{a}_1 = 1$ that the vector $\mathbf{a}_1$ has length 1. This constraint is necessary because the variance could be increased to unlimited values by simply increasing the magnitude of the vector $\mathbf{a}_1$. This constraint is also consistent with our interpretation of $\mathbf{a}_1$ as a direction vector.

Once the variability in the direction $\mathbf{a}_1$ is explained by the first principal component Y_1, we would like to explain the variability in the remaining directions. To achieve this, we define the second principal component as the linear combination $Y_2 = \mathbf{a}_2^\mathsf{T}\mathbf{X} = a_{12}X_1 + a_{22}X_2 + \cdots + a_{p2}X_p$ determined by the direction unit vector $\mathbf{a}_2$ orthogonal to $\mathbf{a}_1$, such that it maximizes the variance $\mathrm{Var}(Y_2)$. The assumption of orthogonality is crucial here. If $\mathbf{a}_2$ was at an acute angle with $\mathbf{a}_1$, then the variability of Y_2 could be maximized by getting $\mathbf{a}_2$ almost collinear with $\mathbf{a}_1$. We could also say that some of the Y_2 variability would be in the $\mathbf{a}_1$ direction. By taking $\mathbf{a}_2$ orthogonal to $\mathbf{a}_1$, we obtain the variability in Y_2 that is uncorrelated with Y_1. This can be explained more formally by taking the vector $\mathbf{a}_2$ represented as a linear combination $\alpha\mathbf{a}_1 + \beta\mathbf{a}_1^\perp$, where $\mathbf{a}_1^\perp$ is a vector orthogonal to $\mathbf{a}_1$. In that case, the variance of Y_2 would be equal to $\alpha^2 \mathrm{Var}(Y_1) + \beta^2 \mathrm{Var}(\mathbf{a}_1^\perp\mathbf{X})$ (see Problem 7.11), and thus it would include the variability already explained by Y_1. Hence, we want $\alpha = 0$, which is the same as the requirement of orthogonality to $\mathbf{a}_1$.

In the same fashion, we define the remaining principal components, where the jth principal component is a linear combination $Y_j = \mathbf{a}_j^\mathsf{T}\mathbf{X}$, $j \le p$, such that $\mathrm{Var}(Y_j)$ is maximized subject to the constraints that $\mathbf{a}_j$ is a vector of length 1, which is orthogonal to all $\mathbf{a}_k$, $k < j$. We can always find a set of exactly p principal components for a given p-dimensional random vector $\mathbf{X}$ because there are p orthogonal directions in $\mathbb{R}^p$.

7.2.2 Finding Principal Components

With the above definition of principal components, it is not immediately clear how to find them as the appropriate linear combinations. Recall that $\mathrm{Var}(Y_j) = \mathbf{a}_j^\mathsf{T}\Sigma\mathbf{a}_j$, $j = 1, 2, \ldots, p$, where Σ is the variance–covariance matrix of $\mathbf{X}$. Hence, the maximization of the variance is equivalent to maximizing the quadratic form defined by the Σ matrix. The following theorem provides the solution.

Theorem 7.1 Let Σ be the variance–covariance matrix of the random vector $\mathbf{X} = \begin{bmatrix} X_1, X_2, \ldots, X_p \end{bmatrix}^\mathsf{T}$ having eigenvalue-normalized-eigenvector pairs $(\lambda_1, \mathbf{e}_1)$, $(\lambda_2, \mathbf{e}_2), \ldots, (\lambda_p, \mathbf{e}_p)$, where the eigenvalues are ordered so that $\lambda_1 \ge \lambda_2 \ge \cdots \ge \lambda_p \ge 0$. Then the jth principal component is given by

$$Y_j = \mathbf{e}_j^\mathsf{T}\mathbf{X} \tag{7.2}$$

and

$$\text{Var}(Y_j) = \mathbf{e}_j^T \Sigma \mathbf{e}_j = \lambda_j, \quad j = 1, 2, \ldots, p. \tag{7.3}$$

Property 7.1 The principal components are uncorrelated since $\text{Cov}(Y_i, Y_j) = \mathbf{e}_i^T \Sigma \mathbf{e}_j = \lambda_j \mathbf{e}_i^T \mathbf{e}_j = 0$ for $i \neq j$.

Proof. See Problem 7.12.

Remark 7.1 In practice, the matrix Σ is unknown, and principal component analysis is performed on the sample estimate $\mathbf{S}$ of Σ. In order to simplify notation, we will mostly denote the eigenvalues and eigenvectors of $\mathbf{S}$ with the same symbols $(\lambda_i, \mathbf{e}_i)$, without emphasizing that these are estimated quantities.

The coordinates of the vector $\mathbf{e}_j$ denoted as $e_{1j}, e_{2j}, \ldots, e_{pj}$ are the coefficients of the jth principal component as a linear combination of the original variables, that is, equation (7.2) can be written as

$$Y_j = e_{1j}X_1 + e_{2j}X_2 + \cdots + e_{pj}X_p. \tag{7.4}$$

The coefficients $e_{1j}, e_{2j}, \ldots, e_{pj}$ are called *loadings* of the jth principal component.

Example 7.1 (cont.) For the radiopaque marker data, we can calculate the estimated variance–covariance matrix

$$\mathbf{S} = \begin{bmatrix} 0.2585 & 0.1269 \\ 0.1269 & 0.1049 \end{bmatrix}, \tag{7.5}$$

which has the eigenvalues $\lambda_1 = 0.330$ and $\lambda_2 = 0.033$ and eigenvectors $\mathbf{e}_1 = [0.871, 0.491]^T$ and $\mathbf{e}_2 = [0.491, -0.871]^T$. Since the screen measurements are assumed to be unbiased, the estimated position of the oscillation axes should go through the sample mean point $\bar{\mathbf{x}}$, and it can be written in the parametric form as $P_1 = \{t \cdot \mathbf{e}_1 + \bar{\mathbf{x}} : t \in R\}$ shown in Figure 7.3. The variability along P_1 is due to the joint effect of the oscillation and measurement error. On the other hand, the variability in the orthogonal direction $\mathbf{e}_2$ (plotted as $P_2 = \{t \cdot \mathbf{e}_2 + \bar{\mathbf{x}} : t \in R\}$ in Figure 7.3) is only due to the measurement error. So, the standard deviation of the measurement error can be estimated as $s_{\text{measure}} = \sqrt{\lambda_2} = 0.18$. If we model the oscillations as being random and independent of the measurement error, we can write that

$$s_{\text{oscillation}}^2 + s_{\text{measure}}^2 = \lambda_1 = 0.330 \tag{7.6}$$

and $s_{\text{oscillation}} = \sqrt{\lambda_1 - \lambda_2} = 0.545$. By using principal component analysis, we are able to estimate the position of the main axis of the marker oscillation, its variability in that direction, and the precision of measurements. □

The geometric interpretation discussed in Example 7.1 and seen in Figure 7.3 can be generalized to the geometry in p dimensions. The eigenvectors $\mathbf{e}_1, \mathbf{e}_2, \ldots, \mathbf{e}_p$ of the variance–covariance matrix define orthogonal directions in which the variability of the random vector $\mathbf{X}$ is sequentially maximized, and the projections on those directions (principal components) are uncorrelated. If we denote by $\mathbf{P}$ the matrix of eigenvectors as columns, the vector $\mathbf{Y}$ of principal components can be written as $\mathbf{Y} = \mathbf{P}^\mathrm{T} \mathbf{X}$. Since $\mathbf{P}^\mathrm{T}$ is an orthogonal matrix representing a rotation (and possibly a permutation of coordinates), PCA can be viewed as a rotation of the data into a more convenient system of coordinates, where the new variables (i.e., principal components) are uncorrelated. We can also express $\mathbf{X}$ in the new system of coordinates $\mathbf{e}_1, \mathbf{e}_2, \ldots, \mathbf{e}_p$ as follows. Since $\mathbf{P}$ is an orthogonal matrix, we have $\mathbf{P}^{-1} = \mathbf{P}^\mathrm{T}$, and we can write $\mathbf{X} = \mathbf{P}\mathbf{Y}$ or

$$\mathbf{X} = Y_1 \cdot \mathbf{e}_1 + Y_2 \cdot \mathbf{e}_2 + \cdots + Y_p \cdot \mathbf{e}_p, \tag{7.7}$$

which shows that $\mathbf{X}$ is a linear combination of the eigenvectors with principal components as random coefficients.

Since the matrix $\boldsymbol{\Sigma}$ describes the variability around $\boldsymbol{\mu} = E(\mathbf{X})$, it is often convenient to use the centered random vector $\mathbf{X}_{\text{centered}} = \mathbf{X} - \boldsymbol{\mu}$. All formulas above can also be written with $\mathbf{X}$ replaced by $\mathbf{X}_{\text{centered}}$. The variance–covariance matrix of $\mathbf{X}_{\text{centered}}$ is still the same matrix $\boldsymbol{\Sigma}$ with the same eigenvectors. In that case, the principal components are also centered and are given by the formula $Y_j^{\text{centered}} = \mathbf{e}_j^\mathrm{T} \mathbf{X}_{\text{centered}}, j = 1, \ldots, p$.

In Chapter 5, we defined the total variability as the sum of variances of all X_i variables, which also turned out to be equal to the sum $\lambda_1 + \lambda_2 + \cdots + \lambda_p$ of eigenvalues (see formula (5.27)). We can now see that the sum is also equal to the total variability in the principal components. This makes sense since the rotation of data does not change its variability.

It is often of interest to calculate the percent of variability accounted for by the ith principal component, which can be calculated from the formula

$$\frac{\lambda_i}{\lambda_1 + \lambda_2 + \cdots + \lambda_p}. \tag{7.8}$$

The following example demonstrates this concept.

Example 7.2 A white colored tile was measured eight times using an X-Rite Series 500 Spectrodensitometer. The resulting 31-band spectral curves are shown in Figure 7.4. This is a subset of Spectrometer Data. The 31 spectral bands (covering the range from 400 to 700 nm wavelength) are treated as random component variables $X_1, X_2, \ldots, X_{31}$, and we have eight observation vectors on those variables. The sample variance–covariance matrix $\mathbf{S}$ is a 31 by 31 dimensional matrix of rank 7 (see inequality (5.24)). We want to investigate the nature of the measurement errors in these measurements. We assume our measurements to be unbiased, and we get the estimates of the measurement errors by subtracting the mean spectral curve $\bar{\mathbf{x}}$ (vector of the components' sample means) from all eight spectral curves. The resulting eight

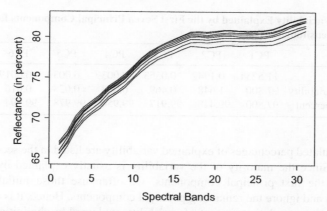

Figure 7.4 Eight reflectance spectra of a white colored tile as discussed in Example 7.2.

31-dimensional vectors are shown in Figure 7.5. The errors range from (-1.22) to 0.98 in percent of reflectance. The variability of measurements as described by the matrix S also describes the variability of the measurement error because the same tile was measured each time.

The first seven eigenvalues of S are listed in Table 7.1. The remaining eigenvalues are zero because of the rank of S equal to 7. The sum of those seven eigenvalues is the total variability in the data set.

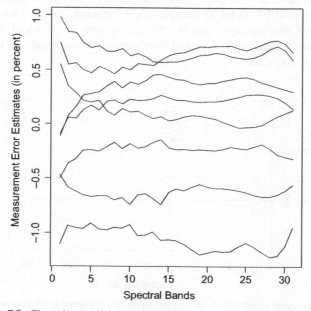

Figure 7.5 The estimates of the measurement errors, as discussed in Example 7.2.

Table 7.1   Variability Explained by the First Seven Principal Components for Example 7.2 Spectrodensitometer Data

	PC1	PC2	PC3	PC4	PC5	PC6	PC7
Eigenvalues	11.5254	0.1942	0.0553	0.0039	0.0033	0.0019	0.0007
Percent of variability	97.800	1.648	0.469	0.033	0.028	0.016	0.006
Cumulative percent	97.800	99.448	99.917	99.950	99.978	99.994	100.000

The calculated percentages of explained variability are listed in the second row of Table 7.1. Since the majority of the variability is usually explained by a certain number of the first principal components, we often use those initial principal components and ignore the remaining principal components. Hence, it is also useful to calculate the cumulative percent of variability explained by the initial principal components. Those numbers are shown in the third row of Table 7.1.

We can see that 97.8% of the measurement error variability is explained by the first principal component. When we also add the second principal component, explaining 1.648%, the first two principal components together explain 99.448% of variability. This means that almost all of the information about the measurement error is contained in one or two variables (principal components). This is a considerable simplification relative to 31 spectral bands.

7.2.3   Interpretation of Principal Component Loadings

Since the first several principal components tend to explain a large amount of variability, it would be useful to interpret their meaning. In order to do that, we need to inspect the loadings, that is, the coefficients of principal components. Figure 7.6 shows the loadings of the first three principal components for Example 7.2 spectrodensitometer data.

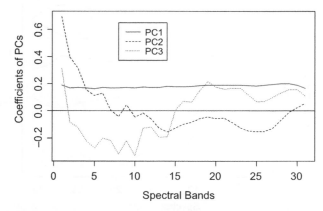

Figure 7.6 Coefficients (loadings) of the first three principal components for Example 7.2 spectrodensitometer data.

The coefficients of the first principal component (PC1) are all positive and are of almost the same magnitude. This means that PC1 is (approximately) proportional to the average over all spectral bands, and it represents an overall magnitude of reflectances in all bands. In the context of variability, PC1 represents the movement of the spectral curve up and down in a parallel way, that is, by approximately the same amount in all bands. To see this, let us investigate what happens when a curve moves up in a parallel way. Since all X variables will increase, the first principal component will also increase. On the other hand, each of the remaining principal components will be almost unchanged because the sum of coefficients for those principal components is close to zero. We can also see this from a different perspective by using equation (7.7). Changing the value of Y_1 will impact all components of $\mathbf{X}$ in approximately the same way because the components of $\mathbf{e}_1$ are approximately the same.

We can see that the spectral curves in Figure 7.4 are largely parallel to each other, which is exactly the variability explained by PC1 in this case. So, it is not surprising that the movement up and down (expressed by PC1) accounts for most of the variability. The same parallel relation is also seen in Figure 7.5.

Figure 7.6 shows that the first three coefficients of the second principal component are significantly larger in magnitude (absolute value) than the remaining coefficients. This means that PC2 is largely dominated by the first three spectral bands, and we would interpret it as explaining variability in the first three spectral bands. To obtain an alternative interpretation, we could contrast the substantially positive coefficients 1–6 with the substantially negative coefficients 13–27.

In order to better interpret the meaning of principal components, it is helpful to introduce the concept of *impact plots*. Formula (7.7) holds for each observation, so it must also hold for the mean of all observations. That is, we can express the sample mean vector $\bar{\mathbf{x}}$ as the function of principal component sample means as follows:

$$\bar{\mathbf{x}} = \bar{y}_1 \cdot \mathbf{e}_1 + \bar{y}_2 \cdot \mathbf{e}_2 + \cdots + \bar{y}_p \cdot \mathbf{e}_p. \tag{7.9}$$

The impact of a given principal component can be measured by how much an observation spectrum vector $\mathbf{x}$ would change due to a change in the principal component value. Let's assume that the jth principal component value is k standard deviation units $\left(= \sqrt{\lambda_j} \right)$ away from the mean, and the remaining principal components have values equal to their means. We want to see how different a hypothetical vector

$$\mathbf{x} = \left(\bar{y}_j + k\sqrt{\lambda_j} \right) \cdot \mathbf{e}_j + \sum_{i \neq j} \bar{y}_i \cdot \mathbf{e}_i \tag{7.10}$$

is from $\bar{\mathbf{x}}$. Based on equations (7.9) and (7.10), we can write that $\mathbf{x} = \bar{\mathbf{x}} + k\sqrt{\lambda_j} \cdot \mathbf{e}_j$. We can say that we start at the point $\bar{\mathbf{x}}$, and move by the distance $k\sqrt{\lambda_j}$ in the direction of $\mathbf{e}_j$. Figure 7.7 shows the values of such $\mathbf{x}$ for $j = 2$, that is, the impact of changes in the second principal component. The mean spectrum $\bar{\mathbf{x}}$ of all spectra is shown as a solid line. The curve of *positive impact* is for $k = 5$ and the *negative impact* is for $k = -5$. We could interpret the second principal component as

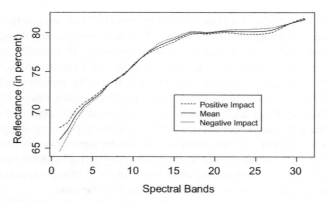

Figure 7.7   The impact of varying the value of the second principal component on the spectral curve based on Example 7.2 data.

explaining variability due to "tilting" in the shape of spectra (as shown in Figure 7.7), where the curve goes up in the lower spectral bands and down in the higher spectral bands for positive k. The tilting goes in the opposite direction for negative k.

The value of $k = 5$ used in Figure 7.7 is rather excessive. A more reasonable value of k would be around 2 or 3 based on the two- and three-sigma rules, but the effect would not be easily seen in the graph due to a much larger variability of reflectance from band to band. To overcome this difficulty, we plot the measurement errors $\mathbf{x} - \bar{\mathbf{x}} = k\sqrt{\lambda_j} \cdot \mathbf{e}_j$ in Figure 7.8. The position of the mean $\bar{\mathbf{x}}$ is represented as the solid line at the zero level. Here we use $k = 2$ in order to reflect the two-sigma limits of the principal component variability. The band variability, calculated as two times the band sample standard deviation, is plotted (dot-dashed lines) as a point of reference to

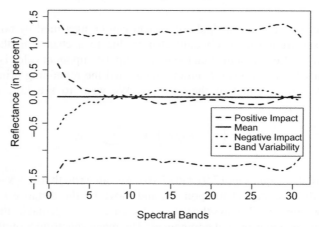

Figure 7.8   The impact of varying the value of PC2 on the measurement errors based on Example 7.2 data (with $k = 2$). The plotted band variability (dot-dashed lines) equals two times the band sample standard deviations.

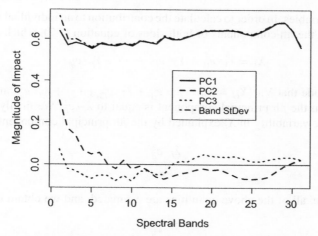

Figure 7.9   The impact plot for the first three principal components in the context of Example 7.2 data.

the variability of the whole data set. We can see a rather small impact of the second principal component in most spectral bands, except for the first three bands.

In order to plot the impact of several principal components, we propose an *impact plot* (see Figure 7.9), where the *impact curves* $\sqrt{\lambda_j} \cdot \mathbf{e}_j$ are plotted. For the $\mathbf{x}$ described in formula (7.10), we have $\mathbf{x} - \bar{\mathbf{x}} = k\sqrt{\lambda_j} \cdot \mathbf{e}_j$. Hence, the impact curve describes the impact on $\mathbf{x}$ of the change in the jth principal component by one standard deviation ($k = 1$). Figure 7.9 can be regarded as a simplified version of Figure 7.8, with only one curve representing each type of variability. The actual variability is described by the plus/minus limits of an impact curve multiplied by k as shown in Figure 7.8 for PC2. Plots similar to Figure 7.8 could be created for the other principal components. However, with some experience, one can draw the same conclusions from Figure 7.9.

The PC1 impact curve is approximately horizontal, which means that the resulting impact on $\mathbf{x}$ will be the up and down movement with respect to $\bar{\mathbf{x}}$ as explained before. The PC3 can be interpreted as a contrast between Bands 4–14 and Bands 18–31, with the tilting similar to that of PC2 but for a different set of spectral bands. The sample standard deviations of spectral bands are plotted as the dot-dashed line, which mostly overlaps with the PC1 line. This reflects the large fraction of variability being explained by the first principal component. The lines do not overlap for the first three bands, where the second principal component contributes most of the remaining variability.

A plot of impact curves is useful when the standard deviations $\sqrt{\lambda_j}$ of principal components are not very different from each other. When they differ a lot, the impact curves with small $\sqrt{\lambda_j}$ are too close to zero level for us to see their variability. In those cases, we can simply plot the vectors $\mathbf{e}_j$ of loadings in order to interpret them.

Another way to evaluate the impact of principal components is to calculate the percent of variability they contribute to individual variables. Formula (7.8) tells us about the contribution to the total variability in all variables jointly, but not for individual variables. In order to calculate the contribution to an individual variable X_i, we can write the ith components of both sides of equation (7.7), which gives

$$X_i = Y_1 \cdot e_{i1} + Y_2 \cdot e_{i2} + \cdots + Y_p \cdot e_{ip}. \tag{7.11}$$

We conclude that $\mathrm{Var}(X_i) = \sigma_{ii} = \lambda_1 \cdot e_{i1}^2 + \lambda_2 \cdot e_{i2}^2 + \cdots + \lambda_p \cdot e_{ip}^2$, and the contribution from the jth principal component is equal to $\lambda_j \cdot e_{ij}^2$. We finally obtain the percent of the variability in X_i explained by the jth principal component as

$$\frac{\lambda_j \cdot e_{ij}^2}{\sigma_{ii}}. \tag{7.12}$$

In practice, all of the above quantities are estimated, and we obtain the formula

$$\frac{\lambda_j \cdot e_{ij}^2}{s_{ii}}, \tag{7.13}$$

where s_{ii} is the estimated variance of X_i, that is, the ith diagonal element of **S**. Applications of these formulas will be demonstrated in the following example.

Example 7.3 This example uses remote sensing data. Some background information about such data can be found in Example 1.3. Here we use data from the Airborne Visible/Infrared Imaging Spectrometer (AVIRIS), which is a sensor collecting spectral radiance in the range of wavelengths from 400 to 2500 nm. It has been flown on various aircraft platforms, and many images of the Earth's surface are available. Figure 7.10 shows a 100 by 100 pixel AVIRIS image of an urban area in Rochester, NY, near the Lake Ontario shoreline. The scene has a wide range of natural and man-made material including a mixture of commercial/warehouse and residential neighborhoods, which adds a wide range of spectral diversity. Prior to processing, invalid bands (due to atmospheric water absorption) were removed, reducing the

Figure 7.10 Color rendering of the AVIRIS scene in Rochester, NY, used in Example 7.3.

Table 7.2   Variability Explained by the First Seven Principal Components for the AVIRIS Image from Example 7.3

	PC1	PC2	PC3	PC4	PC5	PC6	PC7
Eigenvalues	102,453,181	21,445,196	399,655	227,236	117,133	81,133	46,802
Percent of variability	82.06	17.18	0.32	0.18	0.09	0.06	0.04
Cumulative percent	82.06	99.23	99.55	99.73	99.83	99.89	99.93

overall dimensionality to 152 bands. This image has been used in Bajorski et al. (2004) and Bajorski (2011a, 2011b).

When the $p = 152$ spectral bands are treated as variables, and the $n = 10,000$ pixels are treated as observations, one can calculate the resulting eigenvalues of the sample variance–covariance matrix **S** of the whole image. This will tell us the variability explained by principal components. The first seven eigenvalues and the percent variability explained by the first seven principal components are given in Table 7.2.

We can see that 99.23% of the image variability is explained by the first two principal components. Based on research in target detection methods (see Bajorski et al. (2004)), it is known that the remaining principal components still contain important information, and at least 20 principal components should be used for describing this data set. We will also see this later on when discussing residuals in Section 7.5. This means that what would otherwise be considered an overwhelming amount of explained variability is still not sufficient in the context of this spectral data set. This is different from Example 7.2, where a similar percent of explained variability in the measurement error was judged to be sufficient.

Figure 7.11 shows the impact plot for the first three principal components. We can interpret the first principal component as mainly the weighted average of radiance in

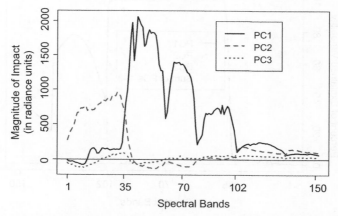

Figure 7.11 The impact plot for the first three principal components for the AVIRIS image discussed in Example 7.3.

the spectral bands 35–100. The remaining coefficients of PC1 are very small, and they have little impact on the PC1 value. The second principal component is mainly the weighted average of radiance in the first 35 bands. In order to see the proportion of variability that each principal component explains in the individual spectral bands, we could also plot values of the bands' standard deviations as we did in Figure 7.9. However, the resulting curve mostly overlaps the PC1 and PC2 curves in the plot, which then becomes difficult to read.

In order to overcome this difficulty, we can calculate the fraction of explained variability based on formula (7.13). For PC1, it becomes $\lambda_1 \cdot e_{i1}^2 / s_{ii}$ for the ith spectral band. The variability in the ith spectral band explained by the first k principal components becomes

$$\frac{1}{s_{ii}} \sum_{j=1}^{k} \lambda_j \cdot e_{ij}^2. \tag{7.14}$$

Figure 7.12 shows the percentages of variability explained by the first three principal components. The first PC explains almost all (above 98%) variability in Bands 39–100. The second PC explains over 86% of the variability in each of the first 34 bands. In bands above 102, both PC1 and PC2 contribute significantly, and PC3 contributes between 3% and 5% in those bands. We can contrast these values with the overall percentages of the explained variability as given in Table 7.2. For example, PC3 explains only 0.32% of the total variability, but as much as 3–5% in some bands.

The cumulative percent variability explained by the first three principal components is plotted as a dot-dashed line in Figure 7.12. We can see that almost all variability is explained by those PCs, but in some bands, up to 10% is still not explained.

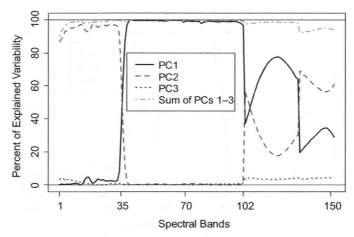

Figure 7.12 Percentages of variability explained by the first three principal components within the spectral bands of the AVIRIS image.

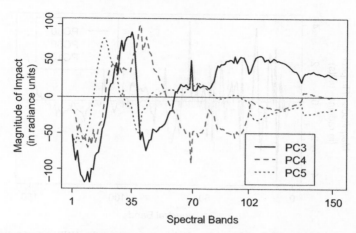

Figure 7.13 The impact plot for the selected principal components for the AVIRIS image discussed in Example 7.3.

The third principal component was difficult to interpret based on Figure 7.11 due to the very small values of impact (because of relatively small $\sqrt{\lambda_3}$). To deal with this obstacle, we could plot the third eigenvector directly without multiplying by $\sqrt{\lambda_3}$. An alternative solution is to create an impact plot of PC3 together with the subsequent principal components, as done in Figure 7.13. Each of the three PCs shown is a contrast between the bands where a given line is above the zero level versus bands where the line is below the zero level.

The concept of a contrast was explained in this section when interpreting loadings in Figure 7.6 and impact curves in Figure 7.9. Here we will show an interpretation of PC3 based on Figure 7.13. We first identify the PC3 impact coefficients $\sqrt{\lambda_3}e_{i3}$ (values shown in Figure 7.13) that are considerably different from zero. We do this based on an arbitrary decision that the coefficients larger than 25 or smaller than -25 are considerable. This leads to an interpretation of PC3 as a contrast between a positive impact of Bands 25–37 and 83–152 versus a negative impact of Bands 1–20 and 39–55. When the value of PC3 increases, the values in Bands 25–37 and 83–152 will also increase (assuming other PCs being constant) and the values in Bands 1–20 and 39–55 will decrease.

Figure 7.14 shows the impact plot for PCs 30, 60, and 100. Those are again contrasts between various spectral bands. The high frequency of oscillations suggests some random behavior and a large impact of image noise. We will provide further interpretations of principal components through their values (called scores) in Section 7.4.

7.2.4 Scaling of Variables

The principal components are defined as linear combination of the original variables $X_1, X_2, \ldots, X_p$. This means an implicit assumption about the comparability of those

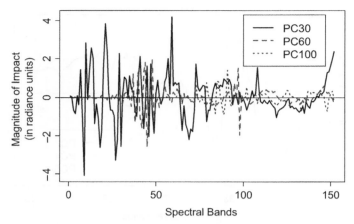

Figure 7.14 The impact plot for selected principal components for the AVIRIS image discussed in Example 7.3.

variables. If one of the variables measures a physical quantity different from those measured by other variables, then combining those variables directly is usually not meaningful. If one variable measured reflectance of a surface in percent, and another variable measured the distance to the surface in meters, then the result of principal component analysis would depend on the specific units used here. If the distance units were changed to centimeters, the variance of that variable would increase 10,000 times, which would significantly change the whole analysis. When the variables are in different physical units, we need to standardize them before performing principal component analysis. In some situations, it might be possible to standardize based on the subject matter knowledge of the variables, for example, by assessing their impact on the process under consideration. In other situations, we need to rely on statistical standardization, where the sample mean is subtracted and the result is divided by the sample standard deviation. Such standardized variables have variance equal to 1, so they can be directly compared. The variance–covariance matrix of the statistically standardized variables is equal to the correlation matrix of the original variables. Hence, we often talk about principal component analysis on the correlation matrix. Most elements of the analysis are the same whether the variance–covariance or correlation matrices are considered. One exception is the statistical inference discussed in Section 7.6. In all other cases, we will refer to the analysis of the estimated variance–covariance matrix $\mathbf{S}$ with the understanding that if the variables are statistically standardized, the matrix $\mathbf{S}$ is the same as the correlation matrix $\mathbf{R}$ of the original variables.

Sometimes, the variables are in the same units, but it is not easy to tell if they should be compared directly. Should the radiance in the visible range of radiation be directly compared to the infrared radiation? If the answer is "no," we should standardize each of the spectral bands before further analysis. This may vary depending on a specific data set. If the signal to noise level is much lower in some spectral bands than in others, then the standardization will amplify the noise from the noisy bands. This will result in an unduly large effect of noise on the analysis. There are also other factors at play, and there is no general agreement on whether spectral data should be standardized within bands. In this book, we do not standardize the spectral data.

7.3 STOPPING RULES FOR PRINCIPAL COMPONENT ANALYSIS

One purpose of PCA is to simplify a larger set of variables into a much smaller set of principal components, which leads to dimensionality reduction of data. We then need to have some stopping rules that would assist us in deciding on the number of principal components to be taken into account. For example, the first k principal components may contain most of the information contained in the whole set of p variables. We can say that we reduce the dimensionality of the data set from p to k, or that the effective or intrinsic dimensionality of the data is equal to k. The decision about a reasonable choice of k will usually depend on the meaning of variability in the specific application, on our ability to interpret principal components, and on other practical considerations. Nevertheless, it is useful to have a set of stopping rules from which to choose.

One such stopping rule is based on the percent of variability explained by the first k principal components. We may decide that once a threshold of 90% or 95% is crossed, the resulting principal components might be sufficient for practical purposes. Using this rule in Example 7.2 would result in one or two principal components, when using a threshold of 95% or 99%, respectively. The percent-of-explained-variability method can be appealing in some applications. For example, when investigating sources of measurement error, we might be satisfied with addressing 99% of variability without worrying about the remaining 1%. On the other hand, in some other imaging applications, this method is not satisfactory. For example, in the context of imaging spectrometer data, the total variability in the image is often very large due to the spectral variability of the materials present in the image scene. Consequently, a small percent of unexplained variability may still play a significant role.

In the context of spectral images, the number k of principal components required to describe the image is usually quite large but still significantly smaller than the actual dimensionality p. The number k is often called the *intrinsic linear dimensionality* because the principal components represent variability in linear directions. They also define an affine subspace in the p-dimensional space of the original observations, which will be formalized in Section 7.5. The dimensionality k can also be called the *intrinsic global linear dimensionality* because it is based on the global variance–covariance matrix estimated from the whole image. This would be in contrast to variance–covariance matrices that can be calculated from subsets of all pixels, for example, only pixels representing certain types of ground cover such as grass areas or water surface.

Example 7.4 Here again, we use remote sensing data. Some background information about such data can be found in Example 1.3. Let us consider the 280 by 800 pixel

Table 7.3   Variability Explained by the First Eight Principal Components for the Cooke City image Data

	PC1	PC2	PC3	PC4	PC5	PC6	PC7	PC8
Percent of variability	88.94	9.08	0.96	0.65	0.16	0.05	0.04	0.03
Cumulative percent	88.94	98.02	98.98	99.64	99.80	99.85	99.89	99.92

Figure 7.15   Color rendering of the 280 by 800 pixel HyMap Cooke City image (see Snyder et al. (2008) for details about the image).

HyMap Cooke City image shown in Figure 7.15, where each pixel is described by a 126-band spectrum (see Snyder et al. (2008) for details about the image). When the $p = 126$ spectral bands are treated as variables, and the $n = 224,000$ pixels are treated as observations, one can calculate the resulting eigenvalues of the sample variance–covariance matrix S, which tell us the variability explained by principal components.

Table 7.3 shows the percentages of the variability explained by the first eight PCs. Even with the threshold of 99.9% of variability, we would opt for only eight principal components, while in fact it is known that the remaining principal components still contain important information. This is analogous to our experience with other spectral images as discussed in Example 7.3 for an AVIRIS image. We will also verify this later on when discussing residuals in Section 7.5. Of course, we could further increase the percentage threshold. However, there is no clear guideline as to where this process should be stopped, and the high percentages no longer give us any intuitive feel for a specific value that should be used.

Another stopping rule for principal components is based on the plot of eigenvalues as shown in Figure 7.16a for the Cooke City image. We are looking for an elbow in the shape of the line, indicating a sudden drop in values. In this case, we would decide on retaining the first two principal components based on the elbow shape. This result is even less satisfactory than the previous selection of eight principal components.

When the eigenvalues differ by orders of magnitude, it is convenient to use the logarithmic scale for better differentiation of the values, as shown in Figure 7.16b. However, in this case, it is still unclear where the threshold for the eigenvalues should be.   □

7.3.1   Fair-Share Stopping Rules

Here we discuss stopping rules based on the fair share of variability expected from a given principal component. If all variability was equally divided into all principal

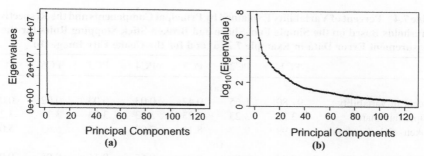

Figure 7.16 A plot of eigenvalues (often called a scree plot) is shown in (a) for the Cooke City image. A logarithmic scale is used in plot (b) for better discrimination of the values that differ by orders of magnitude.

components, the share of each PC would be equal to $(\lambda_1 + \lambda_2 + \cdots + \lambda_p)/p$, where $\lambda_1 \geq \lambda_2 \geq \cdots \geq \lambda_p \geq 0$ are the eigenvalues of the estimated variance–covariance matrix **S**. The *simple fair-share* stopping rule identifies the largest k such that λ_k is larger than its fair share, that is, larger than $(\lambda_1 + \lambda_2 + \cdots + \lambda_p)/p$. We would then use the first k principal components.

If one was concerned that the above method produces too many principal components, a broken-stick rule could be used. The rule is based on the fact that if a line segment of a unit length is randomly divided into p segments, the expected length of the jth longest segment is given by

$$a_j = \frac{1}{p}\sum_{i=j}^{p}\frac{1}{i}, \quad j = 1, \ldots, p, \tag{7.15}$$

which is used as the fair fraction of variability for the jth principal components. Note that $a_1 > a_2 > \cdots > a_p > 0$, so we expect larger variability in earlier principal components. The *broken-stick* stopping rule identifies the principal components with the fraction of variability larger than the fair fraction a_j. The rule tells us to stop at the largest k such that $\lambda_j/(\lambda_1 + \lambda_2 + \cdots + \lambda_p) > a_j$, for all $j \leq k$. The broken-stick stopping rule threshold is usually larger than the simple fair-share threshold, indicating the same or lower intrinsic dimensionality (see Problem 7.7).

We can see from Table 7.4a that both methods would select only the first principal component for the Example 7.2 spectrodensitometer data, which is acceptable because we would explain 97.8% of the measurement error.

However, both rules do not work well for detection of the intrinsic global linear dimensionality of imaging spectrometer data. For the Cooke City image, the two methods would produce intrinsic dimensionalities of 2 or 3 (see Table 7.4b), which are not satisfactory as explained earlier in Example 7.4.

A common scenario when the simple fair-share and the broken-stick stopping rules do not work well is the situation with eigenvalues differing by orders of magnitude. For example, when the first principal component accounts for 90% of variability, not

Table 7.4   Percent of Variability Explained by Principal Components and the Respective Thresholds Based on the Simple Fair-Share and Broken-Stick Stopping Rules for the Measurement Error Data in Example 7.2 (a) and for the Cooke City Image (b)

	PC1	PC2	PC3	PC4	PC5	PC6	PC7
			(a)				
Percent of variability	97.80	1.65	0.47	0.03	0.03	0.02	0.01
Simple fair share	3.23	3.23	3.23	3.23	3.23	3.23	3.23
Broken-stick threshold	12.99	9.77	8.15	7.08	6.27	5.63	5.09
			(b)				
Percent of variability	88.94	9.08	0.96	0.65	0.16	0.05	0.04
Simple fair share	0.79	0.79	0.79	0.79	0.79	0.79	0.79
Broken-stick threshold	4.30	3.51	3.11	2.84	2.65	2.49	2.36

much is left for the remaining principal components and the threshold is crossed very quickly. One could argue that we should consider the magnitude of a given eigenvalue with respect to the remaining lower values (without the impact of the larger values). This leads to the *relative broken-stick* rule, where we analyze λ_j as the first eigenvalue in the set $\lambda_j \geq \lambda_{j+1} \geq \cdots \geq \lambda_p$ of eigenvalues, where $j < p$. This stopping rule is based on the expected length of the longest segment if a line segment of unit length is randomly divided into $(p - j + 1)$ segments, which is equal to a_1 defined in (7.15) with p being replaced by $(p - j + 1)$. This means the threshold is equal to

$$ b_j = \frac{1}{p - j + 1} \sum_{i=1}^{p-j+1} \frac{1}{i}, \tag{7.16} $$

that is, the dimensionality k is chosen as the largest value such that $\lambda_j / (\lambda_j + \cdots + \lambda_p) > b_j$, for all $j \leq k$. The dashed line in Figure 7.17 shows the b_j

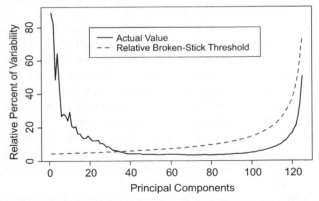

Figure 7.17   The threshold values b_j (the dashed line) based on the relative broken-stick rule calculated for the Cooke City image. The solid line represents the relative percent of explained variability calculated as $\lambda_j / (\lambda_j + \cdots + \lambda_p)$.

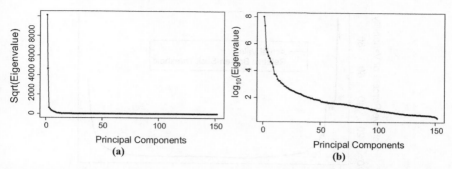

Figure 7.18 A plot of square roots of eigenvalues is shown in (a) for the AVIRIS image used in Example 7.3. A logarithmic scale is used in plot (b) for better discrimination of the values that differ by the orders of magnitude.

threshold values calculated for the Cooke City image. The solid line represents the relative percent of explained variability calculated as $\lambda_j/(\lambda_j + \cdots + \lambda_p)$. The two lines intersect for the k value between 34 and 35. This points to 34 as the intrinsic global linear dimensionality of the Cooke City image based on the relative broken-stick rule, which is an acceptable number based on our earlier discussion in Example 7.4.

Example 7.5 As a continuation of Example 7.3, we would like to identify the intrinsic global linear dimensionality of the AVIRIS image. We could start with plotting the eigenvalues in a scree plot as was done in Figure 7.16a for the Cooke City image. Another possibility is to plot the standard deviation of the principal components, that is, the square roots of eigenvalues, as done in Figure 7.18a. We can see the principal component variability dropping quickly to very small values. Figure 7.18b shows the logarithms of eigenvalues, which follow a steadily decreasing curve without a clear cutoff point.

In order to apply the relative broken-stick rule, we calculated the b_j threshold values using equation (7.16) and plotted them as the dashed line in Figure 7.19. The solid line represents the relative percent of explained variability calculated as $\lambda_j/(\lambda_j + \cdots + \lambda_p)$. The two lines intersect for the k value between 34 and 35. This again points to 34 as the intrinsic global linear dimensionality of the AVIRIS image based on the relative broken-stick rule. We incidentally obtained the same dimensionality as the one for the Cooke City image.

7.3.2 Large-Gap Stopping Rules

Another class of stopping rules is based on a larger gap in eigenvalues. It is convenient to assess the size of gaps based on the proportion $f_j = (\lambda_j - \lambda_{j+1})/\lambda_{j+1} = \lambda_j/\lambda_{j+1} - 1$, which is a relative measure of the size of the gap between λ_j and λ_{j+1} as a fraction of λ_{j+1}. It turns out that in many imaging spectrometer data sets, the proportions f_j have positive values close to zero for large j. That is, here we want to

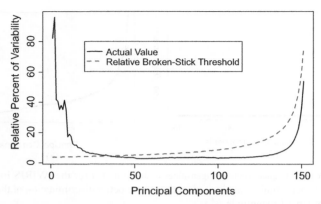

Figure 7.19 The threshold values b_j (the dashed line) based on the relative broken-stick rule calculated for the AVIRIS image. The solid line represents the relative percent of explained variability calculated as $\lambda_j / (\lambda_j + \cdots + \lambda_p)$.

start with large j values and then move backward to the smaller values of j for which f_j tends to get much larger. Moreover, the values f_j, for large j, tend to follow the gamma distribution (see Appendix A for information about the gamma distribution). This motivates the following algorithm that moves backward from the larger to the smaller j as the index of the eigenvalues. The approach was introduced in Bajorski (2011a) in the context of defining the *second moment linear dimensionality* as the intrinsic global linear dimensionality of imaging spectrometer data sets. The following algorithm clarifies the details of calculating the second moment linear dimensionality.

Step 1. Choose an initial set of the "last" proportions $f_{p-m}, \ldots, f_{p-1}$, such that we expect the dimensionality to be smaller than $(p - m)$. We recommend taking $m = [p/3] + 1$, where $[x]$ denotes the integer part of x. Choose the significance level α. We recommend using α equal to $0.05/(p - m - 1)$.

Step 2. Starting with $j = p - m$, fit a gamma distribution to the sample $f_j, \ldots, f_{p-1}$. The easiest way to do this is to use the method of moments, and estimate the scale parameter β using $s^2 / \bar{x}$ and the shape parameter α using $\bar{x}^2 / s^2$, where s^2 is the sample variance.

Step 3. Calculate the p-score of f_{j-1} equal to the upper-tail probability based on the gamma distribution with parameters estimated in Step 2, that is, p-score $= 1 - G(f_{j-1}) = \int_{f_{(j-1)}}^{\infty} g(x)dx$, where G is the cumulative distribution function of the gamma distribution. The p-score tells us how extreme f_{j-1} is in relation to $f_j, \ldots, f_{p-1}$. If the p-score is smaller than or equal to α, then f_{j-1} can be treated as an outlier, which no longer follows the gamma distribution of the other f fractions. The number $(j - 1)$ is called the second moment linear dimensionality at level α. If the p-score is larger than α, continue by going to Step 2 with $j = j - 1$.

Here are some comments regarding the above steps. The algorithm is not particularly sensitive to the choice of m in Step 1. In Step 2, one could also use the maximum likelihood estimates of parameters, but they require a numerical solution of an equation involving special functions. We recommend the simpler method of moments, because there is little difference in the p-scores calculated based on the two methods. Since the algorithm requires a large number of statistical inferences (up to $(p - m - 1)$), we recommend using α equal to $0.05/(p - m - 1)$ in order to account for the joint significance level (based on the Bonferroni method; see Result 6.1). Smaller values of α could also be used, and they will lead to lower dimensionality. Hence, the user can choose the α level depending on the depth of a study. A cursory investigation of an image can be successful with fairly low dimensionality (using small α, that is, less sensitive testing), while a more in-depth investigation may require higher dimensionality (using somewhat larger α, that is, more sensitive testing). A specific choice of α can be supported by consideration of information loss associated with a given dimensionality reduction. Specifics will depend on the type of application. For example, the choice of α can be based on the ability to reconstruct the image or based on the efficiency of target detection in the image. There might be situations where the "last" proportions $f_{p-m}, \ldots, f_{p-1}$ follow a distribution different from the gamma distribution. In that case, the algorithm can identify the first outlier from that distribution.

Example 7.6 As a continuation of Example 7.4, we would like to identify the intrinsic global linear dimensionality of the HyMap image of Cooke City shown in Figure 7.15. We implemented the algorithm for the second moment linear dimensionality. Here, we had $p = 126$, so the algorithm started with $j = p - [p/3] + 1 = 85$. The "last" proportions $f_{85}, \ldots, f_{125}$ turned out to follow a gamma distribution. Then the fraction f_{84} was added, resulting again in a gamma distribution, with slightly different estimated parameters. The procedure then continued until $j = 28$ when the "last" proportions $f_{28}, \ldots, f_{125}$ ($p = 126$) were again confirmed to follow a gamma distribution. Those values are plotted in Figure 7.20 as small circles (excluding the solid circle in the top right-hand corner). The circles line up reasonably well along the straight line $y = x$, showing that the observations follow a gamma distribution. The fraction f_{27} (shown in the top right-hand corner as a solid circle) is significantly larger and can be regarded as an outlier. This was confirmed by the p-score equal to 2.7×10^{-4}. The algorithm stopped here, based on the recommended level $\alpha = 0.05/(126 - 43 - 1) = 0.00061$. This signifies the first (counting from the end) larger gap in eigenvalues between the 27th and 28th eigenvalues, which would suggest retaining the first 27 principal components in that image. This gave the second moment linear dimensionality of 27.

For smaller α values, the algorithm would continue beyond 27, resulting in lower dimensionalities. Table 7.5 shows the second moment linear dimensionalities for such smaller α values. Only a limited choice of lower dimensionalities is available. Clearly, the dimensionality of 2 is too low to be considered for this image, but other values could be used for cursory investigations of the image. □

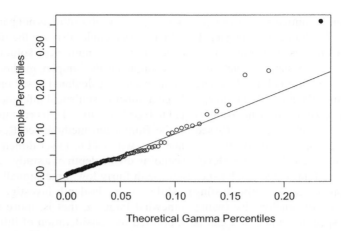

Figure 7.20 A gamma probability plot of the "last" proportions $f_{28}, \ldots, f_{p-1}$ ($p = 126$) for the HyMap image shown in Figure 7.15. The outlier f_{27} is shown in the top right-hand corner as a solid circle.

Table 7.5   The Second Moment Linear Dimensionalities for Various α Levels for the HyMap Image from Figure 7.15

Alpha level	Between 6.1×10^{-4} and 2.7×10^{-4}	Between 2.7×10^{-4} and 9.6×10^{-5}	Between 9.6×10^{-5} and 8.7×10^{-6}	Between 8.7×10^{-6} and 1.9×10^{-6}
SML dimensionality	27	10	5	2

Example 7.5 (cont.) The algorithm for the second moment linear dimensionality was run on the AVIRIS image, and the resulting dimensionalities are shown in Table 7.6. For the recommended level $\alpha = 0.05/(152 - 51 - 1) = 0.0005$, the suggested dimensionality is 18. For smaller α values, a limited choice of lower dimensionalities is available. Those values could be used for cursory investigations of the image when the more detailed information is not needed.

Table 7.6   The Second Moment Linear Dimensionalities for Various α Levels for the AVIRIS Image

Alpha level	Between 5.0×10^{-4} and 4.2×10^{-4}	Between 4.2×10^{-4} and 9.2×10^{-6}	Between 9.2×10^{-6} and 9.9×10^{-10}
SML dimensionality	18	12	9

7.4 PRINCIPAL COMPONENT SCORES

As we have learned earlier in this chapter, each principal component Y_i is a random variable. This means that it will take on a specific value for a given observation vector $\mathbf{x}$. The value would be calculated as $\mathbf{e}_i^T \mathbf{x}$. Prior to this calculation, we usually center each observation vector $\mathbf{x}$ by subtracting the sample mean $\bar{\mathbf{x}}$. The reason is that the variability measured by the estimated variance–covariance matrix $\mathbf{S}$ and considered here is the variability around the mean value $\bar{\mathbf{x}}$. The resulting value is calculated as a linear combination of the coordinates of the vector $(\mathbf{x} - \bar{\mathbf{x}})$ according to the formula

$$y_j = \mathbf{e}_j^T (\mathbf{x} - \bar{\mathbf{x}}). \tag{7.17}$$

The values y_j are called the jth principal component scores. For the ith observation vector $\mathbf{x}_i$, we can calculate its p scores

$$y_{ij} = \mathbf{e}_j^T (\mathbf{x}_i - \bar{\mathbf{x}}), \quad j = 1, \ldots, p. \tag{7.18}$$

It is usually informative to create some scatter plots of scores for pairs of the first few principal components as demonstrated in the following example.

Example 7.7 Upon further inquiries about the spectral measurements discussed earlier in Example 7.2 (and shown in Figure 7.4), we found out that the measurements were in fact taken by two different operators using two X-Rite Series 500 Spectro-densitometers. Some other tiles were also measured within the same experiment, but here we investigate only the white tile measurements (see Appendix B). The operator can be regarded as one factor in this study and the spectrometer as a second factor. The statistical design of the study was the full factorial design with two replicates. The order of runs was generated randomly and is shown in Table 7.7. Although the order of operators might not look random, it was, in fact, generated randomly.

Figure 7.21 shows a scatter plot of scores of the first two principal components for the eight spectral measurements. For $i = 1, \ldots, 8$, the values y_{i2} on the vertical axis are plotted versus y_{i1} on the horizontal axis.

Table 7.7 Order of Runs for Example 7.7 Data

Run Order	Operator	Spectrometer
1	1	1
3	1	1
6	1	2
7	1	2
12	2	1
13	2	2
17	2	1
21	2	2

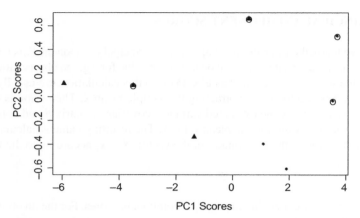

Figure 7.21 A score plot of the Spectrometer Data (Example 7.7). Operator 1 scores are marked with triangles and Operator 2 with dots. Results from Spectrometer 1 are enclosed in circles.

PC1 scores for Operator 1 have a tendency to be lower than those for Operator 2. Recall that PC1 was previously interpreted as being responsible for the movement up and down of the spectral curves, with higher PC1 values representing higher spectra. This means that Operator 1 has a tendency to produce lower spectra versus Operator 2 producing higher spectra. We would need to investigate whether the effect is statistically significant. We can also see a tendency of Spectrometer 1 to have values of PC2 higher than those for Spectrometer 2. This would suggest that the type of tilting observed in Figure 7.7 is different for the two spectrometers. Again, the statistical significance of these conclusions would need to be investigated, especially for such small samples. □

When the principal component scores are calculated for images where the pixels are treated as observations, the scores represent the image pixels, and they can be plotted as an image where the shades of gray indicate the value. There is a tradition in the field of imaging to represent large values in lighter gray and the largest values in white. Unfortunately, this approach often leads to some poor quality displays, where most of the image is almost black and one can see only the spots that are very bright. We believe that in most cases it is a better approach to use black for the largest values and white for the smallest values. This is the approach that we are using throughout this book. The following example shows the principal component scores displayed as an image, so that the spatial information about pixels is preserved.

Example 7.8 This is a continuation of Examples 7.3 and 7.5, where an AVIRIS image of an urban area in Rochester, NY, was used (see Figure 7.10). For each pixel i, where $i = 1, \ldots, n$ and $n = 10,000$, we can calculate the score y_{ij} of the jth principal component given by formula (7.18). The scores are represented as shades of gray in Figure 7.22 with large values shown in darker gray. Each panel shows scores for one principal component. For the PC1 image, we can see a lot of spatial details. Based on

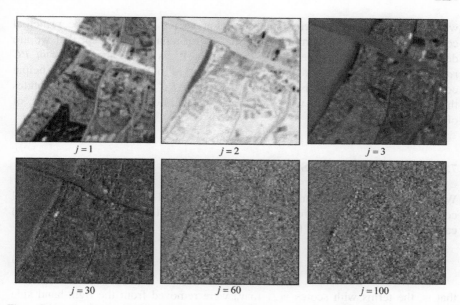

$j = 1$ $j = 2$ $j = 3$

$j = 30$ $j = 60$ $j = 100$

Figure 7.22 The PC scores y_{ij} of the jth principal component given by formula (7.18) are represented as shades of gray with large values shown in darker gray. Each panel shows scores for the principal component with the given value of j.

the interpretation of PC1 in Example 7.3 (see Figure 7.11), these are ground features visible mainly in spectral bands 39–100. The PC2 score image shows ground features visible mainly in the first 34 spectral bands. The third principal component was interpreted in Example 7.3 as a contrast between a positive impact of Bands 25–37 and 83–152 versus a negative impact of Bands 1–20 and 39–55. However, this interpretation assumes scores of other principal components being constant. We can also say that the panel for PC3 in Figure 7.22 reflects the contrast between those bands after the adjustment for other principal components is already made. We can see substantial spatial patterns in the PC3 image, suggesting that important information is included in PC3. The next panel represents PC30 with considerably weaker spatial patterns. This suggests a significant amount of noise in that PC band. The remaining two panels show even more noise in PC bands 60 and 100. We cannot show images of all bands here, but further bands show even more noise, as expected. □

We need to keep in mind that each image of PC scores shows the variability only in one linear direction. We may wonder what information might be contained in several PCs together. For example, is there any important information still included in several PC bands together when the PC bands 101–152 are considered jointly? This question will be answered in Section 7.5 on residuals.

If the scores y_j, $j = 1, \ldots, p$, from equation (7.17) are organized into a vector $\mathbf{y}$, we can write $\mathbf{y} = \mathbf{P}^{\mathrm{T}}(\mathbf{x} - \bar{\mathbf{x}})$ or equivalently $\mathbf{x} = \bar{\mathbf{x}} + \mathbf{P}\mathbf{y}$. This can also be written as

$$\mathbf{x} = \bar{\mathbf{x}} + y_1 \cdot \mathbf{e}_1 + y_2 \cdot \mathbf{e}_2 + \cdots + y_p \cdot \mathbf{e}_p, \qquad (7.19)$$

which clarifies how the original observation vector depends on the principal component scores. This formula is similar to formula (7.7), except that we are dealing with centered scores here. The matrix $\mathbf{P}^{\mathrm{T}}$ is called the *matrix of PC rotation*, because it rotates the original vector of observations into principal components. The matrix $\mathbf{P}$ is the *matrix of inverse PC rotation* because it rotates the principal component values (scores) back into the original vector of observations.

7.5 RESIDUAL ANALYSIS

When PCA is used for dimensionality reduction and only the first k principal components are used to represent the data, the observation $\mathbf{x}$ from equation (7.19) can be approximated by the fitted value defined as

$$\hat{\mathbf{x}} = \bar{\mathbf{x}} + y_1 \cdot \mathbf{e}_1 + y_2 \cdot \mathbf{e}_2 + \cdots + y_k \cdot \mathbf{e}_k, \tag{7.20}$$

that is, the terms with scores y_{k+1} to y_p were removed from the right-hand side of (7.19). At this point, it is critical that the scores are calculated from the centered data as given in formula (7.18). In a data set with n observations $\mathbf{x}_i$, $i = 1, \ldots, n$, we define a fitted value for each observation as

$$\hat{\mathbf{x}}_i = \bar{\mathbf{x}} + y_{i,1} \cdot \mathbf{e}_1 + y_{i,2} \cdot \mathbf{e}_2 + \cdots + y_{i,k} \cdot \mathbf{e}_k, \tag{7.21}$$

where $y_{i,j}$ is the score of the jth principal component for the ith observation. We hope that $\hat{\mathbf{x}}_i$ is close enough to $\mathbf{x}_i$, so that the use of only k principal components is justified. This transformation from the principal component scores to the fitted values $\hat{\mathbf{x}}_i$ can be regarded as the *inverse PC transformation* (but this is not a rotation), because it transforms back to the original space of observations. Note that the inverse PC transformation defined here for $k < p$ is not the exact inverse of the PC rotation, unless the variability in the remaining PCs is zero. Since the information in the principal components numbered from $(k + 1)$ to p was lost, we are not obtaining the exact original values $\mathbf{x}_i$, but only the fitted values $\hat{\mathbf{x}}_i$. The error of this approximation can be assessed by calculating the residual vectors

$$\mathbf{E}_i = \mathbf{x}_i - \hat{\mathbf{x}}_i = y_{i,k+1} \cdot \mathbf{e}_{k+1} + \cdots + y_{i,p} \cdot \mathbf{e}_p. \tag{7.22}$$

As a global measure of approximation, we can use the residual sum of squares defined as

$$\mathrm{RSS} = \sum_{i=1}^{n} \|\mathbf{E}_i\|^2 = \sum_{i=1}^{n} \sum_{j=k+1}^{p} y_{i,j}^2. \tag{7.23}$$

Note that $\sum_{i=1}^{n} y_{i,j}^2/(n-1)$ is the sample variance of the jth principal component (because the sample mean is equal to zero due to centering) and is equal to λ_j calculated from the sample variance–covariance matrix $\mathbf{S}$. Hence,

$$\text{RSS} = (n-1) \sum_{j=k+1}^{p} \lambda_j. \tag{7.24}$$

Note that the fitted values $\hat{\mathbf{x}}_i$ defined in (7.21) belong to a k-dimensional affine subspace going through the mean vector $\bar{\mathbf{x}}$. We may wonder whether there exists a different k-dimensional affine subspace that would give us a better approximation. The answer is *no*, and it is clarified by the following theorem.

Theorem 7.2 Let $\mathbf{z}_i$, $i = 1, \ldots, n$, be a set of points from an affine k-dimensional subspace $L \subset R^p$. Then

$$\sum_{i=1}^{n} \|\mathbf{x}_i - \mathbf{z}_i\|^2 \geq \sum_{i=1}^{n} \|\mathbf{x}_i - \hat{\mathbf{x}}_i\|^2 = \text{RSS}. \tag{7.25}$$

This theorem tells us that the principal components give the best approximation in the sense of the global measure RSS. However, it is worthwhile to check how the approximation works on single observations by inspecting the residual vectors $\mathbf{E}_i$. This approach may not be practical for large data sets, so in those cases we can investigate the lengths $\|\mathbf{E}_i\|$ of the residual vectors. Note that for the p-dimensional residual vector $\mathbf{E} = [1, \ldots, 1]^{\mathrm{T}}$, we would have $\|\mathbf{E}\| = \sqrt{p}$. To make an investigation of residuals more intuitive, we can use an *adjusted norm*

$$\text{Adj.Norm}(\mathbf{E}) = \frac{\|\mathbf{E}\|}{\sqrt{p}}. \tag{7.26}$$

An advantage of the adjusted norm is that it is less dependent on the dimensionality, and for the vector $\mathbf{E} = [1, \ldots, 1]^{\mathrm{T}}$, the standardized norm is 1. In general, it gives us an idea about the average (in the root mean square sense) size of residuals per each variable (e.g., a spectral band). The adjusted norm is the adjusted Euclidean L_2 norm discussed in Section 10.2.1.

Example 7.9 Here we used the eight spectral measurements discussed earlier in Examples 7.2 and 7.7. We decided to approximate the data with $k = 2$ principal components. The resulting eight residual vectors $\mathbf{E}_i$ are shown in Figure 7.23, where Operator 1 values are plotted as solid lines and those for Operator 2 as dashed lines. The black lines indicate Spectrometer 1 and gray lines Spectrometer 2. There is no pattern indicating any specific impact of the two factors (Operator and Spectrometer) on the residuals. The random and rugged shape of the lines is desirable here, indicating

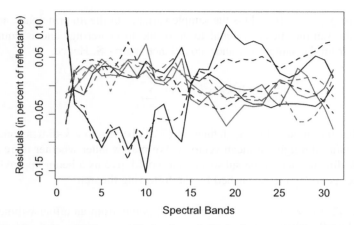

Figure 7.23 The eight residual vectors $\mathbf{E}_i$, $i = 1, \ldots, 8$, for the Spectrometer Data (Example 7.9) are plotted versus spectral band number. Operator 1 residuals are plotted as solid lines and those for Operator 2 as dashed lines. The black lines indicate Spectrometer 1 and gray lines Spectrometer 2.

a random nature of the remaining variability. However, we can also see a pattern where two lines are entirely different from the remaining lines. Hence, there is still some nonrandom effect beyond the first two principal components. All residuals are within the range of (-0.155) and 0.120 in percent of reflectance. Those values should be evaluated in the context of the measurement error estimates shown in Figure 7.5. If the fitted values $\hat{\mathbf{x}}_i$ were used in place of the original values $\mathbf{x}_i$, the measurement error estimates would change by the amounts equal to the residuals. We conclude that the residuals are small enough for most practical purposes.

We can also investigate the adjusted norms of the residual vectors plotted in Figure 7.24. We used a dot plot due to a small sample size. This allowed a representation of the run order of the measurements (the vertical axis). No time patterns can be seen in that plot. Also no pattern can be identified with respect the experiment factors—Operator and Spectrometer. The symbols used here are

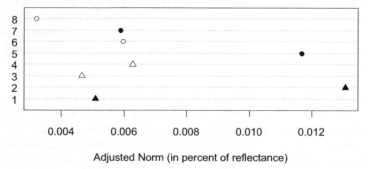

Figure 7.24 A dot plot of the standardized norms of the residual vectors (Example 7.9). The vertical axis indicates the run order of the measurements. Operator 1 values are marked with triangles and Operator 2 with dots. Results from Spectrometer 1 are plotted in solid black.

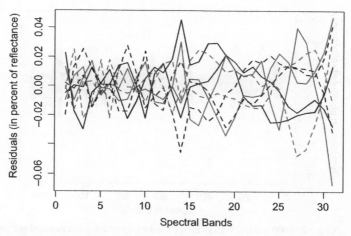

Figure 7.25 A residual plot analogous to Figure 7.23, except that $k = 3$ principal components were used to define the fitted values.

different from those used in Figure 7.21 in order to shown different ways how this can be done.

A similar analysis can be performed with $k = 3$ principal components explaining the variability in the data. The resulting residual vectors $\mathbf{E}_i$ are shown in Figure 7.25. Here we obtain a truly random behavior of the residuals. This means that no more than three principal components are needed to explain the patterns in the measurement error variability. □

The next example shows applications of principal component analysis in a scenario that is more challenging due to a large number of observations, large dimensionality, and complex structures shown in the image.

Example 7.10 In the context of imaging spectrometer data, such as the Cooke City image discussed in Example 7.4, we are interested in the intrinsic global linear dimensionality of the image as discussed in Section 7.3.2, where the dimensionality of 27 was suggested for this image. We want to calculate the residual vectors $\mathbf{E}_i$, assuming that $k = 27$ principal components were used in the calculation of the fitted values in formula (7.21). Figure 7.26 shows all $n = 224,000$ residual vectors. The original image spectra are given in the units of percent of reflectance, so the same units apply to the residuals. Some residuals are as large as 3 in absolute values, which translate to the precision of approximation (by the 27 principal component dimensions) up to $\pm 3\%$ of reflectance. However, most of the remaining values are much smaller. We can see a large range of variability in various spectral bands, which means that the precision of approximation varies significantly from band to band. Due to the large sample size n, it is convenient to use the adjusted norms of the residual vectors. The adjusted norms range from 0 to 0.54, and their histogram is shown in Figure 7.27. Most of the adjusted norms are well below 0.1, but it turns out that as many as 177

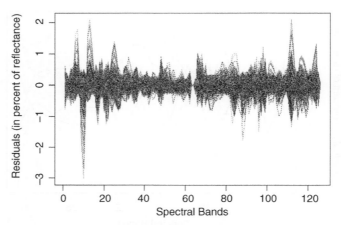

Figure 7.26 All $n = 224{,}000$ residual vectors for the Cooke City image data discussed in Example 7.10.

adjusted norms are above 0.2. The numbers seem small, and the approximation might be suitable for some purposes.

On the other hand, we may want to check if any relevant information is left in the residuals. One way to do that is to look for spatial patterns in residuals. Since we have one residual vectors $\mathbf{E}_i$ for each pixel, we can create an image where the color of a given pixel shows its value of the adjusted norm of $\mathbf{E}_i$. Figure 7.28 shows an image where the darker gray indicates a larger adjusted norm. Most of the image is in a very light shade of gray, reflecting mostly very small adjusted norms in contrast to a small number of large values. In such cases, it is often helpful to rescale the values, as was done in Figure 7.29, where the gradation of gray is proportional to the logarithm of the

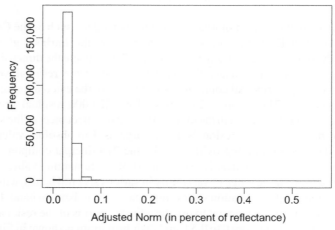

Figure 7.27 A histogram of adjusted norms of the residual vectors shown in Figure 7.26 and discussed in Example 7.10.

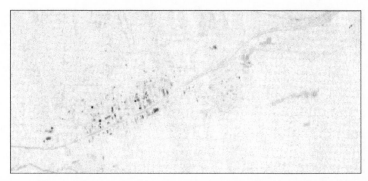

Figure 7.28 The adjusted norms of the residual vectors marked as shades of gray. Darker color indicates larger values.

adjusted norm. In both images, we can see a clear spatial structure, including the town and the road being clearly visible. We conclude that there is still a substantial amount of information in the image beyond the first 27 principal component dimensions. Nevertheless, the first 27 dimensions might be sufficient for some tasks.

In order to check how many principal component dimensions we need for a full representation of the image information, we can inspect the residual images for larger k values. Figure 7.30 shows the image of the adjusted norms when $k = 100$ principal components were used. Recall that the original image dimensionality is $p = 126$, so we are now using most of the original image information. However, we can still see some spatial structure of residuals, mostly in the town area, although it is fairly weak here. Based on these results, we may suspect that the imaging spectrometer data with complex structures and a small amount of noise cannot be precisely approximated by a global linear dimensionality substantially smaller than their original dimensionality. □

In the following example, we want to perform residual analysis on the AVIRIS image considered earlier.

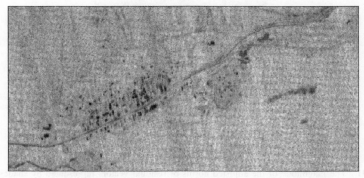

Figure 7.29 The logarithms of the adjusted norms of the residual vectors marked as shades of gray. Darker color indicates larger values.

Figure 7.30 This image is analogous to Figure 7.29, except that here $k = 100$ principal components were used.

Example 7.11 This is a continuation of Example 7.8, where we created images of principal component scores in Figure 7.22. Each of those images represents scores of only one principal component. We observed that each of the principal components numbered from 101 to 152 contained very little, if any, spatial information, which suggested a large amount of noise in those bands. We were then wondering about the amount of spatial information in all PC bands from 101 to 152 when considered jointly.

When we use $k = 100$ principal components in order to approximate the original image information, the residuals describe the joint impact of the remaining PC bands from 101 to 152. Since the residuals are p-dimensional vectors, we use the adjusted norms based on formula (7.26) in order to assess the magnitude of residuals. The adjusted norms are represented in Figure 7.31, where the darker shade of gray indicates larger values. We can see strong spatial structure in the image, mostly in terms of differences among various areas such as the water surface, the sandy shore, and then the remaining areas. The first 100 principal components were not sufficient to approximate precisely all image features.

Figure 7.31 The magnitude of the adjusted norms of residual vectors represented by shades of gray (darker means larger values) for the AVIRIS image discussed in Example 7.11. As many as $k = 100$ principal components were used.

Nevertheless, lower dimensional approximations can still be used for some purposes where less precision is required. $\qquad \square$

7.6 STATISTICAL INFERENCE IN PRINCIPAL COMPONENT ANALYSIS

As discussed in previous chapters, the statistical inference is about drawing conclusions about a population, or its model, based on the available sample. In this process, it is critical to specify the sampling scheme under which a given data set was collected. A classic assumption in the context of principal component analysis is that the multivariate sample of p-dimensional independent and identically distributed vectors $\mathbf{X}_1, \mathbf{X}_2, \ldots, \mathbf{X}_n$ comes from a multivariate distribution with the mean vector $\boldsymbol{\mu}$ and the variance–covariance matrix $\boldsymbol{\Sigma}$. This assumption might be reasonable in some imaging applications, but not in others. In Example 7.2, we considered eight repeated measurements of the same tile. In this context, it is reasonable to assume that $\mathbf{X}_1, \mathbf{X}_2, \ldots, \mathbf{X}_8$ are independent and identically distributed, where the mean vector $\boldsymbol{\mu}$ can be interpreted as the vector of true reflectances. This scenario will be considered in the next subsection. A different scenario will be discussed in Section 7.6.2.

7.6.1 Independent and Identically Distributed Observations

In this subsection, we will assume that the multivariate sample of p-dimensional independent and identically distributed vectors $\mathbf{X}_1, \mathbf{X}_2, \ldots, \mathbf{X}_n$ comes from a multivariate normal distribution $N_p(\boldsymbol{\mu}, \boldsymbol{\Sigma})$ with the mean vector $\boldsymbol{\mu}$ and the variance–covariance matrix $\boldsymbol{\Sigma}$. As usual, the matrix $\boldsymbol{\Sigma}$ is estimated by the sample variance–covariance matrix $\mathbf{S}$. The eigenvalues and normalized eigenvectors of $\mathbf{S}$ will be denoted with hats, that is, as $(\hat{\lambda}_j, \hat{\mathbf{e}}_j)$, $j = 1, \ldots, p$, to emphasize that these are estimated quantities. We have the following asymptotic property.

Property 7.2 Let $\boldsymbol{\Lambda}$ be the diagonal matrix of distinct eigenvalues $\lambda_1 > \lambda_2 > \cdots > \lambda_p > 0$ of the matrix $\boldsymbol{\Sigma}$. Then the distribution of the vector $\sqrt{n}\left(\left[\hat{\lambda}_1, \hat{\lambda}_2, \ldots, \hat{\lambda}_p\right]^{\mathrm{T}} - \left[\lambda_1, \lambda_2, \ldots, \lambda_p\right]^{\mathrm{T}}\right)$ approaches the normal distribution $N_p\left(\mathbf{0}, 2\boldsymbol{\Lambda}^2\right)$ as n tends to infinity.

The proof can be found in Section 13.5.1 in Anderson (2003). This property means that the sample eigenvalues are asymptotically uncorrelated and the distribution of $\hat{\lambda}_j$ can be approximated by the normal distribution $N\left(\lambda_j, 2\lambda_j^2/n\right)$. This allows construction of a large-sample $(1 - \alpha)$ confidence level confidence interval for λ_j, $j = 1, \ldots, p$, given by

$$\frac{\hat{\lambda}_j}{1 + z(\alpha/2)\sqrt{2/n}} \le \lambda_j \le \frac{\hat{\lambda}_j}{1 - z(\alpha/2)\sqrt{2/n}}, \tag{7.27}$$

which can also be written in a more convenient form as

$$1 - z(\alpha/2)\sqrt{2/n} \leq \frac{\hat{\lambda}_j}{\lambda_j} \leq 1 + z(\alpha/2)\sqrt{2/n}, \tag{7.28}$$

or

$$\left| \frac{\hat{\lambda}_j}{\lambda_j} - 1 \right| \leq z(\alpha/2)\sqrt{2/n}, \tag{7.29}$$

which means that $\hat{\lambda}_j$ is different from λ_j by no more than $z(\alpha/2)\sqrt{2/n}$ with $(1 - \alpha)$ confidence.

Since the distribution of $\hat{\lambda}_j$ can be approximated by the normal distribution $N\left(\lambda_j, 2\lambda_j^2/n\right)$, we have $E\left(\hat{\lambda}_j\right) \approx \lambda_j$ and $\text{Var}\left(\hat{\lambda}_j\right) \approx 2\lambda_j^2/n$. The following, more precise, approximations were derived by Lawley (1956):

$$E\left(\hat{\lambda}_j\right) \approx \lambda_j \left[1 + \frac{1}{n}\sum_{k \neq j}\frac{\lambda_k}{\lambda_j - \lambda_k}\right], \quad \text{Var}\left(\hat{\lambda}_j\right) \approx \frac{2\lambda_j^2}{n}\left[1 - \frac{1}{n}\sum_{k \neq j}\left(\frac{\lambda_k}{\lambda_j - \lambda_k}\right)^2\right]. \tag{7.30}$$

This means that the large eigenvalues will be estimated as too large and small ones as too small. In order to correct for the bias, we can use the following adjusted estimates:

$$\hat{\lambda}_j^* = \hat{\lambda}_j\left[1 - \frac{1}{n}\sum_{k \neq j}\frac{\hat{\lambda}_k}{\hat{\lambda}_j - \hat{\lambda}_k}\right]. \tag{7.31}$$

More details about these and some other approximations can be found in Jackson (1991).

7.6.2 Imaging Related Sampling Schemes

The statistical inference discussed in the previous subsection is valid when the multivariate sample consists of p-dimensional independent and identically distributed random vectors. In many applications, this may not be the case. For spectral data representing images of the Earth's surface, like those considered in Examples 7.3 and 7.4, each pixel may have a different distribution. In order to identify what that distribution is, we need to know the population of interest and the sampling process that we want to consider in a given situation. When a global variance–covariance matrix is calculated from the whole image, it incorporates both the spatial variability of the Earth's surface features and other variability such as noise.

In practice, we usually have one image (i.e., one sample) that we want to investigate. A convenient way of thinking about the sampling process we discuss here is to imagine repeated sampling (i.e., repeated images) from the population of interest. For example, imagine that we collect repeated images of the same area under the same conditions. Assume further that within that area, there is a house with horizontal dimensions of 10 m by 20 m, the image pixel size is 5 m by 5 m, and the house is represented by exactly 8 pixels. Assuming the i.i.d. (independent and identically distributed) sampling scheme and an image consisting of 10,000 pixels, we would expect to see anywhere between 3 and 13 pixels of that house in 95% of repeated images (use the binomial distribution with $n = 10^4$ and $p = 8 \times 10^{-4}$). This clearly has no physical justification, and we would conclude that the i.i.d. sampling scheme is not realistic here.

A suitable sampling scheme will depend on a given application and the type of statistical inference we want to perform. The correct sampling scheme should reflect the true relationship between the sample (the image) and the population (the reality the image represents). Here we will make a comparison between two different i.i.d. schemes and one more realistic sampling scheme in order to evaluate their impact on eigenvalues and eigenvectors of the variance–covariance matrix. We will also show a methodology for construction of confidence intervals under those sampling schemes. In a broader context, this is an example of a good practice of using the correct sampling schemes and correct statistical inference in imaging applications. A similar approach can be used in other types of statistical inference such as ROC curves, hypothesis testing, and so on. The results shown in this subsection were published earlier in Bajorski; (2011).

We will use here the following sampling schemes.

Sampling Scheme A. This is an i.i.d. sampling scheme where all spectra x_i's follow a p-dimensional Gaussian (normal) distribution $N_p(\mu, \Sigma)$, where μ and Σ are estimated based on the sample mean $\bar{x}$ and the sample variance–covariance matrix S calculated from the whole image.

Sampling Scheme B. This is an i.i.d. sampling scheme with a discrete uniform distribution on all image spectra x_i, $i = 1, \ldots, n$, with each spectrum having the same probability $1/n$. This sampling scheme is appropriate when we are not willing to make any assumptions about a parametric model for the population distribution (such as those in Scheme A). This type of sampling is used in nonparametric bootstrap methods (see Section 3.9).

Sampling Scheme C. We assume that image spectra follow the so-called *linear mixing model* (as discussed in Schott (2007), Healey and Slater (1999), and Manolakis and Shaw (2002))

$$x_i = \sum_{j=1}^{k} a_{ij} m_j + \varepsilon_i, \tag{7.32}$$

where the term $\sum_{j=1}^{k} a_{ij} m_j$ represents the deterministic component, which would not change in the sampling process. We assume that the constants a_{ij} are such that

$\sum_{j=1}^{k} a_{ij} = 1$, but not necessarily positive. A good choice of p-dimensional vectors $\mathbf{m}_1, \ldots, \mathbf{m}_k$ (representing spectra of material present in the image) should ensure positivity or near positivity of a_{ij}'s. Here we want to make sure that the affine subspace

$$M(\mathbf{m}_1, \ldots, \mathbf{m}_k) = \left\{ \mathbf{x} : \mathbf{x} = \sum_{j=1}^{k} a_j \mathbf{m}_j, \quad \sum_{j=1}^{k} a_j = 1 \right\} \qquad (7.33)$$

gives a good approximation of the image spectra $\mathbf{x}_i$, $i = 1, \ldots, n$. A benefit of this approach is that we do not need to identify the vectors $\mathbf{m}_1, \ldots, \mathbf{m}_k$, but instead we only identify the affine space M based on the first k eigenvectors of the variance–covariance matrix $\mathbf{S}$. We also assume that the error terms ε_i follow the Gaussian distribution $N(0, \Sigma_\varepsilon)$, where Σ_ε is calculated as discussed below.

The deterministic component $\sum_{j=1}^{k} a_{ij} \mathbf{m}_j$ represents the signal in the image, that is, the surface seen in the image as a mixture of some basic materials. Hence, if repeated images of the same area were taken, this deterministic component should not change (if identified correctly). This is why we keep this component constant in this sampling process.

Let $\lambda_1 \geq \lambda_2 \geq \cdots \geq \lambda_p$ be eigenvalues and $e_1, \ldots, e_p$ respective normalized eigenvectors of $\mathbf{S}$. We assume that the variability in the space orthogonal to M is due to noise and is determined by $\lambda_{k+1} \geq \lambda_{k+2} \geq \cdots \geq \lambda_p$. However, we have no information about the noise in the directions within M. We will assume the variance of noise to be λ_{k+1} in all directions within M, which seems to be a reasonable approximation given the lack of any additional information. Let Λ be a diagonal matrix with the first k elements on the diagonal equal to λ_{k+1} followed by $\lambda_{k+1} \geq \lambda_{k+2} \geq \cdots \geq \lambda_p$. Let $\mathbf{P}$ be a matrix of $e_1, \ldots, e_p$ as columns. We now define $\Sigma_\varepsilon = \mathbf{P} \Lambda \mathbf{P}^T$. In order to preserve the total variability in the first k PC dimensions, we need to make a correction to the deterministic part of the model by multiplying the ith PC coordinate of each spectrum (after subtracting $\bar{\mathbf{x}}$) by $b_j = \sqrt{1 - \lambda_{k+1}/\lambda_j}$ for $j = 1, \ldots, k$. Clearly, $b_j \leq 1$ for $j = 1, \ldots, k$, and it can be shown (see Appendix C) that multiplying by b_j is equivalent to reducing the variability of the deterministic part by λ_{k+1}, which is the amount assumed to be due to noise.

In order to investigate a wide range of scenarios for our numerical calculations, we used two very different images, and then three different-size subimages were selected from each of those two images. The first image was a MISI (Modular Imaging Spectrometer Instrument) image of the Lake Ontario shoreline near Russell Station located in Rochester, NY. The data on the 16 NIR (near-infrared) bands (730–985 nm) were collected on September 9, 2001. The MISI instrument was flown at an altitude of 5000 feet with a ground speed of 128 knots. More information about the image can be found in Ientilucci (2003) and about the MISI instrument in Feng et al. (1994). The second image was the AVIRIS image shown in Figure 7.10 and discussed in Example 7.3. With much larger number of spectral bands and being an image of an urban scene, the AVIRIS image is much more complex and very different from the MISI image.

Table 7.8   Numbering of the Six Images Used for Numerical Calculations

Image Size	MISI	AVIRIS
50 by 50 pixels	Image I (5)	Image IV (18)
20 by 20 pixels	Image II (5)	Image V (15)
10 by 10 pixels	Image III (3)	Image VI (12)

The image intrinsic dimensionalities (as calculated in Section 7.6.2.3) are shown in parentheses.

For each of the two images, we selected a 50 by 50 pixel subset ($n = 2500$), and then a 20 by 20 pixel subset ($n = 400$) and a 10 by 10 pixel subset ($n = 100$), each being a subset of the previous subimage. Those choices and the numbering of images are represented in Table 7.8, which can also be used as a reference for interpretation of figures organized in the same fashion.

These images may appear somewhat small to some readers. However, we believe that in modern image analyses, one should not use covariance matrices of the whole large images covering a wide range of different ground covers because such approach does not give a realistic assessment of the pixel-to-pixel variability. Recently more popular and more successful approach is the modeling of individual background components and small multimaterial clusters within a hyperspectral image (for instance, see Schlamm et al. (2008) and Caefer et al. (2008)). For the 50 by 50 pixel images, the PCA inference was largely satisfactory, and for larger images, it would be even more precise. Hence, we were more interested in performance for smaller sample sizes such as $n = 100$ or 400. These smaller images were selected to mimic the context of some of the imaging processing algorithms, where relatively small windows are selected in the process of scanning a large image, for example, in the context of anomaly detection (local RX and other local detectors; see Schaum (2007) and Basener and Messinger (2009)) or in modeling of individual background components. All images were chosen so that their variability is somewhat similar to each other and to the variability of the whole MISI image as measured by eigenvalues. That is, we avoided subsetting of some more uniform areas with small variability. The three images should not be compared directly because they have somewhat different eigenvalues. Nevertheless, they give us some idea about the impact of the sample size n on PCA inference.

For the Sampling Scheme C, one needs to identify the dimensionality k of the affine space M. This decision was based on a simple method of explained variability as discussed in Section 7.3. For all MISI subimages, we used a threshold of 99.3%, and for all AVIRIS subimages, we used 99.98%. The resulting values of k are shown in Table 7.8 in parentheses.

The results of our analysis are presented in the following subsections.

7.6.2.1   Investigation of Eigenvalues

For each of the six images (I–VI) and for each of the sampling schemes (A, B, and C), the following procedure was followed for the resulting 18 scenarios. A number of $N = 10^5$ images of a given size n were generated using the appropriate sampling scheme. For each generated image, PCA was performed, and the sample eigenvalues

$\hat{\lambda}_1 \geq \hat{\lambda}_2 \geq \cdots \geq \hat{\lambda}_p$ were calculated. Then the normalized eigenvalues were calculated as $\bar{\hat{\lambda}}_i / \lambda_i$, where λ_i is the "true" (as assumed by the Monte Carlo simulations) eigenvalue from the given image. For example, the normalized eigenvalue of 1.05 means that the sample value was 5% larger than what it should have been, based on the "true" value λ_i.

For each of the 18 scenarios and each PC direction i $(i = 1, \ldots, p)$, we obtained a distribution of $N = 10^5$ normalized eigenvalues. We then calculated the 2.5th and 97.5th percentiles of those sampling distributions in order to capture 95% of the distribution. Figure 7.32 shows those percentiles for the six images in the layout

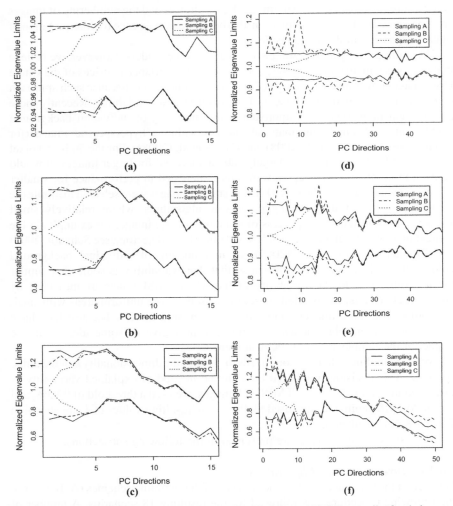

Figure 7.32 The normalized eigenvalue limits based on the 2.5th and 97.5th percentiles for six images. The layout is consistent with Table 7.8, that is, Images I–VI are shown in (a)–(f). The AVIRIS images show only the first 50 PC directions, because they would be mostly of interest.

presented in Table 7.8. Based on Figure 7.32a, 95% of the sampling distribution was approximately within $\pm 6\%$ of the true values, with only a small amount of negative bias for the last several PC directions for the Sampling Schemes A and B in Image I. For the Sampling Scheme C used on the same data set, we can see much smaller sampling variability for the first several PC directions until the variability become the same as for Schemes A and B (where all lines start to overlap). The reason is that the first five PC directions contain significant deterministic components, thus reducing the sampling variability. The first PC direction has the largest deterministic component, which results in the smallest relative variability in the first eigenvalue. Similar behavior is observed in all subpanels in Figure 7.32. This by itself does not mean that Scheme C is better than the other schemes. It simply reflects the lower variability assumed by that sampling scheme.

We can also see that the reduction in the sample size results in larger uncertainty about eigenvalues, reaching levels of about 15–20% limits for $n = 400$ and levels of about 20–40% limits for $n = 100$. For smaller sample sizes, we start seeing a significant amount of negative bias reaching levels of approximately 40% drop for the 50th PC direction for the AVIRIS image with $n = 100$ (Image VI).

7.6.2.2 Investigation of Eigenvectors
The investigation of eigenvectors was performed in a way similar to the one described in the previous subsection when discussing eigenvalues. The difference was that for each generated image, PCA was performed, and the sample eigenvalues $\hat{e}_1, \hat{e}_2, \ldots, \hat{e}_p$ were calculated. Then we calculated the angle between $\hat{e}_i$ and the "true" vector e_i. Large angles indicate lack of precision in estimating the eigenvectors. We calculated the 95th percentile for each sampling distribution of $N = 10^5$ angles.

Figure 7.33 shows those percentiles in a layout analogous to the one in Figure 7.32. For all three MISI images, the results for the three sampling schemes are very similar to each other, except for Sampling Scheme C variability being significantly smaller for the PC directions 2 and 3 in Image III (see Figure 7.33c). We can also notice that starting with the sixth eigenvector, the estimation is highly imprecise. Note that a 90° angle represents the worse-case scenario of an estimate being orthogonal to the true eigenvector. This poor estimation performance for the PC directions 6–16 was expected, because the respective eigenvalues are very close to each other; hence, the directions with maximum variability are difficult to identify. It is known that when several eigenvalues are exactly equal to each other, the eigenvectors are identifiable only up to the appropriate subspace. In the presence of sampling variability, this additional (sampling) variability tends to be assigned to the higher PCs causing positive bias in those higher PCs at the cost of negative bias in lower PCs. We have seen this effect in the classic i.i.d. inference in the context of equation (7.30).

For the three AVIRIS images, the differences among sampling schemes are larger than those for the MISI images. For smaller sample sizes, we observe poor precision in estimation already within the first 10 eigenvectors. This is partially caused by some of the first 10 eigenvalues being close to each other. For both images (MISI and AVIRIS), we can see that the reduction in the sample size consistently results in larger uncertainty about eigenvectors.

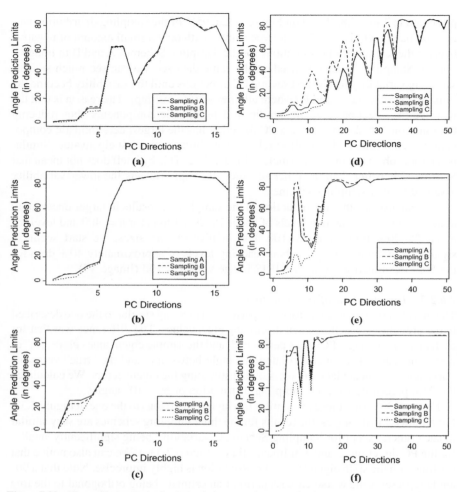

Figure 7.33 The prediction limits for the angle between the true and estimated eigenvectors based on the 95th percentile for six images. The layout is consistent with Table 7.8, that is, Images I–VI are shown in (a)–(f). The AVIRIS images show only the first 50 PC directions, because they would be mostly of interest.

7.6.2.3 Estimation of Image Intrinsic Linear Dimensionality

In Section 7.3, we discussed various stopping rules and introduced the concept of intrinsic linear dimensionality of a spectral image. It is interesting to find out how the intrinsic linear dimensionality might depend on the sampling variability. To this end, we concentrate on the dimensionality defined by the percent of explained variability. As explained earlier, we used a threshold of 99.3% for all MISI subimages, and a threshold of 99.98% for all AVIRIS subimages. This method was applied to the six images, and the resulting values of dimensionality k are shown in Table 7.8 in parentheses. When the simulation is run $N = 10^5$ times, each time we obtain a

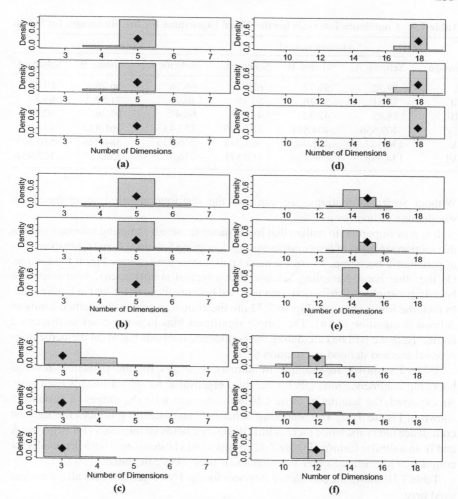

Figure 7.34 Histograms of the sampling distributions of the estimator of the image intrinsic linear dimensionality based on the explained variability. The three histograms within each subpanel are for the Sampling Schemes A (top), B, and C, respectively. The layout is consistent with Table 7.8. The diamonds show the value of the original image intrinsic linear dimensionality (when using the same method).

simulated spectral image and its intrinsic linear dimensionality k can be calculated. The sampling distributions of the dimensionality estimators are shown in Figure 7.34. The precision of estimation is fairly good with some amount of negative bias (not exceeding one dimension) for Images V and VI.

7.6.2.4 Calculation of Confidence Intervals

The simulation results shown in Figure 7.32 can be used for calculation of confidence intervals for the population eigenvalues. Equation (7.27) shows a confidence interval constructed based on a theoretical assumption about the underlying distribution.

Table 7.9 Confidence Intervals for the Third Eigenvalue for the Six Images (in Rows)

Image	Lower Bound			Upper Bound		
	Scheme A	Scheme B	Scheme C	Scheme A	Scheme B	Scheme C
I	59.72	59.72	62.02	66.76	66.76	64.48
II	53.81	53.76	59.19	70.97	71.52	65.47
III	38.83	42.32	43.07	66.47	64.26	60.47
IV	409,656	404,891	432,176	457,453	461,832	434,831
V	436,347	431,940	503,244	576,930	584,724	510,803
VI	134,599	138,384	180,571	236,794	236,155	192,959

Without such assumptions, we can use the distribution information based on simulations or bootstrap.

It is now important to realize that in a broader sense, all sampling schemes used in this section are types of bootstrap sampling. Sampling Schemes A and C are examples of parametric bootstrap, where the model parameters are estimated from the sample. On the other hand, Sampling Scheme B is a typical nonparametric bootstrap. The bootstrap sampling distributions that are used here are those described in Figure 7.32. In fact, the limits shown in Figure 7.32 are the bootstrap percentile method limits as defined in equation (3.48). They show significant bias in some cases as discussed earlier. Here we will be calculating the confidence intervals based on the percentile-reversal method defined in equation (3.51).

Table 7.9 shows an example of numerical results for the percentile-reversal bootstrap confidence intervals for the third eigenvalue for all six images used here. As expected, the lengths of those CIs are consistent with the ranges of variability shown in Figure 7.32. The Sampling Scheme C consistently produces the shortest confidence intervals, while there is little difference between the results for Schemes A and B. In a similar fashion, Figure 7.32 results would be consistent with the results for confidence intervals of other eigenvalues, if they were calculated.

Table 7.10 shows the confidence intervals for the 16th eigenvalue for all six images used here.

Table 7.11 shows the estimated biases of the raw point estimate of the third eigenvalue. All biases are fairly small relative to the magnitudes of the estimates.

Table 7.10 Confidence Intervals for the 16th Eigenvalue for the Six Images (in Rows)

Image	Lower Bound			Upper Bound		
	Scheme A	Scheme B	Scheme C	Scheme A	Scheme B	Scheme C
I	1.98	1.98	1.98	2.17	2.17	2.17
II	1.70	1.71	1.70	2.06	2.06	2.06
III	1.30	1.31	1.31	1.70	1.76	1.70
IV	904	892	905	1005	1019	1001
V	660	651	659	805	813	805
VI	524	539	524	727	737	727

Table 7.11   Raw Point Estimates $\hat{\theta}$ and Their Estimated Biases for the Third Eigenvalue for the Six Images (in Rows)

Image	Point Estimate	Estimated Bias		
		Scheme A	Scheme B	Scheme C
I	63.27	−0.03	−0.03	0.01
II	62.44	−0.26	−0.41	0.06
III	52.03	−1.44	−1.59	−0.02
IV	433,518	−235	−284	8
V	507,076	−1955	−3386	40
VI	187,027	−2275	−2792	220

The biases under the Sampling Scheme C are consistently the smallest relative to the other sampling schemes. Again, this by itself should not be interpreted that the Scheme C is the best. Instead, it means that the bias is actually smaller (as shown by the more realistic Scheme C) than it would be suggested by the unrealistic Sampling Schemes A and B.

Table 7.12 shows the estimated biases of the raw point estimates of the 16th eigenvalue. The biases for Images II and III are fairly large relative to the magnitudes of the estimates. The bias-adjusted estimates can be calculated by subtracting those biases as shown in formula (3.47).

Clearly, there are some general patterns in the results shown here. The reduced sample size results in more variability and some bias in the sampling distributions of eigenvalues. Consequently, the resulting confidence intervals become wider. At the same time, there are also significant differences among the images as well as among different sampling schemes assumed. Hence, researchers should investigate their own case separately, and the appropriate sampling scheme needs to be chosen based on the purpose of a given study. Our purpose was to demonstrate how this should be done. We advocate use of the Sampling Scheme C or a similar type of scheme that would be based on a realistic model for the spectral image under investigation. A similar approach can also be used in other types of statistical inference, such as ROC curves, hypothesis testing, and so on.

Table 7.12   Raw Point Estimates $\hat{\theta}$ and Their Estimated Biases for the 16th Eigenvalue for the Six Images (in Rows)

Image	Point Estimate	Estimated Bias		
		Scheme A	Scheme B	Scheme C
I	2.02	−0.05	−0.05	−0.05
II	1.69	−0.18	−0.19	−0.18
III	1.15	−0.36	−0.39	−0.36
IV	958	3	1	4
V	732	−3	−3	−3
VI	614	−19	−31	−18

7.7   FURTHER READING

More information about principal component analysis can be found in Jackson (1991), Johnson and Wichern (2007), Anderson (2003), and Srivastava (2002).

PROBLEMS

7.1. Refer to the radiopaque marker data shown in Figure 7.1. Assume a point $\mathbf{f} = [f_1, f_2]$ given in the physical coordinates f_1 and f_2. Find the matrix of a linear transformation giving you the mathematical coordinates x_1 and x_2 of that point. Find the algebraic form of that matrix using the vectors v_1 and v_2 without assuming their specific coordinates.

7.2. Refer to the radiopaque marker data shown in Figure 7.1. Assume a point $\mathbf{x} = [x_1, x_2]$ given in the mathematical coordinates x_1 and x_2. Find the matrix of a linear transformation giving you the physical coordinates f_1 and f_2 of that point. Find the algebraic form of that matrix using the vectors v_1 and v_2 without assuming their specific coordinates. If you were also asked to do Problem 7.1, discuss how this solution is different from the one in Problem 7.1 and why.

7.3. The radiopaque marker data provided for Example 7.1 are given in the physical coordinates f_1 and f_2.

    a. Project all points on the X-ray screens defined by the unit vectors $v_1 = [10, 1]/\sqrt{101}$ and $v_2 = [1, 5]/\sqrt{26}$. This will give you the mathematical coordinates x_1 and x_2 of those points. Create a scatter plot of those coordinates and compare it to Figure 7.1b.

    b. Use the mathematical coordinates x_1 and x_2 of those points to project them on the line L_1 described in a parametric form as a set of all vectors $t \cdot \mathbf{b}_1 + \mathbf{d}_1$ parameterized by $t \in R$, where $\mathbf{b}_1 = [-1, 4]^{\mathrm{T}}/\sqrt{17}$ and $\mathbf{d}_1 = [8, 2]/17$. Calculate the variance describing one-dimensional variability of those projections on L_1.

    c. Repeat the process described in point b in order to project on L_2 described by $t \cdot \mathbf{b}_1 + \mathbf{d}_1$ with $\mathbf{b}_2 = [7, -1]^{\mathrm{T}}/\sqrt{50}$ and $\mathbf{d}_2 = 0.07 \cdot [1, 7]$. Calculate the variance describing one-dimensional variability of those projections on L_2.

7.4. Perform principal component analysis on the radiopaque marker data used in Example 7.1. Estimate the standard deviations of the marker's oscillations and the measurement error as demonstrated in Example 7.1.

7.5. Use the infrared astronomy data introduced in Example 3.3. This is a three-dimensional data set ($p = 3$) with bands J, H, and K treated as three variables. We have two groups of multivariate observations, one sample of C AGB stars and one sample of H II regions. Treat all observations as one sample and perform principal component analysis of all data. Create the impact plot for the three principal components like the one in Figure 7.9. Include in the plot the standard deviations of the original bands J, H, and K. Interpret all three principal components.

7.6. Refer to Problem 7.5 and standardize all three bands by subtracting their sample means and dividing by their sample standard deviations. Repeat the tasks of Problem 7.5 for the standardized data. How different are the results? Are they easier to interpret?

7.7. Show that for a_j defined in formula (7.15), we have $a_j \geq 1/p$, for all $j \leq p/3$. This means that if the simple fair-share stopping rule indicates dimensionality k not larger than one-third of p (which is usually the case), then the broken-stick stopping rule will indicate the same or smaller dimensionality.

7.8. For Example 7.2 data, calculate the vectors of principal component coefficients. Your results may have signs opposite to those in Figure 7.6. Explain why this is happening.

7.9. Recreate Figure 7.21.

7.10. Recreate Figure 7.23.

7.11. Consider a p-dimensional random vector $\mathbf{X} = \left[X_1, X_2, \ldots, X_p\right]^{\mathrm{T}}$ and denote by $Y_1 = \mathbf{a}_1^{\mathrm{T}}\mathbf{X}$ the first principal component. Define the vector $\mathbf{a}_2$ as an arbitrary linear combination $\alpha\mathbf{a}_1 + \beta\mathbf{a}_1^{\perp}$, where $\mathbf{a}_1^{\perp}$ is a vector orthogonal to $\mathbf{a}_1$. Show that the variance of Y_2 is equal to $\alpha^2 \operatorname{Var}(Y_1) + \beta^2 \operatorname{Var}(\mathbf{a}_1^{\perp}\mathbf{X})$. *Hint*: Use Theorem 7.1.

7.12. Prove Property 7.1. *Hint*: Use Property 4B.2 and the definition of the eigenvectors of $\mathbf{\Sigma}$.

7.13. * In Example 5.9, we analyzed a 31-dimensional data set consisting of spectral curves as measurements, or observations. For each of 12 calibration tiles, four measurements were taken. Based on those four measurements, a variance–covariance matrix $\mathbf{S}$ was calculated for each tile. Consider only tiles numbered 3, 5, and 8, and the resulting matrices $\mathbf{S}_i$, $i = 3, 5, 8$. We would like to test if the three variance–covariance matrices are statistically significantly different. We would like to use the Box's M-test discussed in Section 6.4. However, one difficulty is that the test requires nonsingular matrices $\mathbf{S}_i$. In order to overcome this difficulty, perform the following tasks.

    a. Pool all 12 observation vectors from the three groups for the three tiles into one data set. The data set can be represented as a 12 by 31 matrix $\mathscr{X}$.

    b. Perform principal component analysis on $\mathscr{X}$. Based on some stopping rules, identify how many PCs you want to retain. The preferred number is three

PCs because this is the rank of each $\mathbf{S}_i$. This step is only for exploratory purposes.

c. Independent of the results of point b, use $k = 3$ principal components for approximations of all 12 observations. Check how precise those approximations are by analyzing residuals.

d. Use the scores of the three principal components calculated in point c, and place them in a 12 by 3 matrix $\mathscr{Y}$ as the data set for further analysis. Split this data set into three groups of observations from the three tiles. For each tile, calculate the matrix $\mathbf{S}_i^*$, $i = 3, 5, 8$. The matrix $\mathbf{S}_i^*$ should approximate the variability in $\mathbf{S}_i$, if the residuals in point c were small enough.

e. Run the Box's M-test for the three matrices $\mathbf{S}_i^*$, $i = 3, 5, 8$, calculated in point d. Draw conclusions.

7.14. Refer to the infrared astronomy data used in Problem 7.5.

a. Create a matrix plot (see Section 5.3) based on the three bands J, H, and K. Mark the C AGB stars and the H II regions with different symbols and/or colors.

b. Create a scatter plot of principal component scores of the first two principal components. Is this plot more informative than the matrix plot from point a in terms of seeing the differences between the two groups of observations? Draw conclusions about the two groups. Draw conclusions about the role of principal components in data analysis.

7.15. For the eight vectors of the estimated measurement errors shown in Figure 7.5, calculate the sample variance–covariance matrix $\mathbf{S}$ and their eigenvalues. We would like to know how precise these estimates are relative to the population eigenvalues. However, we cannot use results of Section 7.6.1 because the sample size is too small. Perform nonparametric bootstrap of your sample using the i.i.d. scheme. Calculate the bias and the confidence intervals for the first three eigenvalues. Interpret your results.

CHAPTER 8

Canonical Correlation Analysis

8.1 INTRODUCTION

Canonical correlation analysis (CCA) deals with two groups of variables and tries to investigate correlations between the two groups. This is primarily used for descriptive purposes, but there are also some predictive applications. One possible application in imaging science is an inverse problem, where the atmospheric effects can be estimated from the observed radiation (see Hernandez-Baquero and Schott (2000)). Another potential application is texture generation, where the interest is in correlating sets of variables (spectral bands and spatial neighboring pixels) and trying to predict one set of variables from the other.

Example 8.1 Here again, we use remote sensing data. Some background information about such data can be found in Example 1.3. Let us consider an example of a 15 by 15 pixel image of grass texture in 42 spectral bands and look at the function plot of all 225 spectra shown in Figure 8.1. Many spectra are parallel to each other, which suggests that the reflectance values from the adjacent spectral bands are highly correlated, with an exception of spectral bands 30 and 31. This can be confirmed numerically by calculating the sample correlations matrix **R** of all 42 spectral bands (represented in Figure 8.2 in shades of gray). Most values in the **R** matrix are very high, which means that most of the 42 spectral bands are highly correlated.

Such correlations play an important role in constructing algorithms for generating textures. The idea is that reflectance values at some spectral bands can be predicted from reflectance values at other spectral bands. It might be even more beneficial to try to establish correlations among sets of spectral bands and their linear combinations. □

Statistics for Imaging, Optics, and Photonics, Peter Bajorski.
© 2012 John Wiley & Sons, Inc. Published 2012 by John Wiley & Sons, Inc.

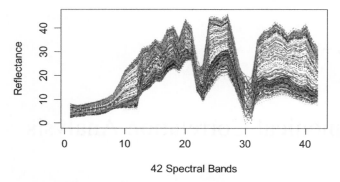

Figure 8.1   Function plot of 225 pixel spectra of grass texture.

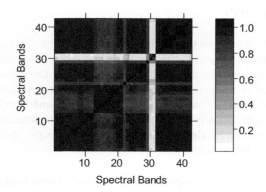

Figure 8.2   The correlations of all 42 spectral bands represented in gray. Larger values are in darker shades.

8.2   MATHEMATICAL FORMULATION

In order to explain the methodology of canonical correlation analysis, we need to introduce some mathematical notation. The two sets of random variables being considered are represented as two random vectors

$$\mathbf{X} = \left[X_1, X_2, \ldots, X_p\right]^{\mathrm{T}} \quad \text{and} \quad \mathbf{Y} = \left[Y_1, Y_2, \ldots, Y_q\right]^{\mathrm{T}}. \tag{8.1}$$

Our goal is to find linear combinations

$$U = \mathbf{a}^{\mathrm{T}}\mathbf{X} = a_1 X_1 + a_2 X_2 + \cdots + a_p X_p \tag{8.2}$$

and

$$V = \mathbf{b}^{\mathrm{T}}\mathbf{Y} = b_1 Y_1 + b_2 Y_2 + \cdots + b_q Y_q \tag{8.3}$$

that maximize the correlation

$$\text{Corr}(U, V) = \frac{\text{Cov}(U, V)}{\sqrt{\text{Var}(U)}\sqrt{\text{Var}(V)}}. \tag{8.4}$$

We will assume that the following variance–covariance and covariance matrices exist:

$$\text{Var}(\mathbf{X}) = \Sigma_{XX}, \quad \text{Var}(\mathbf{Y}) = \Sigma_{YY}, \quad \text{Cov}(\mathbf{X}, \mathbf{Y}) = \Sigma_{XY} \tag{8.5}$$

and the $(p + q) \times (p + q)$ matrix

$$\Sigma = \begin{bmatrix} \Sigma_{XX} & \Sigma_{XY} \\ \Sigma_{XY}^{\text{T}} & \Sigma_{YY} \end{bmatrix} \tag{8.6}$$

has full rank. The following theorem defines the canonical variables and explains how to find them.

Theorem 8.1 Let $m = \min\{p, q\}$ and define $\rho_1^2 \geq \rho_2^2 \geq \cdots \geq \rho_m^2$ as the eigenvalues of the matrix $\Sigma_{XX}^{-1/2} \Sigma_{XY} \Sigma_{YY}^{-1} \Sigma_{XY}^{\text{T}} \Sigma_{XX}^{-1/2}$, and $\mathbf{e}_1, \mathbf{e}_2, \ldots, \mathbf{e}_m$ are the associated normalized p-dimensional eigenvectors. The same quantities $\rho_1^2 \geq \rho_2^2 \geq \cdots \geq \rho_m^2$ are also the m largest eigenvalues of the matrix $\Sigma_{YY}^{-1/2} \Sigma_{XY}^{\text{T}} \Sigma_{XX}^{-1} \Sigma_{XY} \Sigma_{YY}^{-1/2}$ with corresponding normalized q-dimensional eigenvectors $\mathbf{f}_1, \mathbf{f}_2, \ldots, \mathbf{f}_m$. Each eigenvector $\mathbf{f}_k$, $k = 1, \ldots, m$, is proportional to $\Sigma_{YY}^{-1/2} \Sigma_{XY}^{\text{T}} \Sigma_{XX}^{-1/2} \mathbf{e}_k$. The correlation $\text{Corr}(U, V)$ is maximized by the following linear combinations:

$$U_1 = \mathbf{a}_1^{\text{T}} \mathbf{X} = \mathbf{e}_1^{\text{T}} \Sigma_{XX}^{-1/2} \mathbf{X} \quad (\text{i.e.}, \mathbf{a}_1^{\text{T}} = \mathbf{e}_1^{\text{T}} \Sigma_{XX}^{-1/2}) \tag{8.7}$$

and

$$V_1 = \mathbf{b}_1^{\text{T}} \mathbf{Y} = \mathbf{f}_1^{\text{T}} \Sigma_{YY}^{-1/2} \mathbf{Y} \quad (\text{i.e.}, \mathbf{b}_1^{\text{T}} = \mathbf{f}_1^{\text{T}} \Sigma_{YY}^{-1/2}) \tag{8.8}$$

called the *first pair of canonical variables*. The maximum correlation is equal to

$$\text{Corr}(U_1, V_1) = \rho_1 = \sqrt{\rho_1^2} \tag{8.9}$$

and is called the *first canonical correlation*. The kth pair of canonical variables, $k = 2, 3, \ldots, m$,

$$U_k = \mathbf{e}_k^{\text{T}} \Sigma_{XX}^{-1/2} \mathbf{X} \quad \text{and} \quad V_k = \mathbf{f}_k^{\text{T}} \Sigma_{YY}^{-1/2} \mathbf{Y} \tag{8.10}$$

maximizes

$$\text{Corr}(U_k, V_k) = \rho_k = \sqrt{\rho_k^2} \tag{8.11}$$

(called the *kth canonical correlation*) among those linear combinations uncorrelated with preceding $(k-1)$ canonical variables.

Property 8.1 The canonical variables have the following properties:

$$\begin{aligned}
\text{Var}(U_k) &= \text{Var}(V_k) = 1, \\
\text{Corr}(U_k, U_j) &= 0, \quad \text{for } k \neq j, \\
\text{Corr}(V_k, V_j) &= 0, \quad \text{for } k \neq j, \\
\text{Corr}(U_k, V_j) &= 0, \quad \text{for } k \neq j,
\end{aligned} \tag{8.12}$$

for $k, j = 1, 2, \ldots, m$.

In Theorem 8.1, we constructed only m canonical variables, where m is the smaller of the two numbers p and q. Let us assume that $p \geq q$. In that case, we can construct q canonical variables $V_1, V_2, \ldots, V_q$, but we can also construct as many as p canonical variables $U_1, U_2, \ldots, U_p$ by using the formula in (8.10). The additional variables $U_{q+1}, \ldots, U_p$ were not mentioned in Theorem 8.1 because they do not pair with the V variables. Nevertheless, the full set of U variables will be useful. The vectors of canonical variables $\mathbf{U} = \begin{bmatrix} U_1, U_2, \ldots, U_p \end{bmatrix}$ and $\mathbf{V} = \begin{bmatrix} V_1, V_2, \ldots, V_q \end{bmatrix}$ can be expressed as

$$\mathbf{U} = \mathbf{A} \cdot \mathbf{X} \quad \text{and} \quad \mathbf{V} = \mathbf{B} \cdot \mathbf{Y}, \tag{8.13}$$

where $\mathbf{A}$ is a $p \times p$ matrix of $\mathbf{a}_k$ vectors as rows and $\mathbf{B}$ is a $q \times q$ matrix of $\mathbf{b}_k$ vectors as rows. Figure 8.3 shows relationships between the original random components and the canonical variables. Each canonical variable U_j is a linear combination of the X

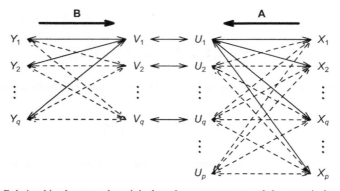

Figure 8.3 Relationships between the original random components and the canonical variables. The matrix **A** transforms **X** to the vector of canonical variables **U**, and **B** transforms **Y** to **V**. Only the canonical variables connected by arrows are potentially correlated.

components, and $\mathbf{A}$ is the matrix of transformation from the components $\mathbf{X}$ to the vector of canonical variables $\mathbf{U}$. In a similar way, $\mathbf{B}$ transforms $\mathbf{Y}$ to $\mathbf{V}$. Only the canonical variables connected by arrows are potentially correlated. We assumed that $p > q$, so the additional canonical variables $U_{q+1}, \ldots, U_p$ are not correlated with any of the remaining canonical variables.

Remark 8.1 In practice, the matrices Σ_{XX}, Σ_{YY}, and Σ_{XY} are unknown, and canonical correlation analysis is performed on the sample estimates $\mathbf{S}_{XX}$, $\mathbf{S}_{YY}$, and $\mathbf{S}_{XY}$, respectively. For simplicity, we are going to use the notation from the above theorem also for the sample canonical correlation analysis, with the only modification that Σ_{XX}, Σ_{YY}, and Σ_{XY} are replaced by $\mathbf{S}_{XX}$, $\mathbf{S}_{YY}$, and $\mathbf{S}_{XY}$.

Remark 8.2 When the canonical variables are calculated based on the above formulas using eigenvectors, they are defined up to a (± 1) factor. Consequently, some of the canonical correlations may have a negative sign. In a sense, we are maximizing the absolute value of the correlations. The negative correlations can be changed to positive ones by changing the signs of some of the canonical variables.

Remark 8.3 Some sources (e.g., Schott, 2007) use a slightly different approach, where a matrix $\Sigma_{XX}^{-1}\Sigma_{XY}\Sigma_{YY}^{-1}\Sigma_{XY}^{T}$ is used instead of $\Sigma_{XX}^{-1/2}\Sigma_{XY}\Sigma_{YY}^{-1}\Sigma_{XY}^{T}\Sigma_{XX}^{-1/2}$ in the definition of canonical variables. That approach is equivalent to our approach, except that the resulting canonical variables can have variances different from 1. Details are explained in Appendix C.

8.3 PRACTICAL APPLICATION

Example 8.1 (cont.) Let's continue the example with a 15 by 15 pixel image of grass texture in 42 spectral bands. We divide all 42 components (spectral bands) into two groups. The first group of components consists of Bands 1–18, and we denote them by $\mathbf{Y}$. The second group of components consists of Bands 19–42, and we denote them by $\mathbf{X}$. In our notation, $p = 24$ and $q = 18$. The first 12 canonical correlations were calculated for this data set, and they are shown in Table 8.1.

The first three canonical correlations are very large, and the subsequent values are also quite large. This means that we found several pairs of linear combinations in both sets of spectral bands that are highly correlated. The result is not surprising when you look at the sample correlations matrix $\mathbf{R}$ of all 42 spectral bands (represented in Figure 8.2). Most values in the $\mathbf{R}$ matrix are very high, which means that most of the 42 spectral bands are highly correlated.

When performing canonical correlation analysis, we hope that the correlation between the two sets of variables can be described by a relatively small set of

Table 8.1 The First 12 Canonical Correlations for the Grass Texture Data

0.9998	0.9941	0.9748	0.8630	0.7866	0.7496	0.6714	0.6192	0.4996	0.4678	0.4626	0.3918

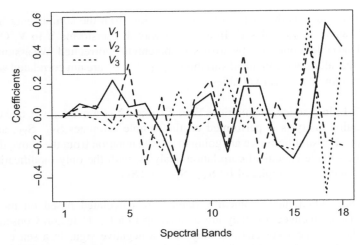

Figure 8.4 Normalized coefficients for the first three canonical variables from the set of the first 18 spectral bands denoted by **Y**.

canonical variables. In our example, the number of highly correlated pairs is relatively large. We should now investigate the canonical variables, try to interpret them, and see how much variability within each set of variables they explain.

Coefficient vectors $\mathbf{a}_k$ and $\mathbf{b}_k$ of the canonical variables are chosen so that the variables have variance 1. Hence, their magnitudes may vary substantially for different k. For interpretation purposes, we can plot the normalized vectors $\mathbf{a}_k$ and $\mathbf{b}_k$. Figure 8.4 shows such normalized values for the first three canonical variables from the first sets of variables. The coefficients alternate between negative and positive, indicating that all three canonical variables are contrasts of spectral bands. However, one cannot see any particular patterns in these values to allow for any more specific interpretation.

Figure 8.5 shows such normalized coefficients for the first three canonical variables from the second sets of variables. The coefficients again alternate between negative and positive, indicating that they are contrasts of spectral bands. We can also see that for all three vectors $\mathbf{b}_k$, the coefficients corresponding to Bands 29, 30, and 31 are almost equal to zero. This was expected from Bands 30 and 31 since they have little correlation with any other bands (see Figure 8.2). Band 29 seems to have larger correlations but apparently is not substantially contributing to correlations with the first 18 bands.   □

8.4   CALCULATING VARIABILITY EXPLAINED BY CANONICAL VARIABLES

We can now calculate how much variability within each set of variables the canonical variables can explain. We will then compare the variability with that of the principal components (which is the largest variability that can be explained). For this purpose, we need the following result.

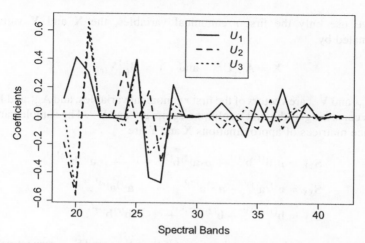

Figure 8.5 Normalized coefficients for the first three canonical variables from the set of the last 24 spectral bands denoted by **X**.

Result 8.1 For any matrix **G** consisting of p columns $\mathbf{g}_j$, $j = 1, \ldots, p$, and matrix **H** consisting of p rows $\mathbf{h}_j^T$, $j = 1, \ldots, p$, we have

$$\mathbf{G} \cdot \mathbf{H} = \mathbf{g}_1 \mathbf{h}_1^T + \mathbf{g}_2 \mathbf{h}_2^T + \cdots + \mathbf{g}_p \mathbf{h}_p^T. \tag{8.14}$$

From equations (8.12) and (8.13), the sample covariance $\text{Cov}(\mathbf{U}, \mathbf{V}) = \mathbf{A} S_{XX} \mathbf{B}^T = \text{diag}_{p \times q}(\rho_1, \rho_2, \ldots, \rho_p)$, sample variance $\text{Var}(\mathbf{U}) = \mathbf{A} S_{XX} \mathbf{A}^T = \mathbf{I}_p$, and sample variance $\text{Var}(\mathbf{V}) = \mathbf{B} S_{YY} \mathbf{B}^T = \mathbf{I}_q$, where $\text{diag}_{p \times q}(\rho_1, \rho_2, \ldots, \rho_p)$ is a $p \times q$ matrix with nonzero elements $\rho_1, \rho_2, \ldots, \rho_p$ on the diagonal only. We can solve for the three **S** matrices to obtain

$$\mathbf{S}_{XY} = \mathbf{A}^{-1} \, \text{diag}_{p \times q}(\rho_1, \rho_2, \ldots, \rho_p) (\mathbf{B}^{-1})^T$$

$$= \rho_1 \mathbf{a}^{(1)} \mathbf{b}^{(1)^T} + \rho_2 \mathbf{a}^{(2)} \mathbf{b}^{(2)^T} + \cdots + \rho_p \mathbf{a}^{(p)} \mathbf{b}^{(p)^T},$$

$$\mathbf{S}_{XX} = \mathbf{A}^{-1} (\mathbf{A}^{-1})^T = \mathbf{a}^{(1)} \mathbf{a}^{(1)^T} + \mathbf{a}^{(2)} \mathbf{a}^{(2)^T} + \cdots + \mathbf{a}^{(p)} \mathbf{a}^{(p)^T},$$

$$\mathbf{S}_{YY} = \mathbf{B}^{-1} (\mathbf{B}^{-1})^T = \mathbf{b}^{(1)} \mathbf{b}^{(1)^T} + \mathbf{b}^{(2)} \mathbf{b}^{(2)^T} + \cdots + \mathbf{b}^{(q)} \mathbf{b}^{(q)^T}, \tag{8.15}$$

where $\mathbf{a}^{(i)}$ and $\mathbf{b}^{(i)}$ denote the ith column of $\mathbf{A}^{-1}$ and $\mathbf{B}^{-1}$, respectively.

Solving equations (8.13), we can recover **X** and **Y** values from the canonical variables using the following formulas:

$$\mathbf{X} = \mathbf{A}^{-1} \mathbf{U} \quad \text{and} \quad \mathbf{Y} = \mathbf{B}^{-1} \mathbf{V}. \tag{8.16}$$

When we use only the first r canonical variables, the $\mathbf{X}$ and $\mathbf{Y}$ values are approximated by

$$\tilde{\mathbf{X}} = \mathbf{A}_{(r)}^{-1} \mathbf{U}_{(r)} \quad \text{and} \quad \tilde{\mathbf{Y}} = \mathbf{B}_{(r)}^{-1} \mathbf{V}_{(r)}, \tag{8.17}$$

where $\mathbf{U}_{(r)}$ and $\mathbf{V}_{(r)}$ are vectors of the first r canonical variables, and $\mathbf{A}_{(r)}^{-1}$ and $\mathbf{B}_{(r)}^{-1}$ are matrices consisting of the first r column vectors $\mathbf{a}^{(i)}$ and $\mathbf{b}^{(i)}$. The sample variance–covariance matrices of approximations $\tilde{\mathbf{X}}$ and $\tilde{\mathbf{Y}}$ are

$$\tilde{\mathbf{S}}_{XY} = \rho_1 \mathbf{a}^{(1)} \mathbf{b}^{(1)^{\mathrm{T}}} + \rho_2 \mathbf{a}^{(2)} \mathbf{b}^{(2)^{\mathrm{T}}} + \cdots + \rho_r \mathbf{a}^{(r)} \mathbf{b}^{(r)^{\mathrm{T}}},$$

$$\tilde{\mathbf{S}}_{XX} = \mathbf{a}^{(1)} \mathbf{a}^{(1)^{T}} + \mathbf{a}^{(2)} \mathbf{a}^{(2)^{T}} + \cdots + \mathbf{a}^{(r)} \mathbf{a}^{(r)^{T}}, \tag{8.18}$$

$$\tilde{\mathbf{S}}_{YY} = \mathbf{b}^{(1)} \mathbf{b}^{(1)^{T}} + \mathbf{b}^{(2)} \mathbf{b}^{(2)^{\mathrm{T}}} + \cdots + \mathbf{b}^{(r)} \mathbf{b}^{(r)^{T}},$$

which can be regarded as approximations of $\mathbf{S}_{XY}$, $\mathbf{S}_{XX}$, and $\mathbf{S}_{YY}$, respectively.

We can now calculate the total variability (as defined in formula (5.26)) explained by the first r canonical variables using the following formulas:

$$\mathrm{tr}(\tilde{\mathbf{S}}_{XX}) = \mathrm{tr}\left(\mathbf{a}^{(1)} \mathbf{a}^{(1)^{T}}\right) + \mathrm{tr}\left(\mathbf{a}^{(2)} \mathbf{a}^{(2)^{T}}\right) + \cdots + \mathrm{tr}\left(\mathbf{a}^{(r)} \mathbf{a}^{(r)^{T}}\right)$$

$$= \left\|\mathbf{a}^{(1)}\right\|^2 + \left\|\mathbf{a}^{(3)}\right\|^2 + \cdots + \left\|\mathbf{a}^{(r)}\right\|^2,$$

$$\mathrm{tr}(\tilde{\mathbf{S}}_{YY}) = \mathrm{tr}\left(\mathbf{b}^{(1)} \mathbf{b}^{(1)^{T}}\right) + \mathrm{tr}\left(\mathbf{b}^{(2)} \mathbf{b}^{(2)^{T}}\right) + \cdots + \mathrm{tr}\left(\mathbf{b}^{(r)} \mathbf{b}^{(r)^{T}}\right)$$

$$= \left\|\mathbf{b}^{(1)}\right\|^2 + \left\|\mathbf{b}^{(2)}\right\|^2 + \cdots + \left\|\mathbf{b}^{(r)}\right\|^2 \tag{8.19}$$

for the first and the second set of variables, respectively.

Example 8.1 (cont.) Table 8.2 summarizes the variability explained by the first 10 canonical variables within the first 18 spectral bands for the grass texture data and compares those values to the variability explained by principal components.

The variability explained by canonical variables is almost as high as that explained by principal components (which, in turn, is the highest possible). This means that in this case the canonical variables achieve two goals at the same time:

1. They give the best correlation with the other set of variables.
2. They give close to the optimal explanation of the variability within the subgroup of variables.

Table 8.3 summarizes similar results on the variability explained by the first 10 canonical variables within the last 24 spectral bands for the grass texture data.

Table 8.2 Variability Explained by the First 10 Canonical Variables and the First 10 Principal Components Within the First 18 Spectral Bands for the Grass Texture Data

		Variables									
		1	2	3	4	5	6	7	8	9	10
Explained Variability	PC	344.349	25.671	0.453	0.261	0.218	0.111	0.040	0.030	0.018	0.014
	CC	337.673	27.284	3.061	0.970	0.232	0.236	0.173	0.195	0.264	0.324
Explained Variability in percent	PC	92.765	6.915	0.122	0.070	0.059	0.030	0.011	0.008	0.005	0.004
	CC	90.967	7.350	0.825	0.261	0.063	0.064	0.047	0.052	0.071	0.087
Cumulative Variability	PC	92.765	99.681	99.803	99.873	99.932	99.962	99.972	99.980	99.985	99.989
	CC	90.967	98.317	99.141	99.403	99.465	99.529	99.575	99.628	99.699	99.786

Table 8.3   Variability Explained by the First 10 Canonical Variables and the First 10 Principal Components Within the Last 24 Spectral Bands for the Grass Texture Data

		Variables									
		1	2	3	4	5	6	7	8	9	10
Explained Variability	PC	1315.29	40.66	10.574	4.482	4.162	2.437	1.013	0.820	0.595	0.271
	CC	1092.24	206.26	22.654	1.261	2.640	1.403	1.195	3.496	1.283	8.079
Explained Variability in percent	PC	95.250	2.944	0.766	0.325	0.301	0.176	0.073	0.059	0.043	0.020
	CC	79.097	14.937	1.641	0.091	0.191	0.102	0.087	0.253	0.093	0.585
Cumulative Variability	PC	95.250	98.195	98.960	99.285	99.586	99.763	99.836	99.896	99.939	99.958
	CC	79.097	94.034	95.674	95.766	95.957	96.059	96.145	96.398	96.491	97.076

Table 8.4   Correlations Between Pairs of Principal Components from the Two Sets of Spectral Bands for the Grass Texture Data

PC1	PC2	PC3	PC4	PC5
0.9475	0.1392	−0.0523	0.0847	−0.4689

This time, we can see much lower variability explained by the first canonical variable, but after including the second canonical variable, the cumulative explained variability is almost as high as that of principal components.

We can also investigate correlations between pairs of principal components from the two sets of spectral bands. Table 8.4 summarizes such correlations for the pairs of the first five PCs. The first correlation coefficient is quite high, but the remaining values are very low and are much lower than the canonical correlations.

When comparing canonical variables with principal components, there is a trade-off between maximizing correlations of the two sets of variables and maximizing the explained variability within those sets. In our example of grass texture data, we can see a relatively good agreement between the two goals. In some cases, it may not work that well. An example when the two approaches give entirely different results is shown in Problem 8.6, where the first principal components from the two groups are uncorrelated and the canonical variables explain little variability.

8.5   CANONICAL CORRELATION REGRESSION

So far, we have been using canonical correlation as a tool to describe the correlations between two sets of variables. It can also be used for predictive purposes in the so-called *canonical correlation regression* (CCR). The idea is to establish a predictive relationship between the sets of canonical variables **V** and **U**.

We are going to perform sample canonical correlation analysis with the following notation. $\mathcal{X}$ and $\mathcal{Y}$ are $n \times p$ and $n \times q$ matrices representing data from the two sets of variables, respectively. All observations have been centered, that is, a mean vector was subtracted from each observation. The values of canonical variables for all n observations can be represented in matrices $\mathcal{U}$ and $\mathcal{V}$ given by the following formulas:

$$\mathcal{U} = \mathcal{X} \cdot \mathbf{A}^{\mathrm{T}} \quad \text{and} \quad \mathcal{V} = \mathcal{Y} \cdot \mathbf{B}^{T}, \tag{8.20}$$

where **A** is a $p \times p$ matrix of $\mathbf{a}_k$ vectors as rows and **B** is a $q \times q$ matrix of $\mathbf{b}_k$ vectors as rows (as defined in the context of formula (8.13)).

The idea of canonical correlation regression is to build a regression relationship between **V** and **U** and then transform it to the relationship between **Y** and **X**. Let us say, we want to predict **Y** based on **X**. This means we need to predict **V** based on **U**. We can do this by using the traditional least-squares method of linear regression.

The resulting fitted values given by an $n \times p$ matrix $\hat{\mathscr{V}}$ can be calculated from the following formula:

$$\hat{\mathscr{V}} = \mathscr{U} \cdot \hat{\beta}_{CC}, \qquad (8.21)$$

where $\hat{\beta}_{CC} = \left(\mathscr{U}^T \mathscr{U} \right)^{-1} \mathscr{U}^T \mathscr{V}$. It turns out that $\hat{\beta}_{CC} = \mathrm{diag}_{p \times q}(\rho_1, \rho_2, \ldots, \rho_m)$, where $m = \min\{p, q\}$. However, some of the correlations on the diagonal can be negative, so they should be calculated directly as the sample correlations between the canonical variables rather than as the square root of ρ_j^2. An alternative computational strategy is to modify the sign of a canonical variable so that only nonnegative correlations are used (as commented in Remark 8.2). Finally, the fitted values for the prediction of $\mathbf{Y}$ can be calculated as

$$\hat{\mathscr{Y}} = \hat{\mathscr{V}} \cdot \left(\mathbf{B}^T \right)^{-1}. \qquad (8.22)$$

By taking together formulas (8.20), (8.21), and (8.22), we obtain

$$\hat{\mathscr{Y}} = \mathscr{X} \cdot \mathbf{C}^T, \qquad (8.23)$$

where $\mathbf{C} = \mathbf{B}^{-1} \cdot \hat{\beta}_{CC}^T \cdot \mathbf{A}$. In order to write down the predictive linear function for canonical correlation regression, we want to operate on the vectors $\mathbf{Y}$ and $\mathbf{X}$. In that case, equation (8.23) can be written as

$$\hat{\mathbf{Y}} = \mathbf{C} \cdot \mathbf{X}, \qquad (8.24)$$

where $\mathbf{C}$ is the matrix of regression coefficients describing the predictive linear transformation between the two sets of variables. The three transformations $\mathbf{A}$, $\hat{\beta}_{CC}^T$, and $\mathbf{B}^{-1}$ are shown in Figure 8.6, so that we can see their chain leading to the transformation $\mathbf{C}$ representing the regression coefficients.

When all canonical variables are used (as shown above), the resulting regression is equivalent to the traditional least-squares regression. Since the initial pairs of canonical variables give the strongest correlations, it makes sense to decide on a smaller number, say $t \leq m$, of canonical variable pairs. In that case, we can still use the above formulas with the matrix β_{CC} equal to $\mathrm{diag}_{p \times q}(\rho_1, \rho_2, \ldots, \rho_t)$ (i.e., the remaining $(m - t)$ correlations on the diagonal are replaced by zeros).

In order to assess the CCR model, we can study residual vectors calculated as rows of the following residual matrix $\mathbf{D}$:

$$\mathbf{D} = \mathscr{Y} - \hat{\mathscr{Y}}. \qquad (8.25)$$

Example 8.1 (cont.) For the grass texture data, a CCR model was fitted for $t = 18$ (the full model). The resulting residual vectors $\mathbf{d}_i$, $i = 1, \ldots, n$ (rows of $\mathbf{D}$), are plotted in Figure 8.7. There are some differences in the residual variability in different spectral bands. This means that the prediction will be more precise in the bands with

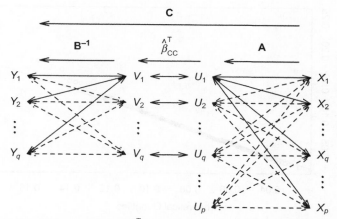

Figure 8.6 The three transformations $\mathbf{A}$, $\hat{\beta}_{CC}^{T}$, and $\mathbf{B}^{-1}$ shown as a chain leading to the transformation $\mathbf{C}$ representing the regression coefficients.

lower residual variability. The distribution of row vectors of $\mathbf{D}$ seems to be well approximated by the p-dimensional normal distribution, which was verified by univariate normal probability plots. It is also confirmed by a chi-squared Q–Q plot of statistical distances shown in Figure 8.8, where most points follow the plotted straight line ($y = x$), except for one outlier.

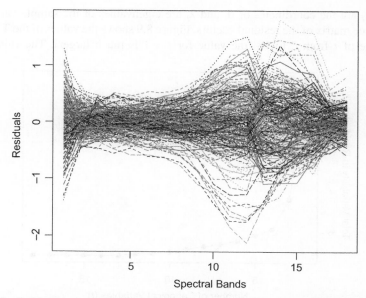

Figure 8.7 Residual vectors (rows of the residual matrix $\mathbf{D}$ defined by formula (8.25)) shown as functions of the spectral band number.

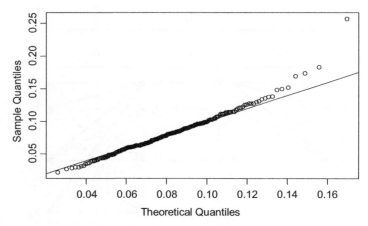

Figure 8.8   A probability plot using the Small's method for the residual vectors calculated as rows of the residual matrix **D** defined by formula (8.25).

It is of interest to investigate the impact of reducing the number k of canonical variables used in CCR. As a one-number summary of variability in residuals, we use the total variability in residuals (TVR)

$$\text{TVR} = \frac{1}{n-1} \sum_{i=1}^{n} \|\mathbf{d}_i\|^2 = \sum_{j=1}^{p} \left(\frac{1}{n-1} \sum_{i=1}^{n} d_{ij}^2 \right) = \sum_{j=1}^{p} \lambda_j, \qquad (8.26)$$

where d_{ij} are the coordinates of $\mathbf{d}_i$ and λ_j are eigenvalues of the sample variance–covariance matrix of the residual vectors. Figure 8.9 shows the values of the TVR for the range of t from 2 to 18 (the value for $t = 1$ is much larger). The values are

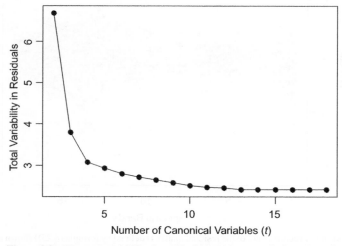

Figure 8.9   The total variability in residuals for the range of t from 2 to 18.

decreasing with the smallest value at $t = 18$, and they are almost constant for large t. This may suggest the use of $t = 18$ as the best choice, or possibly the use of a somewhat smaller t, if one was not concerned with a slight increase in TVR. However, a shortcoming of such analysis is that the TVR will usually be reduced for more complex models such as CCR with a larger value of t.

In general, a drawback of more complex models is that they are more likely to be overfitted to the data. This means that the model gives good predictions (small residuals) for the points from the data set used to fit the model, but it may give poor predictions in future data. Consequently, cross-validation should be performed to compare the models and give a more realistic assessment of prediction in future data. Here we use k-fold cross-validation with several values of k and leave-one-out cross-validation (see Supplement 8A for a general discussion of cross-validation). For each test vector $\mathbf{x}_i$ (realization of the second set of variables $\mathbf{X}$), we calculate the fitted value $\hat{\mathbf{y}}_i = \mathbf{C} \cdot \mathbf{x}_i$. In the leave-one-out cross-validation, each vector from the data set is taken exactly once as a test vector. Hence, we can calculate the estimated test error (ETE) as

$$\text{ETE} = \frac{1}{n} \sum_{i=1}^{n} \|\mathbf{y}_i - \hat{\mathbf{y}}_i\|^2 = \sum_{j=1}^{p} \left(\frac{1}{n} \sum_{i=1}^{n} (y_{ij} - \hat{y}_{ij})^2 \right), \qquad (8.27)$$

where $y_{ij}, j = 1, \ldots, p$, are the coordinates of $\mathbf{y}_i$. The ETE can also be regarded as the total variability of the prediction error around zero (the desired value). For the k-fold cross-validation, each vector from the data set is also taken exactly once as a test vector within one round of cross-validation. Hence, formula (8.27) can also be used for estimation of the test error. The ETE values for the leave-one-out and three types of k-fold cross-validation are plotted in Figure 8.10. The k-fold

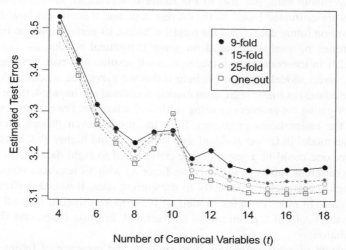

Figure 8.10 The ETE values for the leave-one-out and three types of k-fold cross-validation for a range of the number of canonical variables used in CCR.

cross-validation consisted of 100 rounds each, and consequently the standard errors (due to random sampling) of all estimates are very low (below 0.004). Here, the nine-fold cross-validation leaves out the largest number of vectors ($225/9 = 25$) for testing each time, and hence it is the most challenging case for prediction. This explains why the ETE values are mostly the largest ones for the nine-fold case. For larger k, the k-fold generated ETE values become smaller, with the smallest values for the leave-one-out cross-validation (equivalent to n-fold cross-validation), except for $t = 10$.

Within a given type of cross-validation, the smallest ETE values are observed for $t = 15$ (9-fold and 15-fold) and for $t = 16$ (25-fold and one-out). This suggests slight overfitting in the full CCR model with $t = 18$ and a recommended value of $t = 15$ in CCR. One could also consider $t = 8$, if further simplification is sought, possibly at a cost of an increased prediction error.

8.6 FURTHER READING

Good references for further reading on canonical correlation analysis are Johnson and Wichern (2007) and Jackson (1991). Schott (2007), Hernandez-Baquero and Schott (2000), and Hernandez-Baquero (2001) shows applications of canonical correlation regression in remote sensing.

SUPPLEMENT 8A. CROSS-VALIDATION

In any type of statistical modeling, we are interested in the model fit, not only to the currently available data, but also to the future observations. However, the model parameters are estimated based on the current data, and it is unclear how the model may perform on future data. Once the model is found, its performance on future data can sometimes be evaluated based on some theoretical results, as was done in Section 4.2.6 in the context of regression. Such results, however, assume that we found the correct model. One danger here is that we *overfit* the model to the current data. In the context of linear regression models discussed in Chapter 4, there are some ways of mitigating the problem by using statistical inference. For example, we would eliminate the insignificant predictors from the model, even though they always increase the model fit (lower residual sum of squares and higher R^2). Without such precautions, one could fit a seven-degree polynomial to eight data points perfectly (assuming different values of the only predictor x) with all residuals equal to zero. Even though the model fits perfectly to the current data, it would perform poorly on future data. In more complex models, it is far more difficult, if at all possible, to check analytically if a given model is overfitted. In those cases, one should use cross-validation.

The concept of *cross-validation* is to simulate the presence of future data. We pretend that some of the current data points are the future observations (we call it a *testing sample*), and only the remaining observations (*a learning sample*) are used to fit the model. The fitted model is then evaluated based on the testing sample. We would typically compare different types of models through cross-validation. Once it is decided which model is the most promising one, the specific model parameters are fitted based on the whole data set.

The question that remains is how to choose the testing sample. There are many ways to do this. One popular method is that of k-fold validation, where k is a chosen number such as 5 or 10. The k-fold validation is performed in the following round of steps:

1. The data set is randomly divided into k approximately equal subsets.
2. One of the k subsets is removed as the learning sample, and the model is fitted to the data from the remaining $(k - 1)$ subsets.
3. The model is evaluated based on the removed subset (the testing sample).
4. Steps 2 and 3 are repeated for all k subsets being removed in turn.

In the above steps, each observation is removed exactly once as an element of one of the k subsets. The above round can be repeated several times and the average evaluation results are calculated. The model evaluation in Step 3 involves a measure of prediction error. For continuous vector responses $\mathbf{y}$, we can calculate the *estimated test error* as

$$\text{ETE} = \frac{1}{n}\sum_{i=1}^{n} \|\mathbf{y}_i - \hat{\mathbf{y}}_i\|^2 = \sum_{j=1}^{p}\left(\frac{1}{n}\sum_{i=1}^{n}\left(y_{ij} - \hat{y}_{ij}\right)^2\right), \qquad (8.28)$$

where $\mathbf{y}_i$ is the actual response of the ith observation and $\hat{\mathbf{y}}_i$ is the predicted response for the ith observation based on the learning sample. The values $y_{ij}, j = 1, \ldots, p$, are the coordinates of $\mathbf{y}_i$. The value of ETE can be calculated after one round of cross-validation. If more rounds are performed, the resulting ETE values are averaged. In classification problems, we use different ways to evaluate the model and the resulting classification rule (see Chapter 9 for details).

When k is equal to the sample size n, the k-fold cross-validation becomes the so-called *leave-one-out* method. Here the procedure is repeated n times with exactly one observation removed each time. There is no need to randomize here, so the result is not random, in contrast to the classic k-fold method for smaller k. This is a definite advantage of the leave-one-out cross-validation. On the other hand, the method is limited in predicting some challenges in future data because only one observation is removed at a time. If there are small clusters of outliers in the data, the remaining outliers in the cluster will mask the difficulty of prediction. When we remove more observations at one time, a cluster of outliers might be removed as a whole from the learning sample, resulting in poor, but more realistic, prediction. Another difficulty is that for very large data sets and complex procedures being evaluated, the leave-one-out cross-validation may not be computationally feasible.

PROBLEMS

8.1. Calculate all 18 canonical correlations for the grass data used in Example 8.1. Check your first 12 values with those in Table 8.1. Are some values different only by an opposite sign? Is this a problem? Why?

8.2. Calculate the coefficients for the first 10 canonical variables from both sets of the spectral bands considered in Example 8.1. Also calculate their normalized versions and plot the figures analogous to Figures 8.4 and 8.5. Can you observe any patterns?

8.3. Fit the canonical correlation regression model for all values of k from 1 to 18 for the data used in Example 8.1. Recreate the results of

    a. Figure 8.7.

    b. Figure 8.9.

8.4. Perform leave-one-out cross-validation for the canonical correlation regression model for the data used in Example 8.1. Recreate the results of Figure 8.10 for the leave-one-out cross-validation.

8.5. Perform k-fold cross-validation for the canonical correlation regression model for the data used in Example 8.1. Recreate the results of Figure 8.10 for the k-fold cross-validation.

    a. Use $k = 9$.

    b. Use $k = 15$.

    c. Use $k = 25$.

8.6. Assume that we have two sets of variables written as two random vectors $\mathbf{X}$ and $\mathbf{Y}$, each of dimensionality 2, that is, $p = q = 2$ in our notation. The variance–covariance matrix of the joint vector is given by

$$\mathrm{Var}\left(\begin{bmatrix}\mathbf{X}\\\mathbf{Y}\end{bmatrix}\right) = \begin{bmatrix} 80 & 0 & 0 & 0 \\ 0 & 1 & 0.97 & 0 \\ 0 & 0.97 & 1 & 0 \\ 0 & 0 & 0 & 120 \end{bmatrix}.$$

    a. Find the pairs of canonical variables.

    b. Calculate the fraction of variability explained by each of the first canonical variables within their respective sets of variables.

    c. Find the principal components within each group of variables and how much variability they explain.

    d. Find the correlations among the principal components identified in part c.

    e. Interpret your findings.

8.7. Consider infrared astronomy data used in Example 3.3 and described in Appendix B. Here we want to use only the group of C AGB stars and all four bands J, H, K, and A as variables. The first group of variables, denoted by the random vector **X**, consists of the J and H bands, and second group of variables consists of the remaining bands K and A (described by the random vector **Y**). Perform canonical correlation analysis and see if you can interpret the canonical variables in both groups (use an equivalent of Figure 8.4). Calculate all canonical correlations. Interpret the results.

8.8. Consider the infrared astronomy data used in Problem 8.7.

    a. Find the principal components within each group of variables and how much variability they explain. Try to interpret the principal components by using impact plots, like the one shown in Figure 7.9.

    b. Find the correlations among the principal components identified in part a. Interpret your findings.

    c. If you were also assigned Problem 8.7, compare the results and conclude which set of variables (canonical correlations versus principal components) better summarizes the structure of data.

8.9. Perform tasks of Problem 8.7 for the group of H II regions. If you were also assigned Problem 8.7, compare the results for the two groups.

8.10. Perform tasks of Problem 8.8 for the group of H II regions. If you were also assigned Problem 8.8, compare the results for the two groups. Also compare your results to those in Problem 8.9, if you were assigned that problem as well.

8.11. Consider the infrared astronomy data used in Problem 8.7. Use the canonical correlation regression to predict the vectors of readings in the second group described by the random vector **Y** based on the variables in the first group described by the random vector **X**. See if you can reduce the number of canonical variables used from $t = 2$ to $t = 1$. Base your decision on the residuals and some of their summary measures developed in Section 8.5.

8.12. In the context of Problem 8.11, perform cross-validation of the canonical correlation regression. Use the leave-one-out cross-validation and the k-fold cross-validation for $k = 7$ and $k = 18$.

C H A P T E R 9

Discrimination and Classification – Supervised Learning

9.1 INTRODUCTION

In this chapter, we will describe discrimination and classification analysis, also called supervised learning. The purpose of discrimination analysis is to describe features of observations from several populations that allow differentiating those populations and their observations. On the other hand, classification is a more formal process of assigning new observations to one of the identified populations. In practice, the same methodologies can be used for both discrimination and classification, and we will mostly use the term *classification* to describe those methods.

In the classification problems, we have a data set with observations on one or more variables, and we know which observations come from which populations or groups. This is why the methods for classification are often called supervised learning (we learn how to classify objects into groups), which is different from unsupervised learning (called cluster analysis in statistics) discussed in the next chapter. Our goal is to find a classification rule or procedure that would predict the group membership of future observations based on their observed values of the variables. Here are two examples of classification problems.

Example 9.1 Consider infrared astronomy data described in Example 3.3. Here we want to concentrate on two variables, the J (1.25 μm) and H (1.65 μm) band magnitudes obtained for 126 star objects, of which 67 are C AGB stars and 59 are H II regions. See Appendix B for more details about this data set. Figure 9.1 shows a scatter plot of the two groups of stars based on those two variables. The task of classification is to find a boundary line that would separate the two groups, so that, for example, the observations above the line are classified as C AGB and those below the line as H II. We can see some overlap of the two groups, and it is clear that perfect classification

Statistics for Imaging, Optics, and Photonics, Peter Bajorski.
© 2012 John Wiley & Sons, Inc. Published 2012 by John Wiley & Sons, Inc.

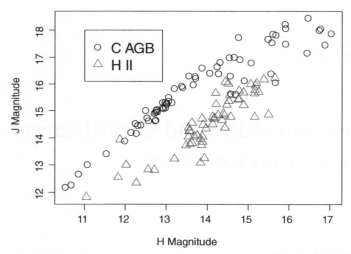

Figure 9.1   A scatter plot of two groups of stars (C AGB and H II) based on the magnitudes in two infrared bands, J and H.

based only on these two variables will not be possible. We will continue using this example throughout the chapter. □

Example 9.2   Consider a 64 by 64 pixel image of grass texture in 42 spectral bands. This means that each pixel is represented by a 42-dimensional vector or spectrum of reflectances in 42 spectral bands. A small area in the top right corner of the image (see Figure 9.2 and Appendix B for the exact definition of that area) is affected by a disease. There are 256 pixels with diseased grass, out of the total of the 4096 pixels in the image.

Our goal is to find out how the diseased grass can be recognized based on the reflectance values in 42 spectral bands. For this purpose, we define two sets of observations (pixels). The first set represents healthy grass, and the second set

Figure 9.2   An area of diseased grass in the top right corner.

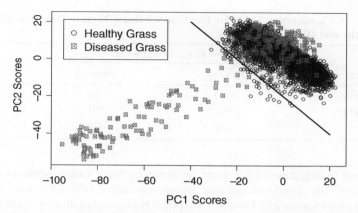

Figure 9.3 Two sets of grass spectra in principal component coordinates (PCA scores), with an example of a possible boundary line for a classification rule.

represents diseased grass, based on known locations of pixels. We will try to characterize the two data sets so that a future observation of a grass pixel can be identified as healthy or diseased based on reflectance values in 42 spectral bands.

A function plot of all 4096 spectra would look somewhat similar to Figure 8.1, where a subset of this data set was used. However, the plot would not be very informative due to a large number of overlapping lines. An alternative solution is to perform principal component analysis and plot values of the first two principal component scores as done in Figure 9.3. The spectra of the diseased grass are represented by crossed squares, and those of the healthy grass by circles. Some of the diseased grass spectra are clearly distinct from those of healthy grass, but there is also some significant overlap of the two groups. It turns out that the first two principal components explain 84.6% of the variability, which is not a lot in the context of spectral images. This suggests significant information contained in the remaining dimensions. Nevertheless, we can use Figure 9.3 for exploratory purposes, and try to construct a classification rule in those two dimensions.

One possibility is to use the straight line shown in Figure 9.3 as the boundary line L for a classification rule. The line goes through two points: $(20, -40)$ and $(-40, 20)$ in the principal component coordinates, or PC scores, that we will denote here by (y_1, y_2). We can classify all observations below the dividing line as representing diseased grass, and those on the line or above as healthy grass. Since the equation of the boundary line is $y_2 = -y_1 - 20$, we can calculate the threshold PC2 value for each observation (y_1, y_2) as $-y_1 - 20$. If y_2 is smaller than the threshold, that is, $y_2 < (-y_1 - 20)$, we would classify the observation as a pixel of diseased grass. However, a simpler approach would be to calculate $y_1 + y_2$ and check if it is below (-20). Note that $y_1 + y_2 = [y_1, y_2] \cdot \mathbf{v}$ is proportional to the projection on $\mathbf{v}$, where $\mathbf{v} = [1, 1]^T$ is a vector orthogonal to L. We can say that the classification is performed based on the projections of points on the vector $\mathbf{v}$. We will return to this general concept later on in this chapter.

Table 9.1   A Classification Table for Example 9.2 Grass Texture Data Classified into Healthy and Diseased Grass

Actual State	Grass Identified as		Total
	Healthy	Diseased	
Healthy grass	3809	31	3840
Diseased grass	149	107	256
Total	3958	138	4096

It is clear from Figure 9.3 that using the above approach leads to some misclassification errors. The classification results are summarized in a *classification table*, or a *confusion matrix*, shown in Table 9.1. We prefer the term classification table and will be using it throughout this book. The number of correctly classified observations is identified on the diagonal of the table or matrix, and the incorrect classifications are outside of the diagonal in the shaded cells.

We might wonder if we can reduce the misclassification rates shown in Table 9.1. When moving the dividing line up, we would reduce the number of misclassified pixels of diseased grass, but at the same time, we would increase the number of misclassified pixels of healthy grass. On the other hand, when moving the dividing line down, we would increase the number of misclassified pixels of diseased grass, but at the same time, we would reduce the number of misclassified pixels of healthy grass. The following questions arise:

- What is the optimum position for the dividing line?
- Is the use of the first two principal components optimal? Should we use the remaining principal components?
- Do we really need principal components, or should we just use the original variables?
- Would it help if we used a dividing line that was not a straight line?

The next section introduces some mathematical notation and results that help in answering these questions.   □

9.2   CLASSIFICATION FOR TWO POPULATIONS

Consider two populations from which two groups, or samples, of observations are drawn. We will denote the two populations by π_1 and π_2. In the context of Example 9.2, π_1 could be the population of healthy grass spectra, meaning not only the grass represented in the image, but also other healthy grass of similar type. The second population, π_2, would then be the population of diseased grass spectra, again not only that represented in the image, but also other diseased grass (although we should probably limit that population to a given type of disease). The two sets of spectra represented in Figure 9.3 (by their PC coordinates only) are two samples drawn from the two populations π_1 and π_2, respectively.

Let **X** be a random vector representing all the variables we observe in order to discriminate between the two populations. In Example 9.2, **X** could be a vector of all 42 spectral bands, or it could be a vector of the first two principal components as discussed in Section 9.1. If an observation comes from the first population π_1, the distribution of **X** is assumed to have a probability density function $f_1(\mathbf{x})$. On the other hand, if an observation comes from the second population π_2, the distribution of **X** is assumed to have a probability density function $f_2(\mathbf{x})$. It is assumed that the densities $f_1(\mathbf{x})$ and $f_2(\mathbf{x})$ are different from each other, since otherwise it would not be possible to distinguish between the two populations.

For each realization **x** of the random vector **X**, we want to decide whether the object comes from the population π_1 or π_2. Let R_1 be the set of **x** values for which we classify the object as coming from π_1, and R_2 be the set of the remaining possible **x** values (for which we classify the object as coming from π_2). The mutually exclusive regions R_1 and R_2 are called *classification regions* for the two populations. In the context of Grass Data from Example 9.2, let us assume that the vector **X** was taken as the vector of the first two principal components. Then R_1 could be the area above the dividing straight line, and R_2 would be the area below the dividing straight line. Table 9.2 shows the four possible events that can happen when an observation is classified into one of the two populations.

We are mainly interested in avoiding the misclassification events. In order to calculate their probabilities, we need to introduce prior probabilities. If we randomly select an object, the probability that it comes from population π_1 is the *prior probability* p_1, and the probability that it comes from population π_2 is the prior probability p_2. Clearly, $p_1 + p_2 = 1$. In practice, the values of p_1 and p_2 might be difficult to estimate. They could be estimated from the proportions of observations in the available data set under the assumption that the sample proportions are reasonable estimates of the population proportions. In many classification problems, this may not be a reasonable assumption since the data are often collected simply to represent the features of the two populations, but the sample proportions are not necessarily representative of the population proportions. In Example 9.2, the proportion of diseased grass shown in Figure 9.2a is highly unlikely to represent the true proportion in the population. If the image were taken as centered at the diseased grass area, those pixels would have larger representation, but again the sampling proportion would most likely be different from the true population proportion.

Table 9.2 Four Possibilities When an Observation is Classified into One of the Two Populations

	Observation Is Identified as Coming from	
Observation Comes from	Population π_1	Population π_2
Population π_1	OK	Misclassified as π_2
Population π_2	Misclassified as π_1	OK

The proportions p_1 and p_2 can sometimes be estimated from some general statistics available independently of the given data set. For example, it might be known that approximately 5% of grass in a given county is affected by a specific disease.

We also need to know some conditional probabilities. The conditional probability $P(2|1)$ of classifying an object as π_2 when, in fact, it is from π_1 is given by

$$P(2|1) = P(X \in R_2|\pi_1) = \int_{R_2} f_1(\mathbf{x})d\mathbf{x}, \qquad (9.1)$$

and similarly, the conditional probability $P(1|2)$ of classifying an object as π_1 when, in fact, it is from π_2 is given by

$$P(1|2) = P(X \in R_1|\pi_2) = \int_{R_1} f_2(\mathbf{x})d\mathbf{x}. \qquad (9.2)$$

We can now write down formulas for misclassification probabilities.

P(observation is misclassified as π_1)

$= P$(observation comes from π_2 and is misclassified as π_1) $\qquad (9.3)$

$= P(X \in R_1|\pi_2)P(\pi_2) = P(1|2)p_2.$

Similarly,

P(observation is misclassified as π_2)

$= P$(observation comes from π_1 and is misclassified as π_2) $\qquad (9.4)$

$= P(X \in R_2|\pi_1)P(\pi_1) = P(2|1)p_1.$

Our goal is to minimize both probabilities of misclassification. In practice, it is usually not possible to find a classification rule that would minimize both probabilities at the same. We could use the following approaches:

1. Fix the probability of one type of misclassification, and minimize the probability of the other type of misclassification (this is similar to the hypothesis testing problem with Type I and Type II errors).
2. Assign costs associated with each type of misclassification, and minimize the total cost of misclassification.

We are going to investigate the second option here. To this end, we can define the misclassification costs as shown in Table 9.3. The cost of correct classification is zero; the cost of misclassification as π_2 is $c(2|1)$; and the cost of misclassification as π_1 is $c(1|2)$.

Table 9.3 The Cost Matrix of Misclassifications

Observation Comes from	Observation Is Identified as Coming from	
	Population π_1	Population π_2
Population π_1	0	$c(2|1)$
Population π_2	$c(1|2)$	0

Definition 9.1 The expected cost of misclassification (ECM) is defined as the average cost of both types of misclassifications and can be calculated from the formula

$$\text{ECM} = c(2|1)P(2|1)p_1 + c(1|2)P(1|2)p_2. \tag{9.5}$$

Theorem 9.1 The classification rule that minimizes ECM is defined by the classification regions R_1 and R_2 given as

$$R_1 = \left\{ x : \frac{f_1(\mathbf{x})}{f_2(\mathbf{x})} \geq \frac{c(1|2)\,p_2}{c(2|1)\,p_1} \right\} \quad \text{and} \quad R_2 = \left\{ x : \frac{f_1(\mathbf{x})}{f_2(\mathbf{x})} < \frac{c(1|2)\,p_2}{c(2|1)\,p_1} \right\}. \tag{9.6}$$

Proof. See Problem 9.2 for a sketch of the proof.

When no information is available about prior probabilities or misclassification costs, they can be assumed to be the same for the two populations. The above classification rule is then simplified to a straightforward rule based on which density has the larger value, that is, $R_1 = \{\mathbf{x} : f_1(\mathbf{x}) \geq f_2(\mathbf{x})\}$ and R_2 otherwise.

The classification rule given by (9.6) requires knowledge of the density functions $f_1(\mathbf{x})$ and $f_2(\mathbf{x})$ for the two populations. Even though one could estimate those densities, it is rather difficult, especially in the multivariate case. Another possibility is to assume a specific form of those densities. We are going to take the latter approach in the next section, and assume normality of observations from both populations.

9.2.1 Classification Rules for Multivariate Normal Distributions

We now assume that $f_1(\mathbf{x})$ and $f_2(\mathbf{x})$ are multivariate normal (Gaussian) densities, the first one with the mean vector $\boldsymbol{\mu}_1$ and the variance–covariance matrix $\boldsymbol{\Sigma}_1$ and the second one with the mean vector $\boldsymbol{\mu}_2$ and the variance–covariance matrix $\boldsymbol{\Sigma}_2$. The classification rules discussed in this section are called Gaussian maximum likelihood rules by some sources, but there is little justification for this name because the maximum likelihood principles are not used here. We will call these rules the *Gaussian classification rules*. The true maximum likelihood rules are more complex and are defined in Anderson (2003) in Section 6.5.5 (equal variance–covariance matrices) and Section 6.10.1 (unequal variance–covariance matrices).

9.2.1.1 *Classification for Normal Populations with Equal Variance–Covariance Matrices*

When $\Sigma_1 = \Sigma_2 = \Sigma$, we can simplify the classification rule defined in equation (9.6). Specifically, one can show that the rule that minimizes the ECM is such that $\mathbf{x} \in R_1$, if

$$(\boldsymbol{\mu}_1 - \boldsymbol{\mu}_2)^{\mathrm{T}} \Sigma^{-1} \mathbf{x} - \frac{1}{2} (\boldsymbol{\mu}_1 - \boldsymbol{\mu}_2)^{\mathrm{T}} \Sigma^{-1} (\boldsymbol{\mu}_1 + \boldsymbol{\mu}_2) \geq \ln \left[\frac{c(1|2) \, p_2}{c(2|1) \, p_1} \right] \qquad (9.7)$$

and $\mathbf{x} \in R_2$ otherwise (the derivation is given as Problem 9.6). In practice, the population parameters are unknown, and they need to be estimated. The variance–covariance matrix Σ is estimated from the pooled (combined) estimate

$$\mathbf{S}_{\text{pooled}} = \frac{(n_1 - 1)}{(n_1 + n_2 - 2)} \mathbf{S}_1 + \frac{(n_2 - 1)}{(n_1 + n_2 - 2)} \mathbf{S}_2, \qquad (9.8)$$

and the classification rule for $\mathbf{x} \in R_1$ becomes

$$(\bar{\mathbf{x}}_1 - \bar{\mathbf{x}}_2)^{\mathrm{T}} \mathbf{S}_{\text{pooled}}^{-1} \mathbf{x} - \frac{1}{2} (\bar{\mathbf{x}}_1 - \bar{\mathbf{x}}_2)^{\mathrm{T}} \mathbf{S}_{\text{pooled}}^{-1} (\bar{\mathbf{x}}_1 + \bar{\mathbf{x}}_2) \geq \ln \left[\frac{c(1|2) \, p_2}{c(2|1) \, p_1} \right] \qquad (9.9)$$

and $\mathbf{x} \in R_2$ otherwise. The rule is called the *linear Gaussian rule*. Since the parameters are estimated here, the rule is no longer guaranteed to minimize the ECM. It is assumed that $\mathbf{S}_{\text{pooled}}$ is a nonsingular matrix. This means that $n_1 + n_2 - 2 \geq p$, where n_1 and n_2 are sizes of the two samples. If this is not the case, we can reduce the dimensionality p of the data set, so that $\mathbf{S}_{\text{pooled}}$ becomes nonsingular, and the classification can be performed in the reduced space. In practice, the sample sizes should be much larger than the minima mentioned here.

The classification rule defined by formula (9.7) or (9.9) is called a linear rule because the classification depends on $\mathbf{x}$ in a linear way. This can be seen from an alternative notation, where inequality (9.9) is written in the form

$$\mathbf{v}^{\mathrm{T}} \mathbf{x} \geq c, \qquad (9.10)$$

where

$$\mathbf{v}^{\mathrm{T}} = (\bar{\mathbf{x}}_1 - \bar{\mathbf{x}}_2)^{\mathrm{T}} \mathbf{S}_{\text{pooled}}^{-1} \quad \text{and} \quad c = \frac{1}{2} \mathbf{v}^{\mathrm{T}} (\bar{\mathbf{x}}_1 + \bar{\mathbf{x}}_2) + \ln \left[\frac{c(1|2) \, p_2}{c(2|1) \, p_1} \right]. \qquad (9.11)$$

This also means that the classification regions R_1 and R_2 are regions in a p-dimensional space separated by a $(p - 1)$-dimensional affine (linear) subspace (a straight line for $p = 2$) orthogonal to the *discriminant vector* $\mathbf{v}$. The boundary subspace is described by the equation $\mathbf{v}^{\mathrm{T}} \mathbf{x} = c$.

We call the direction of the discriminant vector $\mathbf{v}$ a *discriminant direction* because the discrimination (or classification) is performed based on the projection of $\mathbf{x}$ on the vector $\mathbf{v}$. It is worth mentioning that the discriminant direction does not depend on

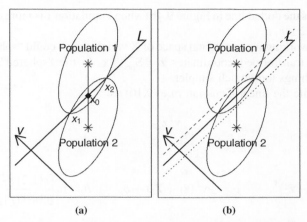

(a)   (b)

Figure 9.4 A scenario of linear classification for two populations following bivariate normal distributions with the same variance–covariance matrix.

the costs of misclassification and the prior probabilities. Only the threshold c applied to the projections on $\mathbf{v}$ depends on those quantities. Figure 9.4 shows graphically the situation for a two-dimensional classification problem ($p = 2$). The ellipses are the contour lines of two bivariate normal distributions with the same estimated variance–covariance matrix $\mathbf{S}_{\text{pooled}}$ and different mean vectors shown as stars. These are the estimated distributions of the two populations. Note that the discriminant direction, marked as $\mathbf{v}$, is not necessarily parallel to the line connecting the two means. Instead, it adjusts to the shape of the distributions, so that the boundary line L, orthogonal to $\mathbf{v}$, approximately minimizes the value of the ECM. The bold boundary line L shown in Figure 9.4 was calculated for the case when $c(1|2)p_2 = c(2|1)p_1$, which could be used when no information about the costs of misclassification and prior probabilities is available. In that case, the boundary line is the line where the two normal densities have the same values, that is, $L = \{\mathbf{x} : f_1(\mathbf{x}) = f_2(\mathbf{x})\}$. This means that L will cross all pairs of points where the two contour lines of densities of equal height intersect (such as points $\mathbf{x}_1$ and $\mathbf{x}_2$ in Figure 9.4a). The line L will also cross the middle point $\mathbf{x}_0$ between the estimated means of the two populations.

Note that $c(1|2)$ is the cost of misclassification per one object from Population 2. If N is the total size of both populations, then $p_2 N$ is the size of Population 2, and $c(1|2)p_2 N$ is the total cost of all misclassifications in Population 2. In the same fashion, $c(2|1)p_1 N$ is the total cost of all misclassifications in Population 1. When $c(1|2)p_2 > c(2|1)p_1$, misclassification in Population 2 will be more costly. This can happen when the objects from Population 2 are more costly when misclassified (i.e., $c(1|2) > c(2|1)$), or when there are more objects in Population 2 ($p_2 > p_1$). In either case, we would want to move the boundary line above L, so that more objects are classified as Population 2, thus reducing misclassifications. Since only the constant c in formula (9.10) is impacted by those changes, the resulting boundary line will be parallel to L, as shown by the dashed line in Figure 9.4b. In the same way, the line will move down (as the dotted line in Figure 9.4b) when Population 1 is more costly (when $c(1|2)p_2 < c(2|1)p_1$).

Instead of working in the original space of $\mathbf{X}$ variables, we could "sphere" all data by using the transformed coordinates $\mathbf{z} = \mathbf{S}_{pooled}^{-1/2}\mathbf{x}$. In the "sphered" space of $\mathbf{z}$ coordinates, things look much simpler.

We can write the linear Gaussian rule (9.10) as

$$\mathbf{v}^{\mathrm{T}}\mathbf{z} \geq c, \tag{9.12}$$

where

$$\mathbf{v}^{\mathrm{T}} = (\bar{\mathbf{z}}_1 - \bar{\mathbf{z}}_2)^{\mathrm{T}}, \qquad c = \frac{1}{2}\mathbf{v}^{\mathrm{T}}(\bar{\mathbf{z}}_1 + \bar{\mathbf{z}}_2) + b, \qquad b = \ln\left[\frac{c(1|2)\,p_2}{c(2|1)\,p_1}\right]. \tag{9.13}$$

The ellipses from Figure 9.4 become circles, and the line connecting the means is parallel to the discriminant direction $\mathbf{v}$ and is orthogonal to the boundary line L, as shown in Figure 9.5.

Errors of Misclassification
When the two costs of misclassification are assumed to be the same ($c(1|2) = c(2|1)$), the ECM criterion is reduced to minimizing the total probability of misclassification (TPM) defined by

$$\mathrm{TPM} = P(2|1)p_1 + P(1|2)p_2. \tag{9.14}$$

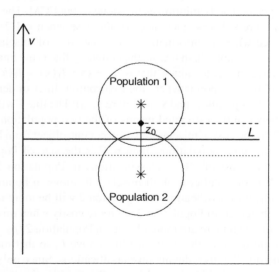

Figure 9.5 A scenario (shown in the sphered $\mathbf{z}$ space) of a linear classification for two populations following bivariate normal distributions with the same variance–covariance matrix.

We will now show how one can estimate both ECM and TPM for the linear Gaussian rule given by (9.10). In a given practical situation, we already know the values for the prior probabilities and the costs of misclassification because they were used in constructing the classification rule. However, we do not know the conditional probabilities. They can be estimated in the sphered $\mathbf{z}$ space shown in Figure 9.5 by using some simple geometric considerations. Note that the Euclidean distance in the $\mathbf{z}$ space is equivalent to the Mahalanobis distance in the original space, and the linear Gaussian rule has the form written in equation (9.12). Let $\Delta = \|\bar{\mathbf{z}}_1 - \bar{\mathbf{z}}_2\|$ be the Euclidean distance between the two sample means, and assume that $\bar{\mathbf{z}}_2$ is at the origin (only for simplicity of explanations). Let $\mathbf{f}$ be a unit vector in the direction of $\bar{\mathbf{z}}_1 - \bar{\mathbf{z}}_2$. Then $\bar{\mathbf{z}}_1 - \bar{\mathbf{z}}_2 = \Delta \cdot \mathbf{f}$ and $(\bar{\mathbf{z}}_1 + \bar{\mathbf{z}}_2)/2 = (\Delta/2) \cdot \mathbf{f}$. Let $\mathbf{z}_0 = h \cdot \mathbf{f}$ be a point on L (this is the case of arbitrary costs of misclassifications, so think of L as the dashed line in Figure 9.5). Hence, $\mathbf{v}^T \mathbf{z}_0 = c$. We also have $\mathbf{v}^T \mathbf{z}_0 = \Delta \cdot h$ and $c = \Delta^2/2 + b$. This leads to $h = \Delta/2 + b/\Delta$. The probability $P(1|2)$ can be approximated as the probability of the area above the dashed line in Figure 9.5 when using the standard bivariate normal distribution with the center at $\bar{\mathbf{z}}_2$. This is only an approximation because the estimate $\mathbf{S}_{pooled}^{-1/2}$ was used instead of the true population matrix $\boldsymbol{\Sigma}^{-1/2}$ in the sphering process. Similar arguments can be used for derivation of $P(2|1)$. Hence,

$$P(2|1) \approx \Phi\left(-\frac{\Delta}{2} + \frac{b}{\Delta}\right) \quad \text{and} \quad P(1|2) \approx \Phi\left(-\frac{\Delta}{2} - \frac{b}{\Delta}\right), \qquad (9.15)$$

where Φ is the cumulative standard normal distribution function,

$$b = \ln\left[\frac{c(1|2)\,p_2}{c(2|1)\,p_1}\right] \quad \text{and} \quad \Delta = \sqrt{(\bar{\mathbf{x}}_1 - \bar{\mathbf{x}}_2)^T \mathbf{S}_{pooled}^{-1}(\bar{\mathbf{x}}_1 - \bar{\mathbf{x}}_2)}. \qquad (9.16)$$

More precise approximations are discussed by Srivastava (2002) in Section 8.4.

Approximations (9.15) require the assumption of normality. For non-normal data, we can still use the linear Gaussian rule, and the conditional probabilities can be estimated from the data. Let us say that a given classification procedure was applied to all observations and the resulting classifications were recorded in a classification table like the one shown in Table 9.4. The table shows the notation that we want to use here. For example, n_{21} is the number of observations classified into Population 2, given they actually come from Population 1. It is clear that

Table 9.4   Classification Table Showing Notation Used in the Text

	Observation Is Identified as Coming from		
Observation Comes from	Population π_1	Population π_2	Total
Population π_1	n_{11}	n_{21}	n_1
Population π_2	n_{12}	n_{22}	n_2
			Grand total n

$P(2|1) \approx n_{21}/n_1$ and $P(1|2) \approx n_{12}/n_2$. Hence, we introduce the *estimated error rate* (EER), defined as

$$\text{EER} = \frac{n_{21}}{n_1}p_1 + \frac{n_{12}}{n_2}p_2, \tag{9.17}$$

as an estimator of TPM. When the prior probabilities are estimated from the sample, we have $p_1 = n_1/n$ and $p_2 = n_2/n$, and $\text{EER} = (n_{21} + n_{12})/n$ reduces to the *apparent error rate* (APER), defined by the same formula $\text{APER} = (n_{21} + n_{12})/n$. APER is often mentioned by other authors and is produced by software. We need to remember that APER should be used only when we are willing to assume the prior probabilities estimated from the sample. Otherwise, it is better to use EER.

Unfortunately, the methods of misclassification error estimation shown here tend to underestimate the true error rates in future samples. The reason is that the classification rule is assessed based on the same data that were used to estimate the rule. This is why it is important to use cross-validation to find more realistic error rates. We will show the use of cross-validation in Section 9.2.2.

Example 9.1 (cont.) The Astronomy Data shown earlier in Figure 9.1 is only mildly non-normal as can be seen in Figure 9.6, where two beta probability plots (see Section 6.5) for the two groups are shown. The H II group shown in Figure 9.6b looks less normal due to two outlying observations. Out of the four tests of univariate normality (for two groups and two variables) only H magnitude variable for H II group shows a significant p-value of 0.0026. This value can be interpreted as a p-value of $4 \times 0.0026 \approx 0.01$ based on the Bonferroni-type arguments (see Result 6.1 in Section 6.2.2) because four inferences were performed here.

In order to test the equality of variance–covariance matrices for the two groups, we can use the Box test (see Section 6.4), and the resulting p-value of 0.0008 tells us that the matrices are statistically significantly different. Consequently, the linear Gaussian

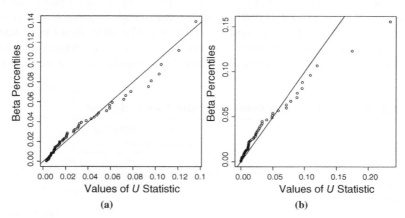

Figure 9.6 Probability plots for the C AGB group (a) and the H II group (b) as discussed in Example 9.1.

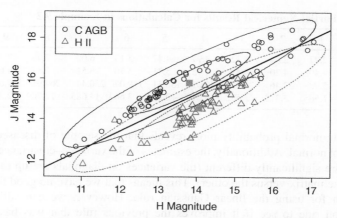

Figure 9.7 The linear Gaussian classification boundary line (the bold line) for Example 9.1 data. The ellipses are the contour lines encompassing 50% and 95% of the assumed bivariate normal distributions. The sample means are marked by the two squares.

rule defined by (9.9) may not be approximately optimal in the ECM sense. Nevertheless, it is worthwhile to try it here. Since we have no information about the frequency of the two types of stars in future applications, we assume $p_1 = p_2 = 0.5$. We also assume equal costs of misclassification. Application of the classification rule (9.9) results in the boundary line shown in Figure 9.7. Table 9.5 shows the classification table with 11 misclassifications. EER is calculated as $(7/67) \times 0.5 + (4/59) \times 0.5 = 0.0861$, and APER as $11/126 = 0.0873$. Based on formula (9.15), both misclassification probabilities are estimated as 0.0781, and formula (9.14) with $p_1 = p_2 = 0.5$ leads to the TPM being estimated as 0.0781. We can see that all three approaches to the estimation of TPM produce somewhat similar results in this case.   □

Example 9.2 (cont.) Let us again consider two populations of grass pixels. The first population π_1 is the population of healthy grass, and the second population, denoted by π_2, is the population of diseased grass. Before we apply the linear Gaussian methodology, we need to check whether the two samples representing our two populations come from multivariate normal (Gaussian) distributions. A simple check

Table 9.5 **Classification Table for the Linear Gaussian Classification Rule Applied to the Example 9.1 Data**

| | Stars Identified as | | |
Actual State	C AGB	H II	Total
C AGB	60	7	67
H II	4	55	59
Total	64	62	126

Table 9.6   Partial Numerical Results for Calculations in Example 9.2

Coordinate	1	2	3	4	5	6	7	8	9	10
$\bar{\mathbf{x}}_1$	2.77	2.84	3.32	3.97	4.47	5.13	6.02	7.49	8.39	9.03
$\bar{\mathbf{x}}_2$	4.40	4.20	4.47	4.95	5.31	5.76	6.51	7.87	9.64	11.49
$\mathbf{S}_{pooled}^{-1}$	5.98	32.46	75.72	140.01	158.51	159.07	230.44	264.91	337.08	259.58
v	-1.505	-3.373	-5.532	-6.610	-2.520	3.663	14.685	-12.709	2.315	0.807

of univariate normal probability plots reveals that the sample of diseased grass is certainly not normal. Additionally, the estimates of the variance–covariance matrices $\mathbf{S}_1$ and $\mathbf{S}_2$ are significantly different (the variances on diagonals are up to 50 times larger in one matrix versus the other). This means that we have no good theoretical justification for using the linear Gaussian rule. However, we can still use this classification rule to see if it improves the previous rule that was based on an arbitrarily chosen straight line in the principal component coordinates as shown in Figure 9.3.

In order to use the classification rule, we need to decide on misclassification costs and prior probabilities. Let us assume that the grass identified as diseased will be treated with inexpensive chemicals that do not bother healthy grass. We further assume that grass identified as healthy will not be treated. Therefore, if the grass is diseased and not treated, there is a danger that it may die. Consequently, the cost $c(1|2)$ is higher than $c(2|1)$. For instance, we can assume that it is three times higher, that is, $c(1|2)/c(2|1) = 3$. Note that we do not need to know the exact costs, but only the proportion of the two costs. In order to establish prior probabilities, we would need to know how much of the healthy grass area we might expect in relation to the diseased grass area. If we have no idea about proportions of the two populations in the general population, it is natural to assume that $p_1 = p_2 = 0.5$. With these assumptions, the right-hand side in (9.9) is equal to $\ln(3) = 1.0986$.

Table 9.6 shows the first 10 coordinates of the $\bar{\mathbf{x}}_1$ and $\bar{\mathbf{x}}_2$ vectors and the first 10 values on the diagonal of $\mathbf{S}_{pooled}^{-1}$. Coordinates of the resulting discriminant vector v are also shown in the same table.

The constant c defined in equation (9.11) is equal to 38.314. The linear Gaussian rule given by (9.10) can now be used to classify all grass pixels. Table 9.7 summarizes the number of correctly and incorrectly classified pixels. The misclassification rates

Table 9.7   Classification Table for the Linear Gaussian Rule Applied to the Grass Data from Example 9.2

Actual State	Grass Identified as		Total
	Healthy	Diseased	
Healthy grass	3837	3	3840
Diseased grass	40	216	256
Total	3877	219	4096

We assumed $c(1|2)/c(2|1) = 3$ and equal prior probabilities.

Table 9.8   Classification Table for the Linear Gaussian Rule Applied to the Example 9.2 Data

Actual State	Grass Identified as		Total
	Healthy	Diseased	
Healthy grass	3838	2	3840
Diseased grass	42	214	256
Total	3880	216	4096

Here we assumed equal misclassification costs and equal prior probabilities.

are much lower than those for the principal component-based rule summarized in Table 9.1.

We may want to investigate the sensitivity of the linear Gaussian rule (9.9) to the assumptions about the misclassification costs and prior probabilities. Assuming $c(1|2)/c(2|1) = 3$, as before, and $p_1 = 0.75$, $p_2 = 0.25$ leads to the right-hand side of (9.9) equal to $\ln(1) = 0$. (The same result is obtained when no prior information is assumed, that is, $c(1|2)/c(2|1) = 1$ and $p_1 = p_2 = 0.5$.) In this case, c is equal to 37.215, and the classification results are summarized in Table 9.8. The misclassification rates are only moderately different from previous ones. A more thorough sensitivity analysis would involve investigation of more alternative values for the misclassification costs and prior probabilities. ☐

9.2.1.2   Classification for Normal Populations with Unequal Variance–Covariance Matrices

When $\Sigma_1 \neq \Sigma_2$, we can also simplify the classification rule defined in equation (9.6). Specifically, one can show that the rule that minimizes the ECM is such that $\mathbf{x} \in R_1$ if

$$-\frac{1}{2}\mathbf{x}^\mathrm{T}\left(\Sigma_1^{-1} - \Sigma_2^{-1}\right)\mathbf{x} + \left(\boldsymbol{\mu}_1^\mathrm{T}\Sigma_1^{-1} - \boldsymbol{\mu}_2^\mathrm{T}\Sigma_2^{-1}\right)\mathbf{x} - k \geq \ln\left[\frac{c(1|2)\,p_2}{c(2|1)\,p_1}\right], \quad (9.18)$$

where

$$k = \frac{1}{2}\ln\left(\frac{|\Sigma_1|}{|\Sigma_2|}\right) + \frac{1}{2}\left(\boldsymbol{\mu}_1^\mathrm{T}\Sigma_1^{-1}\boldsymbol{\mu}_1 - \boldsymbol{\mu}_2^\mathrm{T}\Sigma_2^{-1}\boldsymbol{\mu}_2\right) \quad (9.19)$$

and $\mathbf{x} \in R_2$ otherwise (the derivation is given as Problem 9.8). In practice, the population parameters are unknown, and they are estimated here using the plug-in principle (with the unbiased version of the sample variance-covariance matrix). The classification rule for $\mathbf{x} \in R_1$ becomes

$$-\frac{1}{2}\mathbf{x}^\mathrm{T}\left(\mathbf{S}_1^{-1} - \mathbf{S}_2^{-1}\right)\mathbf{x} + \left(\bar{\mathbf{x}}_1^\mathrm{T}\mathbf{S}_1^{-1} - \bar{\mathbf{x}}_2^\mathrm{T}\mathbf{S}_2^{-1}\right)\mathbf{x} - \hat{k} \geq \ln\left[\frac{c(1|2)\,p_2}{c(2|1)\,p_1}\right], \quad (9.20)$$

where

$$\hat{k} = \frac{1}{2}\ln\left(\frac{|\mathbf{S}_1|}{|\mathbf{S}_2|}\right) + \frac{1}{2}\left(\bar{\mathbf{x}}_1^{\mathsf{T}}\mathbf{S}_1^{-1}\bar{\mathbf{x}}_1 - \bar{\mathbf{x}}_2^{\mathsf{T}}\mathbf{S}_2^{-1}\bar{\mathbf{x}}_2\right), \tag{9.21}$$

and $\mathbf{x} \in R_2$ otherwise. The rule is called the *quadratic Gaussian rule*. It is assumed here that matrices $\mathbf{S}_1^{-1}$ and $\mathbf{S}_2^{-1}$ exist. For this, it is necessary that $n_1 > p$ and $n_2 > p$. If those conditions are not satisfied, we can reduce the dimensionality p of the data set, so that the variance–covariance matrices become nonsingular, and the classification can be performed in the reduced space. In practice, the sample sizes should be much larger than the minima mentioned here.

The quadratic Gaussian rule is more flexible than a linear rule in adjusting to the shapes of distributions of the two groups. When $\mathbf{S}_1$ is close to $\mathbf{S}_2$, the rule should give the results that are similar to those for the linear rule because the first term on the left-hand side of (9.20) will be small. When $\mathbf{S}_1$ is very different from $\mathbf{S}_2$, the quadratic rule should perform much better than the linear one. These properties are based on the distributional assumption of normality, but they often hold for other types of data as well.

The following two examples discuss applications of the quadratic Gaussian rule to the two data sets used earlier.

Example 9.3 As a continuation of Example 9.2, we use the same Grass data set, but we now recognize the fact that the variance–covariance matrices for the two populations are different, and we want to use the quadratic Gaussian rule. As earlier, we assume $c(1|2)/c(2|1) = 3$ and $p_1 = p_2 = 0.5$. Using the quadratic rule, we obtain the classification table shown in Table 9.9.

We can now compare the classifications in Table 9.9 to those in Table 9.7 for the linear Gaussian rule. It may seem surprising at first that the quadratic rule gives a larger total number of misclassifications. However, when using the linear classification rule, we have as many as 40 of the more costly misclassifications. Let us calculate the cost associated with both classification rules. An appropriate way to do so would be to calculate the expected cost of misclassification defined previously. Unfortunately, the assumption of normality was not fulfilled for the Grass Data. Without this assumption we cannot use formulas (9.15), and it is rather difficult to calculate the required conditional probabilities.

Table 9.9   Classification Table for the Quadratic Gaussian Rule Applied to the Example 9.2 Data

Actual State	Grass Identified as		Total
	Healthy	Diseased	
Healthy grass	3793	47	3840
Diseased grass	4	252	256
Total	3797	299	4096

Another possibility is to calculate the observed cost when the two classification rules are applied to the Grass data set. Assuming $c(2|1)$ to be a unit cost per pixel, the total observed misclassification cost for the linear rule is

$$40 \times 3 \text{ units} + 3 \times 1 \text{ unit} = 123 \text{ units}, \tag{9.22}$$

and for the quadratic rule the cost is

$$4 \times 3 \text{ units} + 47 \times 1 \text{ unit} = 59 \text{ units}, \tag{9.23}$$

which is significantly less.   □

Example 9.4 Here we use the star data from Example 9.1 but recognize different variance–covariance matrices as tested earlier. We again assume equal misclassification costs and equal prior probabilities. When using the quadratic rule defined by formula (9.20), we obtain misclassification counts identical to those in Table 9.5 for the linear rule. One reason for this result is the relatively small difference between the variance–covariance matrices of the two groups, even though the difference is statistically significant.   □

9.2.2   Cross-Validation of Classification Rules

Earlier we mentioned that the evaluation of classification rules based on error rates estimated directly from the original data underestimates the true rates. This problem is related to model overfitting. Specifically, the linear and quadratic classification rules developed in the previous section are dependent on the available data set. The rules are then evaluated using the same data. The danger is that a classification rule may be fitted precisely to the data but may not accurately predict future observations. One possible methodology to check for such overfitting is to perform cross-validation as discussed in Supplement 8A. Recall that the idea is to separate a small subset (testing sample) of the original data set, develop a classification rule using the remaining data (learning sample), and then check how well the classification rule works on the testing sample separated at the beginning. In cross-validation, we need to decide on the size of the subset and how it is chosen. Here we want to start with the leave-one-out cross-validation, where we choose a testing sample consisting of only one observation, but we repeat it for all observations in the whole data set. This procedure was performed on the Grass Data for the linear Gaussian classification rule under the assumption that $c(1|2)/c(2|1) = 3$ and $p_1 = p_2 = 0.5$. For each removed observation, the vector **v** and the constant c defined in (9.11) are somewhat different because they are based on a somewhat different data set (if only by one observation). Using the fitted rule, we classify the removed observation as healthy or diseased grass, and then compare it to its actual state as healthy or diseased grass. This gives one count into Table 9.10. If the removed observation was from the diseased grass area and it was classified as healthy, the case would count as

Table 9.10   Leave-One-Out Cross-Validation Results for the Linear Gaussian Rule Applied to the Grass Data Used in Example 9.2

	Grass Identified as		
Actual State	Healthy	Diseased	Total
Healthy grass	3837	3	3840
Diseased grass	41	215	256
Total	3878	218	4096

one of 41 cases shown in the second row of the first column. The process of removing one observation is then repeated 4096 times, and Table 9.10 is filled out with the counts.

The number of misclassifications is just a little bit larger than the one in Table 9.7. This is not surprising because linear rules are typically robust in the sense of not being strongly influenced by a single observation (unless an outlier). Similar calculations were done for the quadratic Gaussian classification rule, and the results are summarized in Table 9.11.

This time, the misclassification rates are not very close to those estimated directly from the whole data set (and shown in Table 9.9). In order to assess an impact of those misclassification rates, it is worthwhile to recalculate the estimate of ECM. Again, assuming $c(2|1)$ being a unit cost per pixel, the total misclassification cost based on the leave-one-out cross-validation for the linear rule is

$$41 \times 3 \text{ units} + 3 \times 1 \text{ unit} = 126 \text{ units}, \tag{9.24}$$

and for the quadratic rule

$$13 \times 3 \text{ units} + 51 \times 1 \text{ unit} = 90 \text{ units}. \tag{9.25}$$

We observe an increase in the estimate of ECM in both cases in relation to the numbers calculated at the end of the previous section (formulas (9.22) and (9.23)). However, the increase is much more significant for the quadratic rule, which shows that the rule is more prone to overfitting. These costs are more realistic estimates of the ECM values in future samples.

Table 9.11   Leave-One-Out Cross-Validation Results for the Quadratic Gaussian Rule Applied to the Grass Data Used in Example 9.2

	Grass Identified as		
Actual State	Healthy	Diseased	Total
Healthy grass	3789	51	3840
Diseased grass	13	243	256
Total	3802	294	4096

Table 9.12   Eight-Fold Cross-Validation Results for the Linear Gaussian Rule Averaged over 50 Random Rounds

Actual State	Grass Identified as		Total
	Healthy	Diseased	
Healthy grass	3836.58	3.42	3840
Diseased grass	41.02	214.98	256
Total	3877.60	218.40	4096

We can now try the k-fold cross-validation also described in Supplement 8A. We will start with $k = 8$, which means eight-fold cross-validation. Here the data set is randomly divided into eight groups, and then each group in turn plays a role of the testing sample. Since there are $4096/8 = 512$ elements in each testing group, we will obtain 512 classification results for each testing sample. Since there are eight testing samples, we will obtain 4096 classification results after the whole round of the eight-fold cross-validation. Those results could be placed into a 2 by 2 classification table like the ones used earlier. Since the result of the whole round of the cross-validation is random here, we can run several rounds and then average the results in order to reduce an impact of spurious variability. Table 9.12 shows the average classification results from 50 random rounds.

The average total misclassification cost for the linear rule is now estimated as $41.02 \times 3 + 3.42 \times 1 = 126.48$, which is only slightly larger than the cost in (9.24) for the leave-one-out cross-validation. Similar calculations were performed for the quadratic Gaussian rule as well, and the results are summarized in Table 9.13. The average cost for the quadratic rule is now estimated as $14.66 \times 3 + 47.38 \times 1 = 91.36$, which is not much larger than the cost in (9.25) for the leave-one-out cross-validation.

Since the results have not changed much between the leave-one-out and the eight-fold cross-validation, we also tried two-fold cross-validation. Here the classification is more challenging because only half of the data are used for the estimation of the classification rule. In this scenario, the misclassification cost of the linear rule went up to 129.74 and that of the quadratic rule went up to 100.24. We conclude that the

Table 9.13   Eight-Fold Cross-Validation Results for the Quadratic Gaussian Rule Averaged over 50 Random Runs

Actual State	Grass Identified as		Total
	Healthy	Diseased	
Healthy grass	3792.62	47.38	3840
Diseased grass	14.66	241.34	256
Total	3807.28	288.72	4096

quadratic rule is less robust to variations in the sample, but it is still a better rule than the linear one for this data set.

The above-described approaches to cross-validation were also implemented on the linear Gaussian rule applied to the Astronomy Data used in Example 9.1. The leave-one-out method produced only one additional misclassification case relative to Table 9.5 results. The six-fold cross-validation resulted in the average misclassifications almost identical to those of the leave-one-out method. Similar results were also obtained for the quadratic Gaussian rule. This shows that both the linear and quadratic Gaussian rules promise to be robust to the variability in the future samples from the same distributions. The future misclassification rates are expected to be only slightly higher than those in Table 9.5.

9.2.3 Fisher's Discriminant Function

The linear Gaussian classification rule can also be derived without the Gaussianity assumption. The approach discussed here was proposed by Fisher (1938). His idea was to look for a linear combination of the observed vector $\mathbf{x}$ that would best separate two samples. Specifically, using a p-dimensional vector of coefficients $\mathbf{a}$, we can define a linear combination $y = \mathbf{a}^{\mathrm{T}}\mathbf{x}$. The y values can be calculated for all observations from the first sample (and will be denoted by $y_{11}, y_{12}, \ldots, y_{1n_1}$) and the second sample (denoted by $y_{21}, y_{22}, \ldots, y_{2n_2}$). The y values can be regarded as projections onto a straight line defined by the vector $\mathbf{a}$ (see Property 4A.1). One can calculate means of the two sets of y values and denote them by $\bar{y}_1$ and $\bar{y}_2$. The separation of the two sets is assessed based on separation of the y values, which, in turn, is measured by the difference between the two means in standard deviation units, that is,

$$\frac{|\bar{y}_1 - \bar{y}_2|}{s_y}, \tag{9.26}$$

where

$$s_y^2 = \frac{\sum_{i=1}^{n_1} (y_{1i} - \bar{y}_1)^2 + \sum_{i=1}^{n_2} (y_{2i} - \bar{y}_2)^2}{n_1 + n_2 - 2} \tag{9.27}$$

is the pooled variance estimated from both samples. The following theorem tells us how to find the direction $\mathbf{a}$ that would maximize the separation between the two samples.

Theorem 9.2 The linear combination $y = \mathbf{a}^{\mathrm{T}}\mathbf{x} = (\bar{\mathbf{x}}_1 - \bar{\mathbf{x}}_2)^{\mathrm{T}}\mathbf{S}_{\mathrm{pooled}}^{-1}\mathbf{x}$ maximizes the ratio

$$\frac{(\bar{y}_1 - \bar{y}_2)^2}{s_y^2} = \frac{[\mathbf{a}^{\mathrm{T}}(\bar{\mathbf{x}}_1 - \bar{\mathbf{x}}_2)]^2}{\mathbf{a}^{\mathrm{T}}\mathbf{S}_{\mathrm{pooled}}\mathbf{a}} \tag{9.28}$$

over all possible coefficient vectors $\mathbf{a}$. The maximum of the ratio is

$$D^2 = (\overline{\mathbf{x}}_1 - \overline{\mathbf{x}}_2)^T \mathbf{S}_{\text{pooled}}^{-1} (\overline{\mathbf{x}}_1 - \overline{\mathbf{x}}_2), \tag{9.29}$$

which is the squared Mahalanobis distance between the two means and can be regarded as a measure of separation between the two samples.

Proof. See Problem 9.5.

The *Fisher discrimination rule* assigns $\mathbf{x}$ to R_1 if its y value is closer to $\overline{y}_1$ than to $\overline{y}_2$, that is, if

$$y = \mathbf{a}^T \mathbf{x} = (\overline{\mathbf{x}}_1 - \overline{\mathbf{x}}_2)^T \mathbf{S}_{\text{pooled}}^{-1} \mathbf{x} \geq (\overline{\mathbf{x}}_1 - \overline{\mathbf{x}}_2)^T \mathbf{S}_{\text{pooled}}^{-1} \frac{(\overline{\mathbf{x}}_1 + \overline{\mathbf{x}}_2)}{2}, \tag{9.30}$$

and $\mathbf{x}$ is assigned to R_2 otherwise. The vector $\mathbf{a} = \mathbf{S}_{\text{pooled}}^{-1}(\overline{\mathbf{x}}_1 - \overline{\mathbf{x}}_2)$ is called the *discriminant vector*. The Fisher discrimination rule is equivalent to the linear Gaussian classification rule (defined by (9.9)) that minimizes the ECM when $c(1|2)p_2 = c(2|1)p_1$. The difference between the two approaches is that Fisher's approach does not require the assumptions of normality, but then it does not guarantee any optimality such as minimizing ECM. Fisher's approach gives a geometric interpretation of the linear Gaussian rule consistent with Figure 9.4, where the vector $\mathbf{v}$ defines the direction of maximum separation between the two samples.

When the data are sphered by applying the transformation $\mathbf{z} = \mathbf{S}_{\text{pooled}}^{-1/2}\mathbf{x}$, the samples should have the variance–covariance matrices close to the identity. The Mahalanobis distance then becomes the Euclidean distance, and the separation between the two samples is maximized along the line connecting the sample means as in Figure 9.5.

Since the discriminant vector points in the direction of the best separation between the two samples, it can be used for plotting the data. For simplicity of presentation, we are now going to use the sphered and centered data after the transformation $\mathbf{z} = \mathbf{S}_{\text{pooled}}^{-1/2}[\mathbf{x} - (\overline{\mathbf{x}}_1 + \overline{\mathbf{x}}_2)/2]$. This means that the middle point between the group means $\overline{\mathbf{z}}_1$ and $\overline{\mathbf{z}}_2$ is the origin of the new system of coordinates. The discriminant direction now becomes $\mathbf{a} = \overline{\mathbf{z}}_1 - \overline{\mathbf{z}}_2$.

Example 9.1 (cont.) Figure 9.8 shows the Astronomy Data from Example 9.1 plotted in the centered and sphered coordinates that were further rotated so that the vertical axis is the normalized discriminant direction. The horizontal axis is orthogonal to the discriminant direction in the sphered coordinates. The horizontal straight line at level zero is the boundary line between the two discrimination regions. Figure 9.8 shows the same information that was shown in Figure 9.7, but it is now easier to appreciate the separation between the two groups. It is also easier to assess normality of data by a comparison to the circles representing contour lines encompassing 50% and 95% of the normal distributions. We can see some mild departures from normality. □

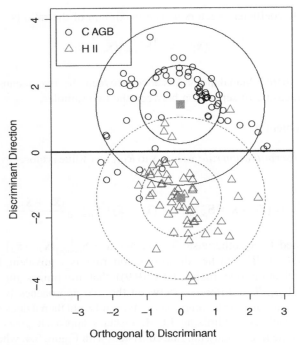

Figure 9.8 The Astronomy Data from Example 9.1 plotted in the centered and sphered coordinates. The vertical axis is the normalized discriminant direction. The contour lines encompass 50% and 95% of the normal distributions. The sample means are marked by two squares.

Example 9.5 For Grass Data from Example 9.2, we can assume equal misclassification costs and equal prior probabilities. In that case, Fisher's rule is equivalent to the linear Gaussian rule, and the classification table for both rules is shown in Table 9.8. For a graphical presentation of the data, we can again use the sphered coordinates with the normalized discriminant vector as the vertical basis vector. However, here the data are 42 dimensional, and there are many possibilities how the horizontal axis can be chosen as the direction orthogonal to the discriminant vector $\mathbf{a}$. Since it is interesting to see the misclassifications in the graph, we chose the direction of maximum variability in the misclassified observations. More specifically, we projected all $k = 44$ centered and sphered spectra of the misclassified observations onto a subspace orthogonal to $\mathbf{a}$ and then placed all those projections into a k by p matrix $\mathbf{W}$ and performed the singular value decomposition (SVD) that gives the representation $\mathbf{W} = \mathbf{U\Lambda V}^{\mathrm{T}}$. The diagonal elements of the $\mathbf{\Lambda}$ matrix describe the variability of data in $\mathbf{W}$ around zero. It is important that the SVD was performed here rather than principal component analysis. The latter would describe the variability around the mean of the misclassified observations, which is not interesting here. Assuming that the diagonal elements are ordered, with the first value being the largest, we choose the first column

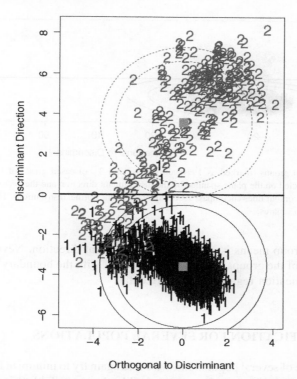

Figure 9.9 Two groups (marked as 1 = healthy grass and 2 = diseased grass) of Grass Data from Example 9.2 in the sphered coordinates of the discriminant direction and an orthogonal direction based on the maximum variability of misclassified observations. The contour lines encompass 99% and 99.9% of the normal distributions. The sample means are marked by two squares.

of the **V** matrix as the horizontal axis. Figure 9.9 shows the Grass Data in those two coordinates. The contour lines encompassing 99% and 99.9% of the normal distributions are plotted this time due to large sample sizes. If the two populations were indeed normal, we would not expect any of the 256 observations from Group 2 (diseased grass) to be outside of the larger circle. In Group 1 (healthy grass) with 3840 observations, we would expect approximately four observations outside of the larger circle. We can say that Group 2 looks more non-normal than Group 1.

We might also be interested in plotting the data in the original space, that is, without the scaling that is done in the process of data sphering. In that case, the discriminant vector $\mathbf{a} = \mathbf{S}_{pooled}^{-1}(\bar{\mathbf{x}}_1 - \bar{\mathbf{x}}_2)$ is not collinear with the vector $\mathbf{v} = \bar{\mathbf{x}}_1 - \bar{\mathbf{x}}_2$ connecting the group means. Hence, we can project all data onto the plane G spanned by $\mathbf{a}$ and $\mathbf{v}$. The normalized vector $\mathbf{a}/\|\mathbf{a}\|$ was again used as the vertical basis vector, and an orthogonal vector in G was chosen as the horizontal axis. Figure 9.10 shows the data projected on G, with the contour lines encompassing 99% and 99.9% of the normal distributions, similarly to Figure 9.9. Note that the data stretch much more in the horizontal direction, and the difference

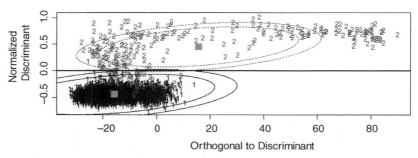

Figure 9.10   Two groups (marked as 1 = healthy grass and 2 = diseased grass) of Grass Data from Example 9.2 projected on the plane spanned by the discriminant direction and the vector connecting the group means. The contour lines encompass 99% and 99.9% of the normal distributions. The sample means are marked by two squares.

between the group means is also much larger in that direction. Nevertheless, the best direction of the groups' separation is vertical, with the boundary line between the two classification regions being horizontal. $\square$

9.3   CLASSIFICATION FOR SEVERAL POPULATIONS

In the presence of several populations, we could again try to minimize ECM, but this would require a specification of a cost matrix (like the one in Table 9.3) with $g(g-1)$ nonzero entries, where g is the number of populations. The appropriate formulas for that approach can be found in Section 11.5 in Johnson and Wichern (2007). Here we take a simplified approach by assuming that all misclassification costs are the same. In that case, minimizing ECM is equivalent to minimizing TPM.

9.3.1   Gaussian Rules

In this subsection, we will assume that the samples come from normal populations. We start with the case of possibly different population variance–covariance matrices. We define the *quadratic Gaussian classification rule* assigning $\mathbf{x}$ to R_i when the quadratic score $d_i^Q(\mathbf{x})$ is the largest of $d_1^Q(\mathbf{x}), d_2^Q(\mathbf{x}), \ldots, d_g^Q(\mathbf{x})$ for g groups (samples), where

$$d_j^Q(\mathbf{x}) = -0.5\left(\mathbf{x} - \overline{\mathbf{x}}_j\right)^{\mathrm{T}} \mathbf{S}_j^{-1}\left(\mathbf{x} - \overline{\mathbf{x}}_j\right) - 0.5\ln|\mathbf{S}_j| + \ln p_j, \quad j = 1, 2, \ldots, g,$$

$$(9.31)$$

and $\overline{\mathbf{x}}_j, \mathbf{S}_j$, and p_j are the sample mean, the sample variance–covariance matrix, and the prior probability for the jth group, respectively.

When all g samples come from normal distributions and their parameters are known exactly (rather than estimated), the above classification rule is optimal in terms of minimizing TPM. The first term in the definition of $d_j^Q(\mathbf{x})$ is the scaled squared Mahalanobis distance between $\mathbf{x}$ and the jth sample mean $\bar{\mathbf{x}}_j$, but the distance is based on a different variance–covariance matrix $\mathbf{S}_j$ for each group.

When all population variance–covariance matrices are equal, that is, $\mathbf{\Sigma}_j = \mathbf{\Sigma}$, $j = 1, 2, \ldots, g$, $\mathbf{\Sigma}$ can be estimated using the pooled estimate

$$\mathbf{S}_{\text{pooled}} = \frac{1}{\sum_{j=1}^{g}(n_j - 1)} \sum_{j=1}^{g}(n_j - 1)\mathbf{S}_j, \tag{9.32}$$

where the sample variance–covariance matrices $\mathbf{S}_j$, $j = 1, \ldots, g$, are from the g samples with their sizes denoted by n_j, $j = 1, \ldots, g$. The classification rule defined by (9.31) simplifies then to the *linear Gaussian classification rule* assigning $\mathbf{x}$ to R_j when the linear score $d_j(\mathbf{x})$ is the largest of $d_1(\mathbf{x}), d_2(\mathbf{x}), \ldots, d_g(\mathbf{x})$ for g groups, where

$$d_j(\mathbf{x}) = \bar{\mathbf{x}}_j^{\mathsf{T}} \mathbf{S}_{\text{pooled}}^{-1} \mathbf{x} - 0.5 \bar{\mathbf{x}}_j^{\mathsf{T}} \mathbf{S}_{\text{pooled}}^{-1} \bar{\mathbf{x}}_j + \ln p_j, \quad j = 1, 2, \ldots, g. \tag{9.33}$$

It is convenient to have the linear form of the scores $d_j(\mathbf{x})$, but we may also relate that rule to distances. When the sample variance–covariance matrices $\mathbf{S}_j$ in (9.31) are replaced by the pooled estimate $\mathbf{S}_{\text{pooled}}$, the resulting rule assigns $\mathbf{x}$ to the closest group based on the adjusted Mahalanobis distance defined as

$$A_j(\mathbf{x}) = D_j^2(\mathbf{x}) - 2 \ln p_j, \tag{9.34}$$

where $D_j^2(\mathbf{x}) = (\mathbf{x} - \bar{\mathbf{x}}_j)^{\mathsf{T}} \mathbf{S}_{\text{pooled}}^{-1} (\mathbf{x} - \bar{\mathbf{x}}_j)$ is the squared Mahalanobis distance based on the common variance–covariance matrix $\mathbf{S}_{\text{pooled}}$. This rule is equivalent to the linear Gaussian classification rule defined by (9.33).

It is often convenient to work on the sphered data obtained by applying the transformation $\mathbf{z} = \mathbf{S}_{\text{pooled}}^{-1/2} \mathbf{x}$. In the new coordinates, the Mahalanobis distance becomes the Euclidean distance, and formula (9.33) simplifies to

$$d_j(\mathbf{z}) = \bar{\mathbf{z}}_j^{\mathsf{T}} \mathbf{z} - 0.5 \|\bar{\mathbf{z}}_j\|^2 + \ln p_j, \quad j = 1, 2, \ldots, g, \tag{9.35}$$

while formula (9.34) simplifies to

$$A_j(\mathbf{z}) = \|\mathbf{z} - \bar{\mathbf{z}}_j\|^2 - 2 \ln p_j. \tag{9.36}$$

Sometimes, we may want to find the boundaries between the classification regions. For the boundary between the jth and kth regions, we need to find all $\mathbf{z}$ satisfying the condition $d_j(\mathbf{z}) = d_k(\mathbf{z})$, which can be written as

$$(\bar{\mathbf{z}}_j - \bar{\mathbf{z}}_k)^{\mathsf{T}} \mathbf{z} = 0.5 \left(\|\bar{\mathbf{z}}_j\|^2 - \|\bar{\mathbf{z}}_k\|^2 \right) - \ln(p_j/p_k). \tag{9.37}$$

The above equation describes a $(p - 1)$-dimensional affine subspace orthogonal to the vector $\bar{\mathbf{z}}_j - \bar{\mathbf{z}}_k$.

The linear and quadratic Gaussian rules are further discussed in the next section, where a numerical example is also shown.

9.3.2 Fisher's Method

Fisher's method discussed earlier in Section 9.2.3 generalizes to the case of several populations. The result will again be equivalent to the linear Gaussian rule (9.33) with equal priors, but the approach will also give us a useful geometric interpretation.

Let us assume that we have g populations with equal population variance–covariance matrices. As usual in such cases, the joint population variance–covariance matrix is estimated by the pooled variance–covariance matrix $\mathbf{S}_{\text{pooled}}$ defined in (9.32). For the remainder of this section, we will work with the sphered data obtained by applying the transformation $\mathbf{z} = \mathbf{S}_{\text{pooled}}^{-1/2}\mathbf{x}$. In the new coordinates, the samples should have the variance–covariance matrices close to the identity and the Mahalanobis distance becomes the Euclidean distance. We denote the sample mean vectors as $\bar{\mathbf{z}}_j$, $j = 1, \ldots, g$, and define the overall mean vector as the weighted mean of the sample means

$$\bar{\mathbf{z}} = \frac{\sum_{j=1}^{g} n_j \bar{\mathbf{z}}_j}{\sum_{j=1}^{g} n_j} = \frac{\sum_{j=1}^{g} \sum_{i=1}^{n_j} \mathbf{z}_{ij}}{\sum_{j=1}^{g} n_j}, \tag{9.38}$$

where $\mathbf{z}_{ij}$ are the observation vectors within the groups. The ability to distinguish between the groups depends on how large the differences $\bar{\mathbf{z}}_j - \bar{\mathbf{z}}$ are. An overall measure of the between-group variability is the matrix $\mathbf{B}$ defined as

$$\mathbf{B} = \sum_{j=1}^{g} n_j (\bar{\mathbf{z}}_j - \bar{\mathbf{z}})(\bar{\mathbf{z}}_j - \bar{\mathbf{z}})^{\mathsf{T}}. \tag{9.39}$$

For the case of two groups, we obtain $\bar{\mathbf{z}}_1 - \bar{\mathbf{z}} = n_2(\bar{\mathbf{z}}_1 - \bar{\mathbf{z}}_2)/(n_1 + n_2)$ and $\bar{\mathbf{z}}_2 - \bar{\mathbf{z}} = n_1(\bar{\mathbf{z}}_2 - \bar{\mathbf{z}}_1)/(n_1 + n_2)$, and $\mathbf{B}$ simplifies to

$$\mathbf{B}_2 = \frac{n_1 n_2}{n_1 + n_2}(\bar{\mathbf{z}}_1 - \bar{\mathbf{z}}_2)(\bar{\mathbf{z}}_1 - \bar{\mathbf{z}}_2)^{\mathsf{T}}, \tag{9.40}$$

which has rank 1 and explains the variability along the line connecting the group means, that is, $\bar{\mathbf{z}}_1 - \bar{\mathbf{z}}_2$ (see Problem 9.1).

As before, Fisher's method is to find a linear combination $y = \mathbf{a}^{\mathsf{T}}\mathbf{z}$ that maximizes separation of samples. The between-group variability in the direction of the coefficient vector $\mathbf{a}$ is described by

$$\mathbf{a}^{\mathsf{T}}\mathbf{B}\mathbf{a} = \sum_{j=1}^{g} n_j \left[\mathbf{a}^{\mathsf{T}}(\bar{\mathbf{z}}_j - \bar{\mathbf{z}})\right]^2, \tag{9.41}$$

where the jth term is proportional to the squared distance between the jth group mean and the overall mean after projection on $\mathbf{a}$. In the Fisher's approach, we want to maximize expression (9.41) subject to the constraint of the unit length of $\mathbf{a}$, that is, $\|\mathbf{a}\| = 1$. For the case of two groups,

$$\mathbf{a}^T \mathbf{B}_2 \mathbf{a} = \frac{n_1 n_2}{n_1 + n_2} \left[\mathbf{a}^T (\bar{\mathbf{z}}_1 - \bar{\mathbf{z}}_2)\right]^2, \tag{9.42}$$

which is maximized for $\mathbf{a}$ being proportional to $\bar{\mathbf{z}}_1 - \bar{\mathbf{z}}_2$. This is illustrated in Figure 9.5, where the discriminant direction $\mathbf{v}$ is parallel to the line connecting the group means. In a general setup, the following theorem provides the solution to maximization of (9.41), and hence to the optimum separation of several samples.

Theorem 9.3 Let $\lambda_1 \geq \lambda_2 \geq \cdots \geq \lambda_s > 0$ be the nonzero eigenvalues of $\mathbf{B}$ and $\mathbf{e}_1, \mathbf{e}_2, \ldots, \mathbf{e}_s$ be the corresponding normalized eigenvectors. The number s is not larger than $\min(g - 1, p)$. Then the vector $\mathbf{a}_1 = \mathbf{e}_1$ maximizes expression (9.41) subject to $\|\mathbf{a}\| = 1$. The linear combination $\mathbf{a}_1^T \mathbf{z}$ is called the sample first discriminant. The sample second discriminant is defined as $\mathbf{a}_2^T \mathbf{z} = \mathbf{e}_2^T \mathbf{z}$, and continuing, we can define the sample kth discriminant as $\mathbf{a}_k^T \mathbf{z} = \mathbf{e}_k^T \mathbf{z}$, $k \leq s$.

The theorem tells us that the sample first discriminant is the best direction for separating the groups, followed by the second discriminant, and so on. This means that we can use a certain number of the discriminants up to the rth discriminant as described by the following classification rule.

Fisher's Classification Rule. Assign $\mathbf{z}$ to R_j when the score $\Delta_j^2(\mathbf{z})$ for the ith group is the smallest of $\Delta_1^2(\mathbf{z}), \Delta_2^2(\mathbf{z}), \ldots, \Delta_g^2(\mathbf{z})$ for all g groups, where

$$\Delta_j^2(\mathbf{z}) = \sum_{m=1}^r \left[\mathbf{a}_m^T (\mathbf{z} - \bar{\mathbf{z}}_j)\right]^2, \quad j = 1, 2, \ldots, g \tag{9.43}$$

and $r \leq s$.

Since the above classification rule uses the first r discriminants or the first r discriminant directions $\mathbf{a}_j$, the classification is performed based on those r dimensions. In most cases, we want to use all s discriminants, that is, $r = s$. We may want to use smaller r when a simplification is needed or when some of the eigenvalues λ_j are very small, suggesting that the respective determinants are not significant.

Remark 9.1 When prior probabilities are all equal and $r = s$, Fisher's classification procedure is equivalent to the linear Gaussian classification rule defined by the linear score (9.33).

The boundary between the jth and kth classification regions can be found from condition (9.37), which can be written as

$$(\bar{\mathbf{z}}_j - \bar{\mathbf{z}}_k)^T \mathbf{z} = 0.5\left(\|\bar{\mathbf{z}}_j\|^2 - \|\bar{\mathbf{z}}_k\|^2\right) \tag{9.44}$$

because equal priors are assumed here. The above equation describes a $(p - 1)$-dimensional affine subspace orthogonal to the vector $\bar{\mathbf{z}}_j - \bar{\mathbf{z}}_k$.

For further consideration, let us assume that all sphered data were further centered at $\bar{\mathbf{z}}$, which becomes the origin in the new, shifted system of coordinates. Let A be the subspace spanned by all group centers $\bar{\mathbf{z}}_j$, and denote by s the dimension of A. For two groups, A is the line connecting the group centers. When p is large, s will generally be equal to $(g - 1)$ because the group centers are not linearly independent (their weighted mean is zero) in the $\bar{\mathbf{z}}$ centered space. The dimension s might also be smaller than $(g - 1)$ when there is more collinearity among the group centers. At the same time, s can never be larger than p. Hence, s is not larger than $\min(g - 1, p)$. Note that $\mathbf{B}$ describes variability in the directions within A, so the rank of $\mathbf{B}$ is s. This is why there are exactly s eigenvalues of $\mathbf{B}$, as described in Theorem 9.3. The s discriminant directions $\mathbf{a}_j$ span the same subspace A. This means that Fisher's rule is entirely determined within the subspace A, that is, if $\mathbf{z} = \mathbf{v} + \mathbf{w}$, where $\mathbf{v} \in A$ and $\mathbf{w}$ is orthogonal to $\mathbf{v}$, then

$$\Delta_j^2(\mathbf{z}) = \Delta_j^2(\mathbf{v}), \quad j = 1, 2, \ldots, g. \tag{9.45}$$

The first two Fisher's discriminants are the most relevant for the group separation, so they are often used for plotting the data in two dimensions. The following example shows an application of discriminants for classification and visualization of data.

Example 9.6   Let us continue with the Grass data set from Example 9.2. When we analyze the diseased grass in the top right corner of our image, it becomes clear that the pixels of diseased grass are not all of the same kind. A small top right corner represents severely diseased grass, while the surrounding area represents less severely diseased grass. This suggests that we can identify three groups of grass pixels. Let us call the healthy grass pixels—Group 1, less-severely diseased grass—Group 2, and severely diseased grass—Group 3. Spatial locations of the three groups are defined in Appendix B and are shown in Figure 9.11a.

We assume equal misclassification costs and equal prior probabilities, so that Fisher's approach can be used. Table 9.14 shows the classification table for Fisher's

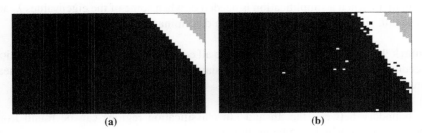

(a)   (b)

Figure 9.11   Spatial locations of the three groups of grass pixels. Healthy grass (Group 1 in black) is the largest group, severely diseased grass (Group 3 in gray) is in the top right corner, and the less severely diseased grass (Group 2 in white) is between the other two groups. Panel (a) shows the actual membership of pixels, and panel (b) shows the results of linear classification summarized in Table 9.14.

Table 9.14   The Classification Table for Fisher's Classification Rule Applied to the Three Groups from Example 9.6

Actual Group	Grass Classified into Groups			Total
	1	2	3	
1	3581	115	0	3696
2	13	285	2	300
3	0	7	93	100
Total	3594	407	95	4096

classification rule applied to the three groups. The classification is largely successful, except for 115 observations from Group 1 that were misclassified to Group 2. The leave-one-out cross-validation produced exactly the same numbers of misclassifications, and the numbers were only slightly larger for the eight-fold cross-validation (less than one additional misclassification in each cell on average). This confirms the usual robust behavior of the linear classification rule.

In this example, the number of groups is equal to $g = 3$, and we obtain $s = 2$ nonzero eigenvalues of the matrix **B**. This means that the two Fisher's discriminant vectors can be used as the basis (system of coordinates) in the subspace A discussed in the paragraph above formula (9.45). Since we are working on the sphered data, the discriminant vectors are orthogonal to each other. Figure 9.12 shows the Grass Data in the coordinates of the first two Fisher's discriminants. The two straight lines show the classification boundaries between Groups 1 and 2 and between Groups 2 and 3. The boundaries were calculated based on formula (9.44). The boundary between Groups 1 and 3 is not shown.

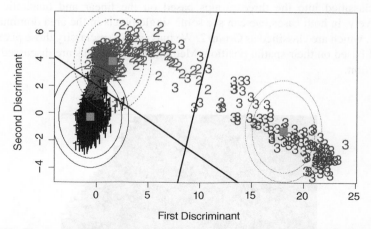

Figure 9.12   Three groups (marked as 1 = healthy grass, 2 = less severely diseased grass, and 3 = severely diseased grass) of Grass Data from Example 9.6 plotted in the coordinates of the first two Fisher's discriminants. The contour lines encompass 99% and 99.9% of the normal distributions. The sample means are marked by three squares.

Table 9.15   Classification Table for the Quadratic Gaussian Classification Rule Applied to the Grass Data from Example 9.6

Actual Group	Grass Classified into Groups			Total
	1	2	3	
1	3601	95	0	3696
2	3	295	2	300
3	0	0	100	100
Total	3604	390	102	4096

Since the variance–covariance matrices of the three groups are significantly different, we should also try the quadratic Gaussian classification rule defined by the scores shown in formula (9.31). The resulting misclassification rates shown in Table 9.15 are significantly lower than those shown in Table 9.14 for the linear Gaussian rule. The spatial positions of the classified pixels are shown in Figure 9.13.

The results of the leave-one-out cross-validation are shown in Table 9.16 with the values somewhat larger than those in Table 9.15. This means that the quadratic rule is not as robust as the linear rule, which is expected. Table 9.17 shows the average classification results from 50 random runs of the eight-fold cross-validation. The results are not much different from the leave-one-out cross-validation results. It is somewhat surprising that there were fewer (96.38) eight-fold misclassifications than leave-one-out ones in Group 1. This was not due to a spurious random fluctuation, since the standard error of each of the results in Table 9.17 was estimated at less than 0.04.

Let us look again at Figures 9.11b and 9.13, which show the spatial positions of pixels classified into the three groups based on the linear and quadratic rules, respectively. In both cases, we can see some single pixels in the area dominated by Group 1, which are classified as Group 2. We may wish to reclassify those pixels into Group 1 based on their spatial position. Methods for doing this are discussed in the next section.   ☐

Figure 9.13   Results of the quadratic Gaussian classification rule. Group 1 is in black, Group 2 in white, and Group 3 in gray.

Table 9.16 Leave-One-Out Cross-Validation Results for the Quadratic Gaussian Rule Applied to the Grass Data from Example 9.6

Actual Group	Grass Classified into Groups			
	1	2	3	Total
1	3596	100	0	3696
2	5	292	3	300
3	0	5	95	100
Total	3601	397	98	4096

Table 9.17 Eight-Fold Cross-Validation Results for the Quadratic Gaussian Rule Applied to the Grass Data from Example 9.6

Actual Group	Grass Classified into Groups			
	1	2	3	Total
1	3599.62	96.38	0	3696
2	6.70	290.58	2.72	300
3	0	6.46	93.54	100
Total	3606.32	393.42	96.26	4096

9.4 SPATIAL SMOOTHING FOR CLASSIFICATION

Sometimes we are able to use additional information in order to improve a given classification procedure. The additional information might be difficult to express as an additional set of **X** variables to be added to previous data. In that case, we may want to fine-tune the classification procedure with a specifically designed follow-up procedure. An example is a situation with spatial data, where the spatial information might be of some help.

Here we show a follow-up fine-tuning procedure that takes classification results of an arbitrary classification rule as an input. We assume here that the classified observations are pixels of an image. We define a *classification image* as an image where each pixel has a value equal to the group number based on the given classification. Figure 9.13 is an example of a classification image, where the groups are represented as colors.

We define the following *spatial smoothing voting procedure* (SSVP) that is applied to each pixel (call it Pixel C) in a classification image.

1. Let j be the group assignment for Pixel C based on the classification image.
2. Find a set F of neighboring pixels that excludes the central Pixel C (here we used a 3 by 3 neighborhood consisting of eight pixels).
3. Find the group assignments of all pixels in F, and find the group number k of the most prevalent group. If there is a tie with Group j, go to Step 5. If the tie is among groups different from j, go to Step 6.

Table 9.18 Classification Table for the Spatially Smoothed Linear Gaussian Classification Rule Applied to the Grass Data from Example 9.6

Actual Group	Grass Classified into Groups			Total
	1	2	3	
1	3592	104	0	3696
2	10	289	1	300
3	0	5	95	100
Total	3602	398	96	4096

4. If the number of pixels in F that are from Group k is at least equal to a threshold m (we used $m = 6$), then assign Pixel C to Group k and stop. Otherwise, go to the next step.

5. Assign Pixel C to Group j and stop.

6. Choose randomly one of the tied groups (call it k). Assign Pixel C to Group k and stop.

The SSVP was applied to the linear Gaussian classification of Grass Data into three groups (see Example 9.6). The classification results are shown in Table 9.18 and in Figure 9.14. The results are much better than those without smoothing shown in Table 9.14. An additional advantage of the smoothed classification is that the spatial boundaries between the groups in the image are smoother than those from the nonsmoothed classification. This might be relevant for some follow-up procedures. In the case of the Grass Data, the diseased grass may need to be treated with some chemicals. Due to the logistics of such treatment, we may not be able to treat small areas represented by single pixels in the middle of the healthy grass area. For such logistical reasons, it would be more convenient to have smoother boundary lines.

We also applied the SSVP to the quadratic Gaussian classification of Grass Data into three groups. The classification results are shown in Table 9.19 and in Figure 9.15. The results are similar to those without smoothing shown in Table 9.15.

Figure 9.14 Results of the spatially smoothed linear Gaussian classification rule. Group 1 is in black, Group 2 in white, and Group 3 in gray.

Table 9.19   Classification Table for the Spatially Smoothed Quadratic Gaussian Classification Rule Applied to the Grass Data from Example 9.6

| Actual Group | Grass Classified into Groups | | | Total |
	1	2	3	
1	3602	94	0	3696
2	4	295	1	300
3	0	0	100	100
Total	3606	389	101	4096

Figure 9.15   Results of the spatially smoothed quadratic Gaussian classification rule. Group 1 is in black, Group 2 in white, and Group 3 in gray.

9.5   FURTHER READING

The classification problem can be formulated as a prediction problem with a categorical response variable (group number). Hence, the logistic regression and some other types of generalized linear models are useful for classification. The logistic regression is about 30% less efficient than the Gaussian rule under the normality assumption, but it often performs much better for non-normal data. The logistic regression boundaries between classification regions are less influenced by the observations lying far away from the boundaries. References for generalized linear models and logistic regression are Montgomery et al. (2006), Kutner et al. (2005), Hosmer (2000), and Agresti (2002).

On the opposite spectrum of classification methods is a highly nonparametric method of k-nearest neighbor, which is based on a majority vote among the closest k neighbors. Other methods include neural networks, support vector machines, and classification trees. All of these methods are prone to overfitting to the sample data, and the cross-validation is very important when applying those tools. A good reference on these and many other classification methods is Hastie et al. (2001). A general reference at a descriptive level is Tso and Mather (2009). Classification in the context of remote sensing is discussed in Canty (2010). Support vector machines for classification are presented in Abe (2005).

PROBLEMS

9.1. For an arbitrary vector $\mathbf{v}$, define a matrix $\mathbf{B} = \mathbf{v}\mathbf{v}^{\mathsf{T}}$. Show that $\mathbf{v}$ is the eigenvector of $\mathbf{B}$ with the eigenvalue equal to $\mathbf{v}^{\mathsf{T}}\mathbf{v}$. Show also that any vector $\mathbf{w} \neq 0$ orthogonal to $\mathbf{v}$ is the eigenvector of $\mathbf{B}$ with the eigenvalue equal to zero. This means that $\mathbf{B}$ has rank 1 and describes variability in the direction of $\mathbf{v}$.

9.2. Prove Theorem 9.1. *Hint*: Use the definition of ECM in equation (9.5) and integrals in (9.1) and (9.2) to obtain

$$\text{ECM} = c(2|1)p_1 \int_{R_2} f_1(\mathbf{x})d\mathbf{x} + c(1|2)p_2 \int_{R_1} f_2(\mathbf{x})d\mathbf{x}. \tag{9.46}$$

Since

$$1 = \int_{R_1 \cup R_2} f_1(\mathbf{x})d\mathbf{x} = \int_{R_1} f_2(\mathbf{x})d\mathbf{x} + \int_{R_2} f_1(\mathbf{x})d\mathbf{x}, \tag{9.47}$$

we can write

$$\begin{aligned}
\text{ECM} &= c(2|1)p_1 \left(1 - \int_{R_1} f_1(\mathbf{x})d\mathbf{x}\right) + c(1|2)p_2 \int_{R_1} f_2(\mathbf{x})d\mathbf{x} \\
&= \int_{R_1} [c(1|2)p_2 f_2(\mathbf{x}) - c(2|1)p_1 f_1(\mathbf{x})]d\mathbf{x} + c(2|1)p_1.
\end{aligned} \tag{9.48}$$

We want to minimize ECM by changing the shape of the region R_1 (everything else is fixed here). This can be achieved by including into R_1 all $\mathbf{x}$ such that the expression under the integral is negative, that is, $c(1|2)p_2 f_2(\mathbf{x}) - c(2|1)p_1 f_1(\mathbf{x}) < 0$ and excluding all $\mathbf{x}$ such that the expression is positive. It does not matter where we place the $\mathbf{x}$ values such that the expression is zero.

9.3. Consider the Astronomy Data used in Example 9.1. Verify the multivariate normality of the data in the two samples.

9.4. Consider the Astronomy Data used in Example 9.1. Test the equality of the population variance–covariance matrices in the two samples.

9.5. Prove Theorem 9.2. *Hint*: Use Property 4A.11 (Maximization Lemma).

9.6. Show how to simplify the classification rule defined in equation (9.6) in order to obtain the rule defined in equation (9.7).

9.7. Apply the linear Gaussian classification rule to the Astronomy Data used in Example 9.1. Assume equal costs of misclassification and equal prior probabilities.

    a. Calculate all entries in the classification table.

    b. Calculate EER and APER.

    c. Use formulas (9.14) and (9.15) to estimate the TPM and compare the value to those obtained in point b.

9.8. Show how to simplify the classification rule defined in equation (9.6) in order to obtain the rule defined in equation (9.18).

9.9. Consider the Astronomy Data used in Example 9.1. Assume equal costs and equal prior probabilities. Perform cross-validation of the linear Gaussian rule to classify observations into the two groups of stars. Use the leave-one-out cross-validation and the k-fold cross-validation for $k = 7$ and $k = 18$.

9.10. Repeat tasks of Problem 9.9 for the quadratic Gaussian rule. What difference do you observe in relation to the results of Problem 9.9? How would you explain it?

9.11. Consider the Astronomy Data used in Example 9.1, but add the third band for three-dimensional data of J, H, and K bands. Find the Fisher discriminant direction **a** and then the Fisher discrimination rule.

    a. Construct the misclassification table and compare it to Table 9.5. Explain differences and similarities.

    b. Plot all data projected onto the plane spanned by **a** and the vector $\mathbf{v} = \bar{\mathbf{x}}_1 - \bar{\mathbf{x}}_2$ connecting the group means. Use different symbols and/or colors for the two groups of observations.

9.12. Repeat tasks of Problem 9.11 for the sphered version of the three-band data. Verify that the Fisher discriminant direction **a** and the vector $\mathbf{v} = \bar{\mathbf{x}}_1 - \bar{\mathbf{x}}_2$ connecting the group means are collinear. Then follow the steps shown in Example 9.5 in order to find the direction **b** of maximum variability in the misclassified observations in the space orthogonal to **a**. Plot all data in the coordinates of two axes in the directions of **a** and **b**. Use different symbols and/or colors for the two groups of observations. If you were also assigned Problem 9.11b, consider if this plot is more helpful than the one created in Problem 9.11b.

9.13. Consider the Astronomy Data used in Example 9.1, but add a third group of stars (oxygen-rich asymptotic giant branch (O AGB)) and two additional bands. We now have four-dimensional data of J, H, K, and A bands. Find the

Fisher discriminant directions and then the Fisher discrimination rule for the three groups.

a. Construct the misclassification table and draw conclusions.

b. Plot all data projected onto the plane spanned by the Fisher discriminant directions. Use different symbols and/or colors for the two groups of observations.

9.14. Consider the scenario of Astronomy Data used in Problem 9.13 with three groups of stars with four bands J, H, K, and A. Perform cross-validation of the Fisher discrimination rule to classify observations into the three groups of stars. Use the leave-one-out cross-validation and the k-fold cross-validation for $k = 7$ and $k = 18$.

9.15. Consider the Astronomy Data used in Problem 9.13 with three groups of stars with four bands J, H, K, and A. Apply the quadratic Gaussian rule to classify observations into the three groups of stars. Construct the misclassification table. If you were also assigned Problem 9.13, compare this table to the one created in Problem 9.13.

9.16. Consider the Astronomy Data used in Problem 9.13 with three groups of stars with four bands J, H, K, and A. Perform cross-validation of the quadratic Gaussian rule to classify observations into the three groups of stars. Use the leave-one-out cross-validation and the k-fold cross-validation for $k = 7$ and $k = 18$. If you were also assigned Problem 9.15, compare your cross-validation misclassification rates with those in the table created in that problem.

CHAPTER 10

Clustering – Unsupervised Learning

10.1 INTRODUCTION

The purpose of cluster analysis is to group objects in such a way that objects in the same group (cluster) are alike, whereas objects in different groups are dissimilar. When a data set is given as a traditional statistical database, objects can be either observations or variables. Sometimes the objects are not directly described, but instead we have information about relations among objects, for instance, in the form of similarity. In cluster analysis, we have no prior information about the grouping of objects. This is why cluster analysis is often called unsupervised learning, which is different from supervised learning (or the classification problem) discussed in Chapter 9.

Clustering is usually done for descriptive or exploratory purposes, so that the relations among objects or the structure of data can be better understood. When we find a cluster of similar variables, we may try to replace the whole group with only one summary variable conveying the same information. When we cluster observations, we may describe the whole data set as consisting of several groups. Each group can then be described by summary statistics. Such descriptions will typically be more precise than global summary statistics of the whole data set.

Finding suitable clusters can be a computationally intensive task. Checking all possible clusters is usually impractical or even impossible. As an example, we are going to find the number of all possible ways to divide the sample of size n into two clusters. For a sample of size n, the number of subsets is 2^n. Any subset (except for the whole data set) and its complement define one clustering, so there are $2^{n-1} - 1$ ways to divide into two subsets. This number can be very large. For $n = 100$, we obtain $2^{99} - 1 \approx 6.3 \times 10^{29}$. If we could check 1 billion clusters in 1 second, it would still take about 2×10^{13} years to check all possible two-cluster solutions, which is over 1000 times longer than the age of the universe. In practice, we do not try to check all

Statistics for Imaging, Optics, and Photonics, Peter Bajorski.
© 2012 John Wiley & Sons, Inc. Published 2012 by John Wiley & Sons, Inc.

possibilities, but instead rely on some heuristic algorithms that search for good clusters. However, there is no guarantee that the clustering found is the best one. An additional difficulty is in defining the best measure to evaluate the quality of a cluster. Such measures are subjective and depend on a given application. In this chapter, we will show various algorithms for clustering and various ways to evaluate the resulting clusters.

In order to group objects into clusters, we need to know how to measure the similarity of the objects, which is discussed in the next section.

10.2 SIMILARITY AND DISSIMILARITY MEASURES

The measures of similarity among observations are usually different from those for comparing variables. Hence, we will discuss them in two separate subsections.

10.2.1 Similarity and Dissimilarity Measures for Observations

Here we assume that a data set consists of n observations on p variables, which means that each observation is represented by a p-dimensional vector. In this context, dissimilarity of observations is usually described by a distance between them. See Section 5.6 for axioms of a distance and other properties. The most popular distance is the Euclidean distance defined as

$$d(\mathbf{x}, \mathbf{y}) = \|\mathbf{x} - \mathbf{y}\| = \sqrt{(\mathbf{x} - \mathbf{y})^{\mathrm{T}}(\mathbf{x} - \mathbf{y})} \tag{10.1}$$

for two p-dimensional vectors $\mathbf{x}$ and $\mathbf{y}$. If the distance is zero, the observations represented by $\mathbf{x}$ and $\mathbf{y}$ are identical. If the observations are far way from each other, they are very different. Like the distance, a similarity measure $s(\mathbf{x}, \mathbf{y})$ is also defined as a real-valued function on pairs of vectors $\mathbf{x}$ and $\mathbf{y}$. A similarity works opposite to dissimilarity, that is, small values indicate little similarity between objects that are very different. Large similarity values indicate similar objects. In general, a similarity may be unbounded with very large values, but we often prefer to assign a value of 1 as the largest possible similarity of identical observations. A similarity can be defined as any decreasing function of a distance. One possibility is the formula

$$s(\mathbf{x}, \mathbf{y}) = \frac{1}{1 + d(\mathbf{x}, \mathbf{y})} \tag{10.2}$$

for any distance d, which gives similarity values between 0 and 1. If we already have a similarity measure, we can define dissimilarity as any decreasing function of the similarity. However, the dissimilarity might not be a distance in the sense of satisfying the axioms of a distance (see Section 5.6).

Assuming n observations in a data set and an $n \times n$ matrix $\mathbf{A} = [a_{ij}]$ of similarities between them with $a_{ij} \leq 1$, we can define a dissimilarity as

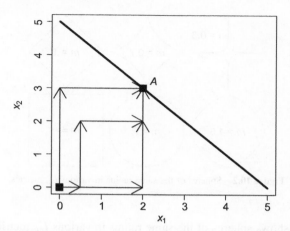

Figure 10.1 Three paths between the origin and Point A, each going along the shortest L_1 distance of 5.

$$d_{ij} = \sqrt{2(1 - a_{ij})}. \tag{10.3}$$

Assume that the largest elements of **A** are its diagonal elements a_{ii} (representing similarities of observations to themselves) all equal to 1. If the matrix **A** is nonnegative definite, then the dissimilarities d_{ij} have the property of a distance. This property will become useful in Section 10.2.2, when defining similarity measures for variables.

In addition to the Euclidean distance, we can define many other types of distances also called metrics. One example is a Minkowski metric often referred to as L_m and defined as

$$L_m(\mathbf{x}, \mathbf{y}) = \left[\sum_{i=1}^{p} |x_i - y_i|^m \right]^{1/m}, \quad m > 0, \tag{10.4}$$

where x_i and y_i are the coordinates of **x** and **y**, respectively. For $m = 2$, formula (10.4) defines the Euclidean distance, also called L_2 metric. For $m = 1$, we obtain L_1 called the city-block distance or Manhattan distance because the distance is measured along the directions parallel to the axes as shown in Figure 10.1. Note that all three paths from the origin to point A show the shortest L_1 distance of 5. Any point on the bold line in Figure 10.1 is in the L_1 distance of 5 from the origin. When m tends to infinity, the L_m distance gets close to the Chebyshev distance L_∞ defined as

$$L_\infty(\mathbf{x}, \mathbf{y}) = \max_{1 \le i \le p} |x_i - y_i|. \tag{10.5}$$

It is often convenient to define the distance between the point **x** and the origin as the norm of **x**. The L_m norm corresponding to the L_m distance is defined as

$$\|\mathbf{x}\|_m = \left[\sum_{i=1}^{p} |x_i|^m \right]^{1/m}. \tag{10.6}$$

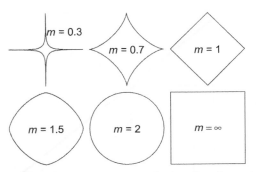

Figure 10.2 Spheres of the same radius in various L_m metrics.

Figure 10.2 shows spheres of the same radius in various L_m metrics. Note that a unit length vector in the direction of an axis of the system of coordinates has the length of 1 in any of the L_m metrics. This might be a desirable property in some applications, but not in the context of spectral curves. Instead of treating a spectral curve as a vector, we may want to treat it as a function of the range of spectral wavelengths that the curve represents. In that case, we would calculate the distance between two spectral functions f and g as the appropriate L_m norm expressed as the following integral:

$$\left(\int_R \left| f(\lambda) - g(\lambda) \right|^m d\lambda \right)^{1/m}, \tag{10.7}$$

where R is the range of the spectral wavelengths under consideration. In practice, the spectral curves are discretized as vectors, and it is convenient to use the equivalent of (10.7) in the form of an *adjusted L_m metric* defined as

$$L_m^{\text{adj}}(\mathbf{x}, \mathbf{y}) = \left[\frac{1}{p} \sum_{i=1}^{p} \left| x_i - y_i \right|^m \right]^{1/m} = \frac{1}{p^{1/m}} L_m(\mathbf{x}, \mathbf{y}) \tag{10.8}$$

and the adjusted L_m norm defined as

$$\|\mathbf{x}\|_m^{\text{adj}} = \left[\frac{1}{p} \sum_{i=1}^{p} \left| x_i \right|^m \right]^{1/m} = \frac{1}{p^{1/m}} \|\mathbf{x}\|_m. \tag{10.9}$$

Formula (10.8) can be viewed as an approximation of integral (10.7), where the interval R is divided into p equal subintervals.

The adjusted L_m metric has a desirable property in that its value does not depend on the dimensionality p in the following sense. Assume that two reflectance spectral

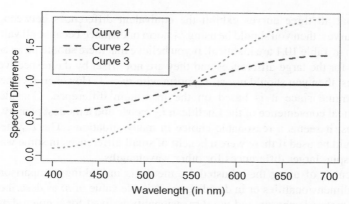

Figure 10.3 An example of three difference curves. Each curve is a difference between two hypothetical vector spectra.

curves described by the vectors **x** and **y** are being compared, and their distance has been calculated as $L_m(\mathbf{x}, \mathbf{y})$. We can now divide each spectral band into two halves and assign the original reflectance to both of those narrower spectral bands. This results in a spectral curve practically identical to the one described by **x** that is now described by $\mathbf{x}^*$ with twice the number of spectral bands. In the same way, we can denote by $\mathbf{y}^*$ a $(2p)$-dimensional vector corresponding to **y**. Note that $L_m(\mathbf{x}^*, \mathbf{y}^*) = 2^{1/m} L_m(\mathbf{x}, \mathbf{y})$ even though both $(\mathbf{x}^*, \mathbf{y}^*)$ and $(\mathbf{x}, \mathbf{y})$ describe the same pair of spectral curves. On the other hand, we have an equality between the corresponding adjusted L_m distances $L_m^{\text{adj}}(\mathbf{x}^*, \mathbf{y}^*) = L_m^{\text{adj}}(\mathbf{x}, \mathbf{y})$.

Another property of the adjusted L_m metric can be explained by considering a difference $\Delta = \mathbf{x} - \mathbf{y}$ between hypothetical vector spectra represented by the vectors **x** and **y**. Figure 10.3 shows three scenarios for such difference vectors. The difference vectors are shown as hypothetical symmetric curves, that is, each curve extends above the level of 1 on the right-hand side of the plot by the same amount that it is below the level of 1 on the left. The adjusted L_m norm does not depend on m for a constant-difference curve like Curve 1 in Figure 10.3. Table 10.1 shows values of adjusted L_m norms for the three curves from Figure 10.3. These values may help you in deciding which L_m metric to use in a given context. This will depend on which column in Table 10.1 best reflects the nature of differences described by the three curves. If you

Table 10.1 The Adjusted L_m Norms for the Three Curves from Figure 10.3

	Value of m					
	0.3	0.7	1	1.5	2	∞
Curve 1	1.00	1.00	1.00	1.00	1.00	1.00
Curve 2	0.97	0.99	1.00	1.02	1.04	1.40
Curve 3	0.81	0.93	1.00	1.10	1.19	1.90

believe that all three curves exhibit the equivalent differences between pairs of spectral curves, then you should be using L_1 norm or distance because all values in the L_1 column in Table 10.1 are identical. If you believe that there should be an additional "penalty" for the large differences and they are not offset by the respective smaller differences, then you should use an L_m norm with $m > 1$. However, L_∞ seems to be rather extreme since it is based on the maximum difference. Considering the mathematical convenience of the Euclidean L_2 metric and a modest penalty for large differences, it seems a reasonable choice in many situations. The L_m metric with $m < 1$ could be used if there were a benefit of small differences in some wavelength ranges despite larger differences for other wavelengths.

The benefit of using the adjusted L_m metric is in making comparisons across different dimensionalities or in deciding the suitable value of m as described above. Once the metric is chosen and the dimensionality is fixed for a given data set, the adjusted L_m metric is entirely equivalent to the classic L_m metric, and either of the two can be used.

A distance that takes into account variability is the Mahalanobis distance used in previous chapters and defined again here as

$$d_M(\mathbf{x}, \mathbf{y}) = \sqrt{(\mathbf{x} - \mathbf{y})^T \mathbf{S}^{-1} (\mathbf{x} - \mathbf{y})}. \tag{10.10}$$

The variance–covariance matrix $\mathbf{S}$ can be either a global statistic calculated from the whole data set or a more local statistic calculated from a cluster or from some other available information. It is often practical to sphere the data by using the transformation $\mathbf{S}^{-1/2}$ and then use a suitable metric such as L_m in the sphered coordinates. When L_2 is used in the sphered coordinates, it is equivalent to using the Mahalanobis distance in the original coordinates. This approach is often used in images with pixels described by spectral curves in order to adjust for the inherent structure of the data as described by $\mathbf{S}$.

If the original variables are in different physical units, then they should be standardized so that the variables are comparable. This will preserve the correlations among variables (unlike the sphering, which results in uncorrelated variables). If the original variables are in the same physical units but their variances are very different, then the decision about standardization depends on whether or not we want to preserve those differences in variability.

Different measures of similarity are needed in the case of categorical variables. A categorical variable takes on labels or names as values. The simplest case is that of a binary variable, where observations can be characterized by the presence or absence of a certain characteristic. A binary variable can be coded with 0 and 1 representing the only two values taken by the variable. Here we will discuss only binary variables.

Let us assume that we have p binary variables. Each observation is then represented by a p-dimensional vector of 0 and 1 values. A similarity between the ith observation, represented by a vector $\mathbf{x}$, and the kth observation, represented by $\mathbf{y}$, can be defined by counting the number of matches and mismatches at the same positions. The counts can

Table 10.2   Frequencies of Matches and Mismatches Between the ith and kth Observations

Observation i	Observation k	
	1	0
1	a	b
0	c	d

be summarized in a contingency table like the one in Table 10.2, where the counts of mismatches are shown in the shaded cells. For example, with $\mathbf{x} = [0, 1, 1, 1, 0, 0]^{\mathrm{T}}$ and $\mathbf{y} = [0, 0, 1, 1, 0, 0]^{\mathrm{T}}$, we obtain $a = 2$, $b = 1$, $c = 0$, and $d = 3$. Note that $a + b + c + d = p$.

A simple measure of similarity can be the fraction of matches out of all comparisons, that is, $f = (a + d)/p$. We can also calculate the Euclidean distance of the two vectors, which turns out to be equal to $\sqrt{b + c}$, as a measure of dissimilarity. Note that the fraction f of matches is a strictly decreasing function of the Euclidean distance, that is,

$$f = 1 - \frac{(\text{Euclidean Distance})^2}{p}. \tag{10.11}$$

The fraction f treats all matches equally, but sometimes matches on 1 might be more important than matches on 0. In order to adjust for different levels of importance, we can assign the weight of w_{11} to a match on 1 and the weight of w_{00} to a match on 0. Similar weights can be assigned to mismatches as shown in Table 10.3.

A weighted measure of similarity can be calculated as

$$w = \frac{w_{11}a + w_{00}d - w_{10}b - w_{01}c}{w_{11}a + w_{00}d + w_{10}b + w_{01}c} = 2\frac{w_{11}a + w_{00}d}{w_{11}a + w_{00}d + w_{10}b + w_{01}c} - 1. \tag{10.12}$$

We can assume that the weights are nonnegative, so that $-1 \leq w \leq 1$, with values close to 1 indicating strong similarity and those close to -1 strong dissimilarity. If all weights are equal to 1, then $w = 2f - 1$.

Table 10.3   Weights for Importance of Matches and Mismatches Between the ith and kth Observations

Observation i	Observation k	
	1	0
1	w_{11}	w_{10}
0	w_{01}	w_{00}

10.2.2 Similarity and Dissimilarity Measures for Variables and Other Objects

Sometimes, we may want to cluster variables into groups of highly correlated variables, so that they can be summarized by only one or two new variables. The new summary variables could be constructed based on the particular knowledge about those variables or through some statistical methods such as principal component analysis performed on the cluster of variables. A correlation coefficient can be used here as a measure of similarity. Some clustering procedures require an input in the form of a dissimilarity (or even a distance) matrix. Since the correlation matrix is nonnegative definite, we can use formula (10.3) to define a distance between variables.

For binary variables (i.e., having only two possible values), one can simply treat variables as observations and observations as variables, and use the similarity measures developed in Section 10.2.1 for binary data. We can also define a distance between such binary variables, especially the Euclidean distance, which is related to one of the discussed similarities as shown in equation (10.11).

Another situation is when information is not available in the form of a statistical database with observations and variables. Instead, we have objects and some similarities among objects are directly available. For example, in an experiment evaluating similarity among colors, several judges can be asked to rate the similarity for each pair of colors on a scale from 0 to 100. The similarity measure is then the average calculated over all judges.

10.3 HIERARCHICAL CLUSTERING METHODS

We start a discussion of clustering with hierarchical methods, which often provide more clustering information than other methods. There are two types of hierarchical clustering algorithms: divisive hierarchical algorithms and agglomerative hierarchical algorithms. Divisive hierarchical algorithms start with placing all objects in one cluster, and try to divide it into the two most meaningful clusters. Next, one of the clusters is further divided, and the process continues until a desired number of clusters are achieved. Agglomerative hierarchical algorithms work in the opposite direction. They start with each object forming a separate cluster, and then try to merge the most similar clusters until all objects are in a single cluster.

We are going to discuss agglomerative hierarchical algorithms and will start with three different linkage methods. These algorithms, as many other clustering methods, can use either a dissimilarity or similarity measure. For simplicity of presentation, we will refer to a distance as an example of dissimilarity. When a similarity measure is used, some obvious modifications to the algorithms need to be made, such as taking a maximum instead of a minimum, for instance. All linkage methods follow these steps

1. Assuming a set of n objects, start with n clusters, each cluster containing one single object. Calculate an $n \times n$ symmetric matrix of distances among the objects.
2. Search for the most similar pair of clusters. (One needs to find the smallest element in the distance matrix, excluding the zero distances on the diagonal.) Let us call these two clusters U and V.
3. Merge clusters U and V into a new cluster called (UV). Update the distance matrix by
   a. Deleting the rows and columns corresponding to clusters U and V.
   b. Adding a row and column for cluster (UV) specifying the distance between cluster (UV) and the remaining clusters.
4. Repeat Steps 2 and 3 a total of $(n-1)$ rounds, until all objects are in a single cluster.

(10.13)

After each round in the algorithm, the number of clusters is reduced by one. This means that we actually have n different clustering solutions, each with a different number of clusters from 1 to n. Of course, the extreme solutions with 1 or n clusters are not interesting, but we have many other clustering solutions from which to choose. The appealing property of the hierarchical algorithms is that they do not require *a priori* determination of the number of clusters, but instead provide the flexibility of choosing the number after we see all potential choices.

The three linkage algorithms we are going to present differ in the method of calculating the distance between two clusters as discussed in the following subsections.

10.3.1 Single Linkage Algorithm

In the *single linkage algorithm* (also called *minimum distance* or *nearest neighbor algorithm*), the distance between clusters is calculated as the distance between the two closest elements from the two clusters. This rule would suggest that we need to use the original distances at each step of the algorithm. However, it turns out that there is a computationally simpler approach that can be described as follows (the explanation why the two approaches are equivalent is given as Problem 10.1). When two clusters U and V are merged (in Step 3 above), the distance between (UV) and any other cluster W is calculated as

$$d_{(UV)W} = \min\{d_{UW}, d_{VW}\}, \tag{10.14}$$

where d_{UW} and d_{VW} are distances between clusters U and W and clusters V and W, respectively. We would use the maximum for similarities. Since the operations of the minimum and maximum are invariant to a strictly increasing transformation of distances, the single linkage clustering is also invariant to a strictly increasing transformation of distances (or similarities) in the sense that the same clustering is produced based on the original distance and the transformed distance (see Problem 10.2).

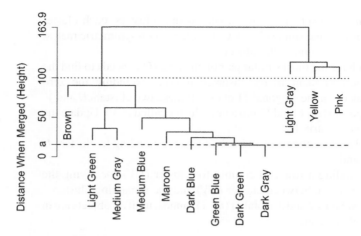

Figure 10.4 A dendrogram showing clusters of 12 tiles when using the single linkage method as discussed in Example 10.1.

Example 10.1 Let us consider the Tile Data from Example 5.1. The data set consisted of 4 measurements for each of the 12 monochromatic (uniformly colored) tiles for a total of 48 observations in 31 spectral bands (variables). Here we want to find spectral similarities among the tiles. We calculate averages of the 4 measurements for each of the 12 tiles and treat those means as 12 observations, each representing one tile. The mean spectral reflectance curve for a given tile describes the inherent property of how the tile reflects light at different wavelengths. The distance between the spectral curves will measure how different those properties are for different tiles. This is not the same as the difference between the colors of the tiles as perceived by humans. For a meaningful analysis of similarity between colors, one would need to transform the spectral reflectance curves into coordinates in a color space that approximates human vision, such as the CIELAB (see Berns (2000)). This transformation depends on the type of the illuminant used, that is, the colors may look different in daylight than under artificial light indoors.

The single linkage algorithm with Euclidean distance between spectra has been applied to those 12 observations, and the clustering results are represented in the form of a dendrogram, or tree diagram, shown in Figure 10.4. The branches in the tree represent clusters. The merging of clusters is shown as two branches coming together at nodes. The vertical position of nodes (Height) indicates the value at which the merging occurred, that is, the distance between the two clusters. Let's read the dendrogram from the bottom up. At the bottom, you can see "dark green" merging with "dark gray." This first merging occurs at the level of $a = 15.7$, where a horizontal dashed line is drawn. This means that the Euclidean distance between the spectra of the dark green and dark gray tiles is 15.7.

In the second round of the algorithm, the cluster consisting of "dark green" and "dark gray" was merged with "green blue." The subsequent merging occurs at higher distances, sometimes between clusters and sometimes between two observations, such as "light green" and "medium gray." The final merging of two clusters into the whole data set is done at the highest level shown in the graph at 163.9 of the Euclidean distance.

We can say that the dendrogram provides the hierarchical structure of clusters. The dendrogram can assist us in deciding on a suitable number of clusters. One way to do this is to decide on a threshold level in the dendrogram. For example, if we believe that 100 is a large enough distance to decide that two clusters are different, then we can draw a horizontal line at that level and identify the branches below the line as clusters. Based on Figure 10.4 with the threshold of 100, we would identify four clusters—three one-element clusters of "light gray," "yellow," and "pink," and the fourth cluster containing all remaining tiles. In the next example, we will discuss some other criteria for finding a suitable number of clusters.

It seems that the tile brightness plays an important role in the clustering results. Some hues, such as green, are found in quite different clusters. Clustering analysis using the CIELAB color space will be done in Example 10.3 in order to reflect human color vision. □

Example 10.2 Consider the RGB image from an Eye Tracking experiment that was discussed in Example 2.5. A scatter plot of intensity of Red versus intensity of Green for all 16,384 pixels was shown in Figure 2.15. Here we selected a sample of 196 pixels randomly from all pixels. Our goal was to cluster those pixels into distinct groups based on the red and green intensity values. Those distinct groups can be characterized by some descriptive statistics. The whole set of pixels can then be described as consisting of several groups with given characteristics. The analysis would ultimately be done on the whole image in all three RGB channels, but we show here a simplified scenario for educational purposes. Figure 10.5 shows a scatter plot of color intensity values for the 196 pixels used in this example.

The single linkage algorithm with Euclidean distance has been applied to the set of 196 pixels, and the results are shown in the form of the dendrogram in Figure 10.6. The graph may seem busy and confusing. However, for a large data set, the purpose of a dendrogram is not to see all details of clustering, but instead to see major patterns. Here our practical goal is to identify several large clusters that encompass most pixels, but not necessarily all of them. This means that we can also leave some of the pixels as unclassified. If we use the threshold level of Height = 0.081 (the solid horizontal line in Figure 10.6 cutting the tree into branches positioned below the threshold), then we obtain three large clusters and the remaining pixels form one-element clusters or some small clusters. We could also use Height = 0.072 (the dashed horizontal line in Figure 10.6), which results in four large clusters. We will further evaluate those clusters later on in this section. □

It turns out that the single linkage clustering is closely related to the graph theory concept of the minimum spanning tree. In mathematics, a *graph* is a representation of a set of objects with *vertices*, where some pairs of the objects are connected by *links* or *edges*. When planning an optical fiber network connecting a number of sites, we may initially consider some connections between them as feasible and other as infeasible. Figure 10.7 shows six sites with feasible connections plotted as edges of the graph.

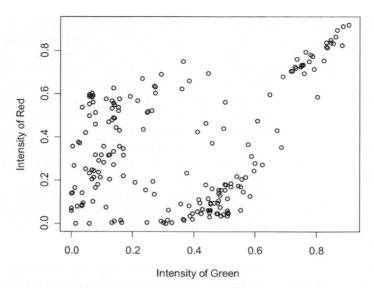

Figure 10.5   A scatter plot of color intensity values for the 196 pixels used in Example 10.2.

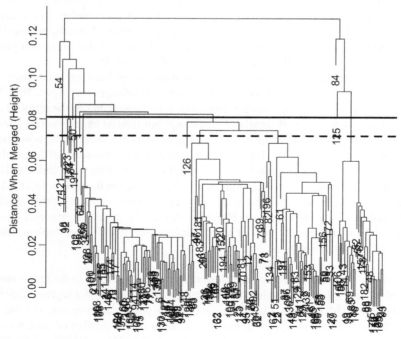

Figure 10.6   A dendrogram describing clustering results of the single linkage algorithm applied to the set of 196 pixels discussed in Example 10.2.

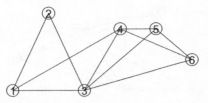

Figure 10.7 A graph with six sites as vertices and feasible connections plotted as edges or links.

Note that the positions of vertices may or may not be associated with the actual geographic locations of the sites, depending on our preference.

In the next step of planning the network, we realize that there is no need for multiple paths between sites. All we need is to have a connection between any two sites by a simple path. Considering the cost of burying the optical fiber into the ground, it is not cost efficient to make any unnecessary links, even if the connection needs to go around through some other sites. A graph where each two vertices are connected by exactly one simple path is called a tree. There are no cycles in a tree, such as the cycle connecting Sites 1, 2, and 3 in Figure 10.7. An example of a tree is shown in Figure 10.8 as a potential optical fiber network. This is a *spanning* tree because it connects all vertices in the graph.

In further planning for the network, we would also need to take into account the costs associated with each link. In order to minimize the cost, we would need to find a tree with the smallest total cost equal to the sum of all *weights*, or costs, associated with the tree edges. This tree is called the *minimum spanning tree*. There could be more than one minimum spanning tree for a given graph. However, if all weights of edges are different, then the solution is unique.

In the context of clustering, we typically assume that each object (vertex) can be connected with any other object. So, the initial graph may look like the one in Figure 10.9, which means that all connections are feasible. This is what we will assume from now on, and we will define the weights of the links as equal to the distances between the sites.

The minimum spanning tree for this configuration of objects, or sites, is shown in Figure 10.10. This is the optimal solution for the optical fiber network, if we assume that the cost is proportional to the actual distances shown in the graph.

In order to find the clustering of sites, we need to drop the longest link from the minimum spanning tree resulting in two trees, each connecting elements of one cluster as seen in Figure 10.11. The graph shown in Figure 10.11 is called a *forest*

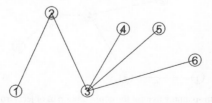

Figure 10.8 An example of a tree as a potential optical fiber network.

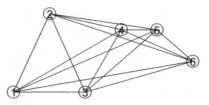

Figure 10.9 A graph of objects in the context of a typical clustering problem, where dissimilarity is defined for all pairs of objects.

because it consists of multiple trees. By removing the next longest link at each step, we can construct further divisions of all sites into more clusters. This process results in a clustering equivalent to the single linkage algorithm discussed earlier (the proof is given as Problem 10.7). We call the forest shown in Figure 10.11 the *first-level minimum forest* because it is obtained at the first level of clustering, counting from the top of the dendrogram. Moving down the tree, we obtain the *second-level minimum forest* and so on. At each level, each tree is the minimum spanning tree of a cluster.

The results of the single linkage algorithm applied to the six sites are shown in Figure 10.12. We can again see the first split into two clusters, one consisting of Sites 1 and 3 and the remaining sites being in the other cluster. We can now apply the concept of the minimum spanning tree to larger data sets such as the one discussed in Example 10.2.

Example 10.2 (cont.) The clustering solution shown in Figure 10.6 for 196 pixels discussed in Example 10.2 can now be represented as the minimum spanning tree shown in Figure 10.13. In order to evaluate the clustering proposed in Figure 10.6 by cutting the tree at the level of 0.081, we can find the equivalent k-level minimum forest shown in Figure 10.14. Here k turns out to be equal to 10. We can then plot the resulting clusters as shown in Figure 10.14. The three large clusters are marked in different colors and symbols, and the remaining clusters, including single observations, are regarded as unclassified and are marked with circles.

Each of the three large clusters can be described by a two-dimensional mean vector and a 2 by 2 variance–covariance matrix. In Figure 10.14, we plotted ellipses encompassing 90% of the normal (Gaussian) distributions with the means and variance–covariance matrices of the clusters. We conclude that the clusters conform

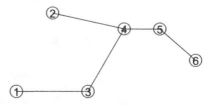

Figure 10.10 The minimum spanning tree for the configuration of objects from Figure 10.9 with weights given by Euclidean distances.

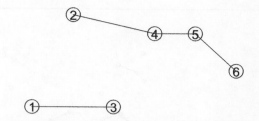

Figure 10.11 A forest consisting of two trees as clusters.

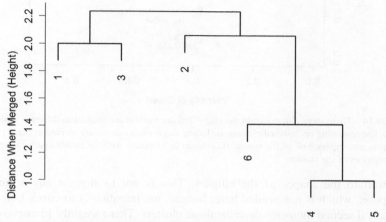

Figure 10.12 A dendrogram describing clustering results of the single linkage algorithm applied to the sites shown in Figure 10.11.

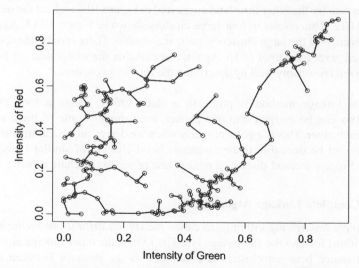

Figure 10.13 The minimum spanning tree equivalent to the clustering solution shown in Figure 10.6 for 196 pixels discussed in Example 10.2.

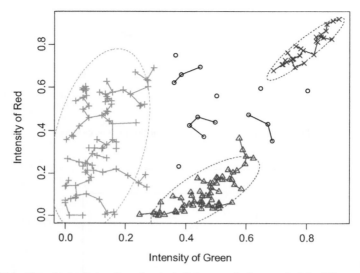

Figure 10.14   Three large clusters from the single linkage method are marked in different colors and symbols. The remaining unclassified clusters, including single observations, are marked as black circles. The ellipses encompass 90% of the normal (Gaussian) distributions with the means and variance–covariance matrices of the clusters.

quite well to the shapes of the ellipses. This is not to suggest any normality assumption, which is not needed here. Instead, the intention is to check how well the first and second moments describe those clusters. There are only 15 unclassified pixels (less than 10% of all pixels). It seems that the whole data set has been characterized reasonably well by describing the three large clusters.

We can also try the other threshold suggested in Figure 10.6 and cut the tree at the level of 0.072. This results in four large clusters shown in Figure 10.15. Again, the characterization of the large clusters is quite reasonable. There is only one additional unclassified pixel for a total of 16. Again, it seems that the whole data set has been characterized reasonably well by describing the four large clusters.   □

The single linkage method is prone to a chain effect, where a long chain of observations can be merged into one cluster, even though some of them are very far from each other. This is a good property when we design an optical fiber network, but it may not be desirable if the emphasis is on clusters of similar objects. The complete linkage method discussed next is less prone to the chain effect.

10.3.2   Complete Linkage Algorithm

The *complete linkage* algorithm (also called *maximum distance* or *farthest neighbor* algorithm) follows the same steps (see (10.13)) as the other linkage algorithms, but the distance between clusters is calculated as the distance between the two farthest elements from the two clusters. Again, it turns out that there is a computationally simpler approach that can be described as follows (the explanation

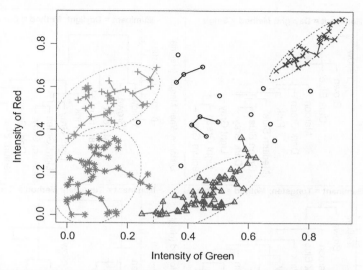

Figure 10.15 Four large clusters from the single linkage method with the 90% probability ellipses characterizing the normal distributions of the clusters.

why of the two approaches are equivalent is given as Problem 10.3). When two clusters U and V are merged in Step 3, the distance between (UV) and any other cluster W is calculated as

$$d_{(UV)W} = \max\{d_{UW}, d_{VW}\}, \tag{10.15}$$

where d_{UW} and d_{VW} are distances between clusters U and W and clusters V and W, respectively. We would use the minimum for similarities. Again, the complete linkage clustering is invariant to a strictly increasing transformation of distances (or similarities) in the sense that the same clustering is produced based on the original distance and the transformed distance (see Problem 10.2).

The complete linkage method has a tendency to produce tight clusters because any new observation merged with a cluster cannot be too far from all observations already in that cluster.

Example 10.3 This is a continuation of Example 10.1, where we were clustering mean spectra of 12 tiles. For a different perspective that takes into account color perception, we transformed the 12 spectra into three-dimensional CIELAB color space coordinates. See Appendix B for details. This transformation depends on the type of the illuminant used, that is, the colors may look different in daylight than under artificial light indoors. We used two illuminants, one representing the noon daylight with overcast sky (Illuminant D_{65}) and the other representing the incandescent or tungsten light source found in homes (Illuminant A). In both cases, the Euclidean distance in the three dimensions was used as a dissimilarity measure. The results of the single and complete linkage algorithms for both illumination conditions are shown as dendrograms in Figure 10.16. One can compare the methods by inspecting the panels

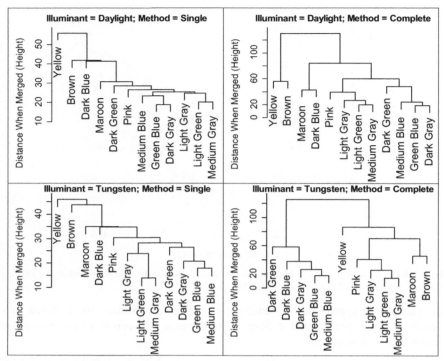

Figure 10.16 The dendrograms of the single and complete linkage (in the two columns of panels) clustering results for two illumination conditions (in rows of panels).

in the same row since they are for the same illumination. For both illuminants, the two methods show different results, although they seem to be more similar under daylight conditions.

The panels in columns are based on the same method and the colors under different illuminations can be compared. According to the single linkage method, the colors cluster only somewhat differently. Based on the complete linkage, the difference is much more significant. For example, in daylight, the brown tile is most similar to the yellow one, but in the tungsten light, the brown tile is most similar to the maroon one. □

Example 10.4 This is a continuation of Example 10.2, where 196 pixels were clustered based on bivariate values of red and green color intensity. The complete linkage algorithm with Euclidean distance was applied to the data, and the results are shown as the dendrogram in Figure 10.17. The dendrogram looks very different from the one for the single linkage algorithm. There are no single observations merged at the large distances (Height). Instead, all observations merge into some clusters early in the clustering process. We can again define three clusters with the threshold of Height $= 0.8$ (the solid horizontal line in Figure 10.17). For each cluster, we used the minimum spanning tree of that cluster only for the purpose of graphical presentation

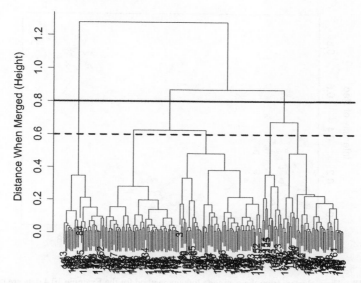

Figure 10.17 A dendrogram describing clustering results of the complete linkage algorithm applied to the set of 196 pixels discussed in Example 10.4.

as one group. The clusters are very different from the three clusters in Figure 10.14 that were based on the single linkage method.

One can also use the threshold of Height = 0.6 (the dashed horizontal line in Figure 10.17) in order to define five clusters shown in Figure 10.19. Those clusters look quite appealing, except perhaps the fifth cluster marked with circles. The fifth cluster is rather small (13 pixels) and is spread "thinly" over a relatively large area. This clustering is similar to the one in Figure 10.15, except that unclassified pixels in Figure 10.15 are now gathered into one cluster. For the purpose of a sparse representation of the whole set of pixels with summary statistics of clusters, we could either use five clusters from Figure 10.19 or prefer a solution with some observations being unclassified and choose the solution from Figure 10.15. The latter solution might work better for dealing with outliers in cases when they spread out into other areas where they will not merge into the fifth cluster. □

10.3.3 Average Linkage Algorithm

The *average linkage* algorithm (also called *average distance*) follows the same steps as the other linkage algorithms (see (10.13)), but the distance between clusters is calculated as the average distance over all possible pairs of elements from the two clusters. When two clusters U and V are merged (in Step 3), the distance between (UV) and any other cluster W is calculated as

$$d_{(UV)W} = \frac{\sum_i \sum_k d_{ik}}{n_{(UV)} n_W}, \qquad (10.16)$$

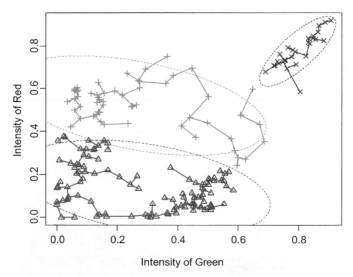

Figure 10.18   Three clusters from the complete linkage method for the Example 10.4 data are marked in different colors and symbols. The ellipses encompass 90% of the normal (Gaussian) distributions with the means and variance–covariance matrices of the clusters.

where d_{ik} is the distance between object i in the cluster (UV) and the object k in the cluster W and $n_{(UV)}$ and n_W are the numbers of objects in clusters (UV) and W, respectively. Unlike the previous two linkage methods, the average linkage is not invariant to a strictly increasing transformation of distances or similarities (see Problem 10.2).

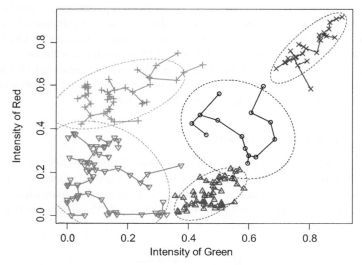

Figure 10.19   Five clusters from the complete linkage method with the 90% probability ellipses characterizing the normal distributions of the clusters.

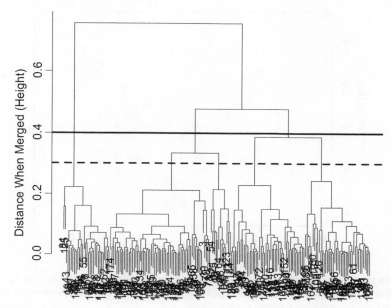

Figure 10.20   A dendrogram describing clustering results of the average linkage algorithm applied to the set of 196 pixels discussed in Example 10.5.

The *average linkage* method can be considered a compromise between the single and complete linkage methods since the average distance is always between the shortest and the longest distances.

Example 10.5   This is a continuation of Examples 10.2 and 10.4, where 196 pixels were clustered based on bivariate values of red and green color intensity. The average linkage algorithm with Euclidean distance was applied to the data, and the results are shown as the dendrogram in Figure 10.20. The dendrogram looks very similar to the one for the complete linkage shown in Figure 10.17. However, it is difficult to see from the dendrogram the actual cluster membership in a large data set like this one. Figure 10.21 showing three clusters based on the threshold of Height = 0.4 reveals that the clusters are very different from the results of the complete linkage (Figure 10.18). In this case, the average linkage results are more similar to the single linkage results (Figure 10.14), except that the unclassified pixels are now included in one of the three large clusters.

Figure 10.22 shows five clusters based on the threshold of Height = 0.3. These results are very similar to those for the complete linkage in Figure 10.19. Consistency of results among the three linkage methods with five groups (or four groups with the fifth group of unclassified pixels for the single linkage) points to the stability of this solution.   □

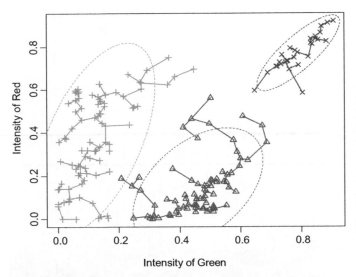

Figure 10.21 Three clusters from the average linkage method for the Example 10.5 data are marked in different colors and symbols. The ellipses encompass 90% of the normal (Gaussian) distributions with the means and variance–covariance matrices of the clusters.

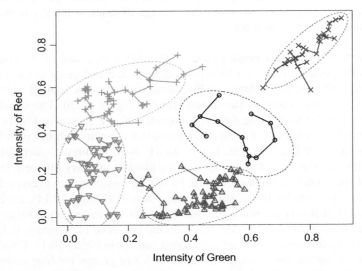

Figure 10.22 Five clusters from the average linkage method with the 90% probability ellipses characterizing the normal distributions of the clusters.

10.3.4   Ward Method

The Ward method, introduced in Ward (1963), is an agglomerative hierarchical clustering procedure, but unlike the linkage methods, it does not define a distance between clusters. Instead, the method is based on the *error sum of squares* defined for the jth cluster as the sum of squared Euclidean distances from points to the cluster mean

$$\text{ESS}_j = \sum_{i=1}^{n_j} \left\| \mathbf{x}_{ij} - \bar{\mathbf{x}}_j \right\|^2, \tag{10.17}$$

where $\mathbf{x}_{ij}$ is the ith observation in the jth cluster and $\bar{\mathbf{x}}_j$ is the mean of all observations in the jth cluster. The *error sum of squares for all clusters* is the sum of the ESS_j values from all clusters, that is,

$$\text{ESS} = \text{ESS}_1 + \text{ESS}_2 + \cdots + \text{ESS}_k, \tag{10.18}$$

where k is the number of clusters. The algorithm starts with each observation forming its own one-element cluster for a total of n clusters, where n is the number of observations. In each one-element cluster, the mean is equal to that one observation and $\text{ESS}_j = 0$, $j = 1, \ldots, n$. Consequently, $\text{ESS} = 0$. In the first round of the algorithm, we try to merge two elements into one cluster in a way that would increase ESS by the smallest possible amount. This can be done by merging the two closest observations within the data set. Up to this point, the Ward algorithm produces the same result as any of the three linkage methods discussed earlier. At each subsequent round, we find merging that produces the smallest increase in ESS. This minimizes the distances between the observations and the centers of the clusters. The process is continued until all observations are in one cluster.

One can show (see Problem 10.8) that formula (10.17) can be replaced with the following one:

$$\text{ESS}_j = \frac{1}{2n_j} \sum_{i=1}^{n_j} \sum_{m=1}^{n_j} \left\| \mathbf{x}_{ij} - \mathbf{x}_{mj} \right\|^2. \tag{10.19}$$

This means that all calculations can be based on an $n \times n$ matrix of distances between observations. This matrix is sometimes the only available information, while the vectors $\mathbf{x}_i$ of the original observations are unknown. Many computer software implementations of clustering procedures use the matrix of distances as the only input. This works with the Ward method and the linkage methods discussed earlier, but some other clustering methods require the vectors $\mathbf{x}_i$ of the original observations.

The Ward method was used to cluster the 196 pixels from Example 10.5. The results were almost identical to those obtained in that example when using the average linkage method (Figures 10.21 and 10.22). Only one pixel was assigned to a different cluster in both cases of three and five selected clusters. The two methods often give similar results. The Ward method tends to give clusters of circular or elliptical shapes.

10.4   NONHIERARCHICAL CLUSTERING METHODS

Nonhierarchical clustering methods identify a specific number of clusters rather than building the whole structure of clusters (as do the hierarchical methods). The number of clusters can be either determined by the user or calculated within the algorithm. The most popular among nonhierarchical clustering methods is the K-means algorithm discussed in the next subsection.

10.4.1   *K*-Means Method

We now look more broadly at the error sum of squares and define it for any set C of k clusters and any set of centroids $\mathbf{m}_1, \ldots, \mathbf{m}_k$ that may not be directly dependent on the clusters. We define a general error sum of squares as

$$\text{ESS}(C, \mathbf{m}_1, \ldots, \mathbf{m}_k) = \sum_{j=1}^{k} \sum_{i=1}^{n_j} \left\| \mathbf{x}_{ij} - \mathbf{m}_j \right\|^2, \tag{10.20}$$

where $\mathbf{x}_{ij}$ is the ith observation in the jth cluster. Given a set of observations, our global minimization problem is to find the set C of k clusters and a set of centroids $\mathbf{m}_1, \ldots, \mathbf{m}_k$ that would minimize $\text{ESS}(C, \mathbf{m}_1, \ldots, \mathbf{m}_k)$. Even for a fixed number (k) of clusters, it is difficult to find the global minimum, but we can find some local ones, and this is what the K-means method does.

For a given assignment C of observations into clusters, the minimum of $\text{ESS}(C, \mathbf{m}_1, \ldots, \mathbf{m}_k)$ is achieved by $\mathbf{m}_j$ equal to the cluster mean $\bar{\mathbf{x}}_j$ for each $j = 1, \ldots, k$ (the proof is given as Problem 10.6). When $\mathbf{m}_j = \bar{\mathbf{x}}_j$, the criterion $\text{ESS}(C, \mathbf{m}_1, \ldots, \mathbf{m}_k)$ becomes ESS defined in equation (10.18). So, the minimization criterion is essentially the same here as the one in the Ward method discussed in Section 10.3.4. The Ward method minimizes ESS over all clustering solutions that can be obtained by merging two out of $(k + 1)$ clusters obtained in the previous round of the algorithm. This allows building the hierarchical structure of the Ward method solution. The K-means method is not limited by those choices of new clusters. Instead, once the centroids are fixed, the algorithm reassigns the observations into potentially entirely new clusters in each round, so that $\text{ESS}(C, \mathbf{m}_1, \ldots, \mathbf{m}_k)$ is minimized. An easy way to do this is to assign each observation to the cluster defined by the closest centroid. If the observation were assigned to a different cluster, the longer distance to a different centroid would increase $\text{ESS}(C, \mathbf{m}_1, \ldots, \mathbf{m}_k)$. More specifically, the classic K-means algorithm follows these steps:

1. Identify k initial clusters.
2. Calculate centroids (mean vectors) for all k clusters (minimization of $\text{ESS}(C, \mathbf{m}_1, \ldots, \mathbf{m}_k)$ given C).
3. Reassign each observation into the cluster whose centroid is nearest (minimization of $\text{ESS}(C, \mathbf{m}_1, \ldots, \mathbf{m}_k)$ given $\mathbf{m}_1, \ldots, \mathbf{m}_k$).
4. Repeat Steps 2 and 3 until no more reassignments need to be made in Step 3.

We can say that the K-means algorithm finds the local minimum of $ESS(C, \mathbf{m}_1, \dots, \mathbf{m}_k)$, but not necessarily the global minimum. Since for a given clustering C, we would always prefer to use the means as centroids, the criterion $ESS(C, \mathbf{m}_1, \dots, \mathbf{m}_k)$ is essentially the same as ESS defined in equation (10.18), but the notation used here helps in a better understanding of the alternating minimization performed by the K-means algorithm.

Some implementations of the K-means algorithm attempt to "jump out" of the local minimum by investigating some other reassignments into clusters. For example, Hartigan and Wong (1979) implementation checks if a single switch of one observation from one cluster to another would reduce ESS. If yes, then the switch is made even though the observation was closer to a different cluster mean before the switch. Note that the reduction in ESS is possible only because the switch will change the cluster means as well (we change both C and $\mathbf{m}_1, \dots, \mathbf{m}_k$ in one step, rather than one of them at a time as in the classic implementation of the algorithm). The switched observation will be closer to its new cluster mean than to any other cluster mean (otherwise, it would not minimize ESS).

Another way to deal with suboptimal local minima is to run the K-means algorithm several times, each time with a different set of initial clusters in Step 1. We then choose the solution with the lowest ESS.

We still need to discuss how the initial clusters are identified in Step 1. This is often done by a random assignment of observations into clusters. An alternative solution is to run a different clustering algorithm and use its results as a starting point of the K-means algorithm. This is especially meaningful when the preprocessing algorithm is used for determination of the number of clusters k. For example, a hierarchical clustering algorithm can be run first. We can then use the hierarchy to help us in deciding on a suitable value of k. The resulting clusters are then used as the initial clusters in the K-means algorithm. One problem with this approach is that the hierarchical algorithms tend to be computationally intensive, while the advantage of the K-means algorithm is that it is cheaper computationally. So, a different strategy is needed for very large data sets.

For the initialization of the K-means algorithm, Step 1 can be skipped altogether as long as we specify the initial centroids in Step 2. One promising example of this approach is the K-means++ algorithm proposed by Arthur and Vassilvitskii (2007). For the set of observation vectors $\mathbf{x}_i$, $i = 1, \dots, n$, the algorithm finds the initial centroids $\mathbf{m}_j$, $j = 1, \dots, k$, by following these steps:

1. $\mathbf{m}_1$ is chosen randomly as one of the vectors $\mathbf{x}_i$, $i = 1, \dots, n$, each with the same probability of $1/n$.
2. For each $\mathbf{x}_i$, calculate the squared distance $d^2(\mathbf{x}_i)$ between $\mathbf{x}_i$ and the closest centroid that has already been chosen.
3. Calculate the weights $p_i = d^2(\mathbf{x}_i) / \sum_{j=1}^{n} d^2(\mathbf{x}_j)$.
4. Choose a new centroid randomly as one of vectors $\mathbf{x}_i$, $i = 1, \dots, n$, where each has the probability p_i of being chosen.
5. Repeat Steps 2, 3, and 4 until all k centroids are found.

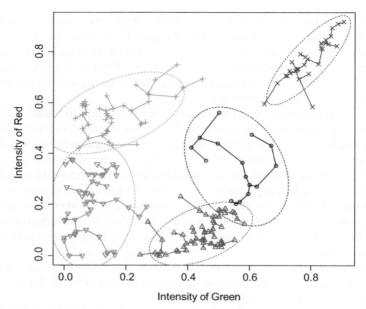

Figure 10.23 Five clusters from the K-means algorithm with the 90% probability ellipses characterizing the normal distributions of the clusters.

The algorithm favors the candidates for a new centroid that are far away from the previous centroids. However, due to its random nature, it is not locked into the farthest observations that could be outliers. This means that the K-means algorithm initialized by K-means $++$ algorithm still should be run several times, as usual, in order to deal with local minimum issue, but it is more likely to give a better initial configuration of centroids than the standard randomization.

Example 10.6 Here we continue the previous examples, where 196 pixels were clustered based on bivariate values of red and green color intensity. We first want to evaluate the previous five-cluster solutions in light of the error sum of squares ESS criterion given in formula (10.18). The ESS was calculated as 2.769 for the Ward method solution (not shown in a separate graph), 2.924 for the complete linkage (shown in Figure 10.19), and 2.727 for the average linkage (shown in Figure 10.22). The Ward and the average linkage solutions were different only by one observation assignment to a different cluster, and the average linkage happened to be slightly better based on the ESS criterion.

We then ran the Hartigan and Wong (1979) implementation of the K-means algorithm. When the initial clustering was the result of the Ward method or the complete linkage, the algorithm would converge to a clustering with ESS = 2.6321. When starting with the clustering result of the average linkage, there was only a slight improvement with ESS = 2.6320. The same solution was obtained as the best one when trying many random initializations. This solution is shown in Figure 10.23, and it seems to provide the global minimum for ESS in this data set, but we cannot be sure about it. These results are meant as examples of calculations that can be replicated by the reader for educational purposes rather than an evaluation of the performance of the algorithms in general. □

10.5 CLUSTERING VARIABLES

When performing cluster analysis of variables, we usually use the correlation coefficients as similarities as discussed in Section 10.2.2. These similarities can be transformed to a distance measure by using formula (10.3). Once we have a similarity or a dissimilarity measure, any of the linkage algorithms can be used. In order to use the K-means method discussed in Section 10.4, we would need to have vectors representing variables. This could be done by using an n-dimensional vector of all values of a given variable. However, this approach would not be useful in most applications. For example, two variables could be highly or even perfectly correlated, but magnitudes of their values could be very different, resulting in a large distance between the variables. This may or may not reflect the intended dissimilarity in a given application.

For binary variables (i.e., having only two possible values), one can simply treat variables as observations and observations as variables as discussed in Section 10.2.2. We can then use the similarity measures developed in Section 10.2.1 for binary data. In this context, the approach with using an n-dimensional vector of all values of a given variable would work because binary variables can become highly correlated only through a large number of matches of values. Note also that the Euclidean distance is meaningful here, and it is, in fact, related to one of the discussed similarities as shown in equation (10.11). This means that we could use the K-means method for binary variables. The centroids would not be binary variables, but this should not create any problems.

Example 10.7 In Example 9.2, we considered a 64 by 64 pixels image of grass texture in 42 spectral bands. As is usually the case for spectral images, the spectral bands, treated here as variables, are highly correlated. However, we may want to check which pairs of variables are more correlated with each other than with other variables. This might be helpful in band selection procedures, where one is interested in a subset of all spectral bands that would carry most of the information available from all bands. This is relevant for more efficient planning of specialized remote sensing devices. If a band is highly correlated with other bands, it may not be carrying much information other than what is already contained in the other bands.

The 42 by 42 correlation matrix was used as a similarity matrix. The similarities were then transformed to distances by using formula (10.3), and the three linkage algorithms were run. Figure 10.24 shows the dendrogram obtained with the single linkage algorithm. We can see that many spectral bands merge at the levels very close to zero, which is equivalent to the correlation coefficient close to one. Band 1 merges with any of the clusters very late in the process. It merges at the level of the distance

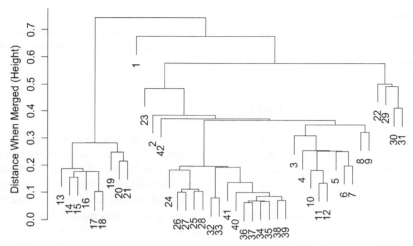

Figure 10.24   A dendrogram showing clusters of 42 spectral bands as obtained with the single linkage algorithm.

equal to 0.674. Since formula (10.3) can be solved for similarity to produce $a_{ij} = 1 - d_{ij}^2/2$, we conclude that the highest correlation coefficient of Band 1 with any other variable is $1 - (0.674)^2/2 = 0.773$. Bands 13–21 are somewhat different from all other bands. They merge with the remaining variables in the last round of the algorithm at the level of 0.7423, which means that the highest correlation between any of the Bands 13–21 and the remaining bands is 0.7245. The vertical axis in Figure 10.24 could show the values of the correlation coefficient directly to make these interpretations easier.

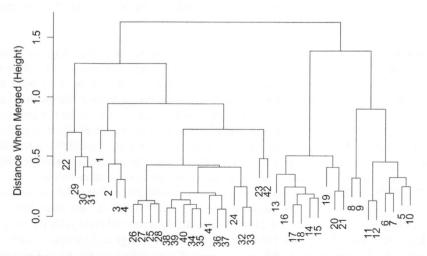

Figure 10.25   A dendrogram showing clusters of 42 spectral bands as obtained with the complete linkage algorithm.

The hierarchical clustering is very useful in this example because we are not looking for a particular number of clusters, but instead we are trying to understand the correlations among variables for further investigation. We also need to keep in mind that correlations describe the pairwise relations between pairs of variables without taking into account more complex multicollinearity structures. In further analysis, we may want to utilize canonical correlation analysis (see Chapter 8) to see if some bands already contain the information included in linear combinations of other variables.

The dendrogram based on the complete linkage algorithm is shown in Figure 10.25. Since this algorithm merges clusters based on the farthest distance, it can be interpreted as the worse-case scenario. Bands 13–21 again form a fairly tight cluster, and they merge with bands 5–12 at the level of 1.393. This means that the least correlated pair of bands, one from the range 13–21 and the other from 5–12, has the correlation coefficient of only 0.030, which practically means no correlation.

The results of the average linkage algorithm are not shown. They are more difficult to interpret, especially that the algorithm produces different results depending on whether the correlations are used directly as a similarity measure or the distances defined by formula (10.3) are used. □

10.6 FURTHER READING

Clustering, or unsupervised learning, is a broad topic and many other methods are discussed in Hastie et al. (2001) and Clarke et al. (2009). Clustering in the context of remote sensing is discussed in Canty (2010).

PROBLEMS

10.1. In the single linkage algorithm, the distance between clusters is calculated as the distance between the two closest elements from the two clusters. Show that this distance is equal to the one defined by formula (10.14).

10.2. Assume that a distance $d(\mathbf{x}, \mathbf{y})$ is defined for all pairs of objects $\mathbf{x}$ and $\mathbf{y}$. Define a new distance $d_{new}(\mathbf{x}, \mathbf{y}) = F(d(\mathbf{x}, \mathbf{y}))$, where $F(\cdot)$ is a strictly increasing function. Show that

   a. $d(\mathbf{x}_1, \mathbf{y}_1) < d(\mathbf{x}_2, \mathbf{y}_2)$ if and only if $d_{new}(\mathbf{x}_1, \mathbf{y}_1) < d_{new}(\mathbf{x}_2, \mathbf{y}_2)$.

   b. The clustering obtained by applying the single linkage algorithm with the distance $d(\mathbf{x}, \mathbf{y})$ is identical to the one obtained with the distance $d_{new}(\mathbf{x}, \mathbf{y})$.

   c. The clustering obtained by applying the complete linkage algorithm with the distance $d(\mathbf{x}, \mathbf{y})$ is identical to the one obtained with the distance $d_{new}(\mathbf{x}, \mathbf{y})$.

d. The clustering obtained by applying the average linkage algorithm with the distance $d(\mathbf{x}, \mathbf{y})$ is not necessarily the same as the one obtained with the distance $d_{\text{new}}(\mathbf{x}, \mathbf{y})$.

10.3. In the complete linkage algorithm, the distance between clusters is calculated as the distance between the two farthest elements from the two clusters. Show that this distance is equal to the one defined by formula (10.15).

10.4. In Example 10.7, we used a 64 by 64 pixels image of grass texture in 42 spectral bands. Use that data set to

   a. Run the average linkage algorithm to cluster variables using the correlations as a similarity measure (if your software accepts only dissimilarities, use $1 - r_{ij}$, where r_{ij} is the correlation; explain why that is equivalent to using r_{ij} as similarities).

   b. Run the average linkage algorithm to cluster variables using the distances defined as $d_{ij} = \sqrt{2(1 - r_{ij})}$ (see formula (10.3)). Are the results the same as those from part a? Explain why.

10.5. Prove that for a given set of points $\mathbf{x}_i$, $i = 1, \ldots, n$, the mean squared error of prediction $\text{MSE}(\mathbf{m}) = \sum_{i=1}^{n} \|\mathbf{x}_i - \mathbf{m}\|^2 / n$ is minimized by $\mathbf{m}$ equal to the sample mean $\bar{\mathbf{x}} = \sum_{i=1}^{n} \mathbf{x}_i / n$. This is a multivariate version of the property in Problem 4.5 written here for the sampling distribution.

10.6. Prove that for a given assignment C of observations into clusters, the minimum of $\text{ESS}(C, \mathbf{m}_1, \ldots, \mathbf{m}_k)$ given by formula (10.20) is achieved by $\mathbf{m}_j$ equal to the cluster mean $\bar{\mathbf{x}}_j$ for each $j = 1, \ldots, k$. Use the lemma written as Problem 10.5.

10.7. Assume that we constructed a minimum spanning tree for a given data set based on a distance matrix. We can find two clusters by removing the longest link from the tree. This results in the *first-level minimum forest*. In the second step, we can remove the second longest link from the forest, which will result in three clusters. We can continue this process until all links are removed and all observations form their own clusters. Prove that this process results in a clustering equivalent to the single linkage algorithm.

10.8. Show that formula (10.17) is equivalent to formula (10.19). *Hint*: Add and subtract the cluster mean inside $\| \cdot \|$ in formula (10.19).

10.9. Recreate Figure 10.3 and the results in Table 10.1. Give an example of an application where you would like to calculate distances of multivariate observations. Based on the results here, decide which L_m would be most appropriate for your application. Explain why.

10.10. For the Tile Data in the CIELAB color space (for both illumination conditions) as used in Example 10.3, perform clustering by using the average linkage algorithm. Create dendrograms and compare them to the dendrograms shown in Figure 10.16.

10.11. For the Tile Data in the CIELAB color space (for both illumination conditions) as used in Example 10.3, perform clustering by using the Ward method. Create dendrograms and compare them to the dendrograms shown in Figure 10.16.

10.12. Replicate the result of Example 10.6, including initialization of the K-means algorithm from the clustering of the other algorithms and also calculating the resulting error sums of squares ESS.

10.13. Use the Ward method to cluster the 196 pixels from Example 10.5. Create a dendrogram and then consider two cases:
   a. Find three clusters based on the dendrogram and create a plot analogous to Figure 10.21.
   b. Find five clusters based on the dendrogram and create a plot analogous to Figure 10.22.

APPENDIX A

Probability Distributions

A.1 INTRODUCTION

The purpose of this appendix is to define the most often used probability distributions and present some of their properties. We provide the probability mass function or a density function for each distribution, but a cumulative distribution function is provided only when it has a simple form. Otherwise it needs to be calculated as an integral of the density function, that is, $F(x) = \int_{-\infty}^{x} f(u)du$ for the continuous distribution or a summation for a discrete distribution. The values of the cumulative distribution function for the normal distribution are tabulated in Section A.18. We also show the notation used throughout this book for the upper percentiles of some of these distributions. The upper percentiles for some distributions are tabulated in Section A.18.

More details about these and some other distributions can be found in Forbes et al. (2010), Balakrishnan and Nevzorov (2003), and Johnson et al. (1994, 1995).

A.2 BETA DISTRIBUTION

The distribution of X is called the *standard beta distribution* with parameters $\alpha > 0$ and $\beta > 0$ if its probability density function is given by

$$f(x) = \frac{1}{B(\alpha, \beta)} x^{\alpha-1}(1 - x)^{\beta-1} \quad \text{for} \quad 0 \leq x \leq 1 \quad \text{(A.1)}$$

and zero otherwise, where $B(\alpha, \beta)$ is the beta function defined as follows:

$$B(\alpha, \beta) = \int_{0}^{1} u^{\alpha-1}(1 - u)^{\beta-1} \, du. \quad \text{(A.2)}$$

Statistics for Imaging, Optics, and Photonics, Peter Bajorski.
© 2012 John Wiley & Sons, Inc. Published 2012 by John Wiley & Sons, Inc.

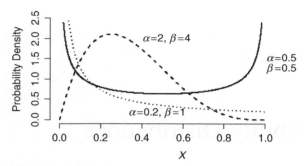

Figure A.1   Three examples of density functions of the standard beta distributions.

The kth moment of X (about zero) is given by

$$E(X^k) = \prod_{i=0}^{k-1} \frac{\alpha + i}{\alpha + \beta + i} = \frac{B(\alpha + k, \beta)}{B(\alpha, \beta)}.$$ (A.3)

We also have $E(X) = \alpha/(\alpha + \beta)$ and $\mathrm{Var}(X) = \alpha\beta(\alpha + \beta)^{-2}(\alpha + \beta + 1)^{-1}$. Three examples of beta density functions are shown in Figure A.1.

A random variable Y following a general *beta distribution* on an interval $[a, b]$ can be defined as $Y = a + (b - a)X$, that is, its probability density function is given by

$$f(x) = \frac{1}{B(\alpha, \beta)} (y - a)^{\alpha - 1} (b - y)^{\beta - 1} \quad \text{for} \quad a \le y \le b.$$ (A.4)

When $\beta = 1$, the distribution is sometimes called a power function distribution.

A.3   BINOMIAL DISTRIBUTION

Let us say, we observe a sequence of n independent trials with the outcomes of the trials represented by random variables X_i, $i = 1, \ldots, n$, such that $P(X_i = 1) = p$ and $P(X_i = 0) = 1 - p$, where $0 \le p \le 1$. The outcome $X_i = 1$ can be called a success and $X_i = 0$ can be called a failure. Such trials are called Bernoulli trials. The number of successes in n trials, equal to $X = \sum_{i=1}^{n} X_i$, follows the *binomial distribution* with parameters $0 \le p \le 1$ and $n \ge 1$. The probability mass function of the binomial distribution is given by

$$b(k; n, p) = \binom{n}{k} p^k (1 - p)^{n-k} \quad \text{for} \quad 0 \le k \le n$$ (A.5)

and zero otherwise, where $\binom{n}{k} = \frac{n!}{k!(n-k)!}$ is the number of combinations of size k from an n-element set. We have $E(X) = np$ and $\mathrm{Var}(X) = np(1 - p)$. When $n = 1$, the distribution is called a *Bernoulli distribution*.

A.4 CAUCHY DISTRIBUTION

The distribution of X is called the *standard Cauchy distribution* if its probability density function is given by

$$f(x) = \frac{1}{\pi(1 + x^2)} \quad \text{for} \quad x \in \mathbb{R}. \tag{A.6}$$

A random variable Y following a general *Cauchy distribution* can be defined as $Y = \theta + \beta X$, where θ is the location parameter and $\beta > 0$ is the scale parameter, that is, its probability density function is given by

$$f(y) = (\pi\beta)^{-1} \left[1 + \left(\frac{y - \theta}{\beta} \right)^2 \right]^{-1} \quad \text{for } y \in \mathbb{R} \tag{A.7}$$

and its cumulative distribution function is

$$F(y) = \frac{1}{2} + \pi^{-1} \arctan\left(\frac{y - \theta}{\beta} \right) \quad \text{for } y \in \mathbb{R}. \tag{A.8}$$

The Cauchy distribution has no finite moments. For example, the mean $E(X)$ does not exist. The Cauchy distribution is symmetric around its mode θ.

The Cauchy distribution has the following interpretation. Let us say, photons are sent from point A shown in Figure A.2 in random directions on the plane (uniformly with respect to the angle). Some of those photons reach the horizontal line L drawn in Figure A.2. Assume that (a) the photon reaches the line at a random point X, (b) the distance between A and the line is equal to β, and (c) the point on the line L that is closest to A has the coordinate θ. The angle γ is between the line connecting A with θ and the line of the photon (the angle is negative when $X < \theta$). Note that $\tan(\gamma) = (X - \theta)/\beta$. We can assume that γ is random, and it follows the uniform distribution on the interval $(-\pi/2, \pi/2)$ since for other angles the photon will not reach the line. The random variable X has the Cauchy distribution with parameters θ and β.

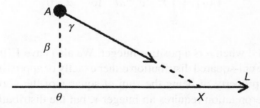

Figure A.2 A photon is sent from point A and it hits the line at point X.

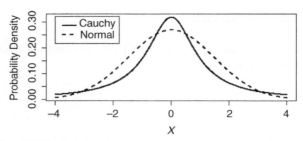

Figure A.3 The probability density function of the standard Cauchy distribution (solid line) in comparison to the normal distribution with $\sigma = 1/z_{0.25} = 1.4826$, where $z_{0.25}$ is the 25th upper percentile from the standard normal distribution.

The standard Cauchy distribution is the same as the t-distribution with one degree of freedom (see Section A.15). Figure A.3 shows the probability density function of the standard Cauchy distribution (solid line) in comparison to the normal distribution with $\sigma = 1/z_{0.25} = 1.4826$, where $z_{0.25}$ is the 25th upper percentile from the standard normal distribution. The two distributions have the same first and third quartiles equal to ± 1. The Cauchy distribution is more peaked than the normal distribution in the center, and it has heavier tails. The kurtosis cannot be calculated because the moments are not finite.

A.5   CHI-SQUARED DISTRIBUTION

Consider v independent random variables X_i, $i = 1, \ldots, v$, each following the standard normal distribution. Define their sum of squares as $X = \sum_{i=1}^{v} X_i^2$. The distribution of X is called the *chi-squared distribution* with parameter v called the number of degrees of freedom. Its probability density function is given by

$$f(x) = \frac{1}{2^{v/2}\Gamma(v/2)} x^{(v-2)/2} \exp(-x/2) \quad \text{for} \quad x \geq 0 \tag{A.9}$$

and zero otherwise, where $\Gamma(\cdot)$ is the gamma function defined by

$$\Gamma(\alpha) = \int_0^\infty t^{\alpha-1} e^{-t} \, dt \quad \text{for} \quad \alpha > 0 \tag{A.10}$$

and $\Gamma(n) = (n - 1)!$ when n is a positive integer. We also have $\Gamma(0.5) = \sqrt{\pi}$. Many statistics follow the chi-squared distribution either exactly or approximately, due to the earlier mentioned property of being the sum of squares of the normally distributed variables. This interpretation requires an integer v, but the distribution is defined for any $v > 0$.

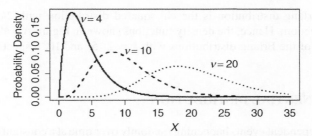

Figure A.4 Three examples of chi-squared density functions for various numbers of degrees of freedom ν.

The chi-squared distribution is a special case of the gamma distribution with the shape parameter $\alpha = \nu/2$ and the scale parameter $\beta = 1$. The kth moment of X (about zero) is given by

$$E(X^k) = 2^k \prod_{i=0}^{k-1} \left(i + \frac{\nu}{2} \right) = 2^k \frac{\Gamma(k + \nu/2)}{\Gamma(\nu/2)}. \qquad (A.11)$$

We also have $E(X) = \nu$ and $\mathrm{Var}(X) = 2\nu$. The coefficient of skewness is $\gamma_1 = 2^{3/2}\nu^{-1/2}$ and kurtosis is $\mathrm{Kurt}(X) = 3 + 12/\nu$. Three examples of chi-squared density functions are shown in Figure A.4. Based on the central limit theorem (see Section 2.7), the chi-squared distribution with large ν (degrees of freedom) can be approximated by the normal distribution.

The upper (100α)th percentile from the chi-squared distribution with ν degrees of freedom is denoted by $\chi_\nu^2(\alpha)$ and is tabulated in Section A.18.

A.6 ERLANG DISTRIBUTION

Consider k independent random variables X_i, $i = 1, \ldots, k$, each following the exponential distribution with the same scale parameter $\beta > 0$. Define their sum as $X = \sum_{i=1}^{k} X_i$. The distribution of X is called the Erlang distribution with the shape parameter $k \geq 1$ and the scale parameter $\beta > 0$. Its probability density function is given by

$$f(x) = \frac{(x/\beta)^{k-1}}{\beta(k-1)!} \exp(-x/\beta) \quad \text{for} \quad x \geq 0 \qquad (A.12)$$

and zero otherwise. The cumulative distribution function is given by

$$F(x) = 1 - \exp(-x/\beta) \sum_{i=0}^{k-1} \frac{(x/\beta)^i}{i!} \quad \text{for} \quad x \geq 0 \qquad (A.13)$$

and zero otherwise. The Erlang distribution is a special case of the gamma distribution, and some of its properties can be found in Section A.10. For the scale parameter

$\beta = 1$, the Erlang distribution is the chi-squared distribution with $2k$ number of degrees of freedom. Hence, the density functions shown in Figure A.4 also represent the densities of the Erlang distributions with $k = 2$, 5, and 10, respectively.

A.7 EXPONENTIAL DISTRIBUTION

Consider independent events happening randomly over time at a constant average rate as discussed in Section A.14 in the context of the Poisson distribution. In this context, the length of time between events follows the exponential distribution. The distribution of a random variable X is called the *exponential distribution* with the scale parameter $\beta > 0$ if its probability density function is given by

$$f(x) = \frac{1}{\beta} \exp(-x/\beta) \quad \text{for} \quad x \geq 0 \tag{A.14}$$

and zero otherwise (Figure A.5). The cumulative distribution function is given by

$$F(x) = 1 - \exp(-x/\beta) \quad \text{for} \quad x \geq 0 \tag{A.15}$$

and zero otherwise. The exponential distribution is sometimes parameterized by the rate parameter $\lambda = 1/\beta$. This is the only continuous distribution having the property of "lack of memory" that can be described as follows. If the waiting time between events is modeled by the exponential distribution, and we are waiting for an event that has not happened yet, the remaining waiting time follows the same exponential distribution. That is, our waiting time so far does not change the chances of the event happening, no matter how long we are already waiting.

The kth moment of X about zero is given by $E(X^k) = \beta^k k!$. We also have $E(X) = \beta$ and $\text{Var}(X) = \beta^2$. The coefficient of skewness is $\gamma_1 = 2$ and kurtosis is $\text{Kurt}(X) = 9$. The exponential distribution is a special case of the gamma distribution discussed in Section A.10.

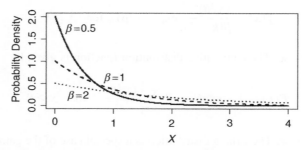

Figure A.5   Three examples of exponential density functions for various values of the scale parameter β.

Table A.1 Some Special Cases of the Exponential Power Distribution

Distribution	Shape Parameter α
Normal	$\alpha = 2$
Laplace	$\alpha = 1$
Uniform	$\alpha \to \infty$

A.8 EXPONENTIAL POWER DISTRIBUTION

The distribution of X is called the *exponential power distribution* (also called *general error distribution*) with the location parameter μ, the scale parameter $\beta > 0$, and the shape parameter $\alpha > 0$ if its probability density function is given by

$$f(x) = \frac{\alpha}{2\beta\,\Gamma(1/\alpha)} \exp\left[-\left(\frac{|x-\mu|}{\beta}\right)^{\alpha}\right] \quad \text{for} \quad x \in \mathbb{R}, \qquad (A.16)$$

where $\Gamma(\cdot)$ is the gamma function defined in equation (A.10). The distribution is symmetric with respect to the location parameter μ, which is also the distribution mean $E(X) = \mu$. The kth moment of X about the mean for even k is given by

$$E\left[(X-\mu)^{k}\right] = \beta^{k}\,\frac{\Gamma((k+1)/\alpha)}{\Gamma(1/\alpha)}. \qquad (A.17)$$

In particular, we have $\text{Var}(X) = \beta^{2}\Gamma(3/\alpha)/\Gamma(1/\alpha)$. For k odd, we have $E\left[(X-\mu)^{k}\right] = 0$ from the symmetry about μ.

Table A.1 shows some special cases of the exponential power distribution.

When the shape parameter α is less than or equal to 1, the exponential power density function has a peak at $x = 0$ and is not differentiable at that point. Figure A.6 shows three examples of such densities. The case of $\alpha = 1$ is the Laplace distribution. When $\alpha > 1$, the density has a derivative equal to zero at $x = 0$. Figure A.7 shows three examples of such density functions. The case of $\alpha = 2$ is the normal distribution. In both figures, the scale parameter β for each distribution is chosen so that the resulting

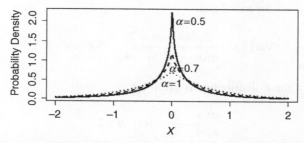

Figure A.6 Three examples of the exponential power density functions with the shape parameter $\alpha \leq 1$. The scale parameter β for each distribution is chosen so that the resulting variance is equal to 1 in each case.

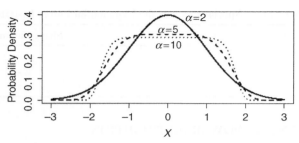

Figure A.7   Three examples of the exponential power density functions with the shape parameter $\alpha > 1$. The scale parameter β for each distribution is chosen so that the resulting variance is equal to 1 in each case.

variance is equal to 1 in each case. A comparison of the Laplace distribution to the normal distribution can be seen in Figure A.10.

A.9   F-DISTRIBUTION

Consider two independent random variables V and W following the chi-squared distributions with n and m degrees of freedom, respectively. The random variable

$$X = \frac{V/n}{W/m} \tag{A.18}$$

follows the F-distribution with n numerator degrees of freedom and m denominator degrees of freedom. This is a distribution of many statistics, where estimates of two different variances are being compared, each having a distribution proportional to the chi-squared distribution. The kth moment of X about the origin is finite only for $k < m/2$ and is given by

$$E[X^k] = \left(\frac{m}{n}\right)^k \frac{\Gamma(n/2 + k)\Gamma(m/2 - k)}{\Gamma(n/2)\Gamma(m/2)}, \tag{A.19}$$

where $\Gamma(\cdot)$ is the gamma function defined in equation (A.10). In particular, we have $E(X) = m/(m - 2)$ for $m > 2$ and

$$\text{Var}(X) = \frac{2m^2(n + m - 2)}{n(m - 2)^2(m - 4)} \quad \text{for} \quad m > 4. \tag{A.20}$$

The mode of the F-distribution is given by

$$\frac{m(n - 2)}{n(m + 2)} \quad \text{for} \quad n > 2, \tag{A.21}$$

which is always less than 1, but is close to 1 for large degrees of freedom n and m.

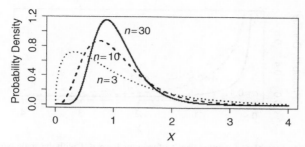

Figure A.8 Three densities of the F-distributions with various numerator degrees of freedom n and the fixed $m = 30$ denominator degrees of freedom.

Figure A.8 shows three densities of the F-distributions with various numerator degrees of freedom n shown in the graph and the fixed $m = 30$ denominator degrees of freedom. The (100α)th upper percentile from the F-distribution with n and m degrees of freedom is denoted by $F_{n,m}(\alpha)$ and is tabulated in Section A.18.

A.10 GAMMA DISTRIBUTION

The distribution of X is called the *gamma distribution* with the shape parameter $\alpha > 0$ and the scale parameter $\beta > 0$ if its probability density function is given by

$$f(x) = \frac{1}{\beta^{\alpha}\Gamma(\alpha)} x^{\alpha-1} \exp(-x/\beta) \quad \text{for} \quad x \geq 0 \tag{A.22}$$

and zero otherwise, where the gamma function $\Gamma(\alpha)$ is defined in (A.10). Table A.2 shows some distributions as special cases of the gamma distribution.

The kth moment of X about zero is given by

$$E(X^k) = \beta^k \frac{\Gamma(\alpha + k)}{\Gamma(\alpha)}. \tag{A.23}$$

We also have $E(X) = \alpha\beta$ and $\text{Var}(X) = \alpha\beta^2$. The coefficient of skewness is $\gamma_1 = 2\alpha^{-1/2}$ and kurtosis is $\text{Kurt}(X) = 3 + 6/\alpha$.

In order to understand the shapes of various gamma distributions, we can assume the scale parameter $\beta = 2$. For integer α, the gamma distribution is the same as the Erlang distribution, which in turn is the same as the chi-squared distribution (since $\beta = 2$) with even number of degrees of freedom v. The shapes of such distributions are shown in Figure A.4. For non-integer values of $\alpha > 1$, the shapes are similar to those shown in

Table A.2 **Some Special Cases of the Gamma Distribution**

Distribution	Shape Parameter α	Scale Parameter β
Chi-squared	$\alpha = v/2$	$\beta = 2$
Erlang	Any positive integer	Any β
Exponential	$\alpha = 1$	Any β

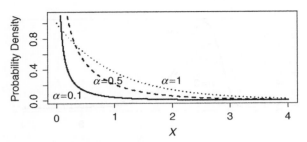

Figure A.9   Three examples of gamma density functions for $\alpha \le 1$ with the scale parameter $\beta = 1$.

Figure A.4. Note that for any $\alpha > 1$, the density at the point zero is equal to zero, that is, $f(0) = 0$. For $\alpha = 1$, that is, the exponential distribution, we have $f(0) = 1$ (or $f(0) = 1/\beta$ for an arbitrary $\beta > 0$), and for $\alpha < 1$, we have $\lim_{x \to 0+} f(x) = \infty$. Figure A.9 shows three examples of gamma density functions for $\alpha \le 1$.

Marchetti et al. (2002) discuss tests for testing the gamma distribution.

A.11   GEOMETRIC DISTRIBUTION

Let us say, we observe a sequence of Bernoulli trials (see Section A.3) and denote by X the number of failures before the first success. The random variable X follows the *geometric distribution*, which has the probability mass function given by

$$P(X = k) = p(1 - p)^k \quad \text{for integer} \quad k \ge 0. \tag{A.24}$$

The cumulative distribution function is given by

$$F(k) = P(X \le k) = 1 - (1 - p)^{k+1} \quad \text{for integer} \quad k \ge 0. \tag{A.25}$$

The mean is equal to $E(X) = (1 - p)/p$, and the variance is $\text{Var}(X) = (1 - p)/p^2$. The geometric distribution is the only discrete distribution having the property of "lack of memory" (the exponential distribution is the only continuous distribution with this property). That is, when we start our waiting time for a success (counted in the number of failures) at a given time, the distribution of that waiting time is always the same, independent of the events that happened up to that point.

A.12   LAPLACE DISTRIBUTION

The distribution of X is called the *Laplace distribution* (or double exponential distribution) with the location parameter μ and the scale parameter $\beta > 0$ if its probability density function is given by

$$f(x) = \frac{1}{2\beta} \exp\left[-\frac{|x - \mu|}{\beta} \right] \quad \text{for} \quad x \in \mathbb{R}, \tag{A.26}$$

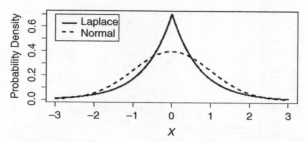

Figure A.10 The density function of the Laplace distribution with $\beta = 1/\sqrt{2}$ (so that its variance is 1) in comparison to the standard normal distribution.

where $\Gamma(\,\cdot\,)$ is the gamma function defined in equation (A.10). This is a special case of the exponential power distribution (see Section A.8). The distribution is symmetric with respect to the location parameter μ, which is also the distribution mean $E(X) = \mu$. The kth moment of X about the mean for even k is equal to $E\left[(X - \mu)^k\right] = k!\beta^k$. In particular, we have $\mathrm{Var}(X) = 2\beta^2$. For k odd, we have $E\left[(X - \mu)^k\right] = 0$ from the symmetry about μ. Figure A.10 shows the density function of the Laplace distribution with $\beta = 1/\sqrt{2}$ (so that its variance is 1) in comparison to the standard normal distribution.

A.13 NORMAL (GAUSSIAN) DISTRIBUTION

The distribution of X is called the *normal distribution* (or *Gaussian distribution*) with the location parameter μ and the scale parameter $\sigma > 0$ if its probability density function is given by

$$f(x) = \frac{1}{\sigma\sqrt{2\pi}} \exp\left[-\frac{1}{2}\left(\frac{x-\mu}{\sigma}\right)^2\right] \quad \text{for} \quad x \in \mathbb{R}. \qquad (A.27)$$

The important role of the normal distribution stems mainly from the central limit theorem (see Section 2.7). The normal distribution is symmetric with respect to the location parameter μ, which is also the distribution mean $E(X) = \mu$. The kth moment μ_k of X about the mean for even k is given by

$$\mu_k = E\left[(X - \mu)^k\right] = \sigma^k(k-1)(k-3)\cdots 3 \cdot 1, \qquad (A.28)$$

which gives a simple recursive formula $\mu_k = \sigma^2(k-1)\mu_{k-2}$ for even $k \geq 4$ and $\mu_2 = \mathrm{Var}(X) = \sigma^2$. For k odd, we have $E\left[(X - \mu)^k\right] = 0$ from the symmetry about μ.

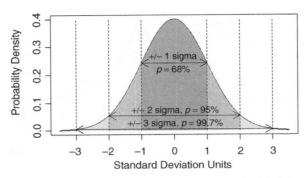

Figure A.11   The shape of the normal density function shown in the standard deviation units. One-, two-, and three-sigma rules are shown by highlighting areas under the normal density curve.

The normal distribution with $\mu = 0$ and $\sigma = 1$ is called the *standard normal distribution*. Its probability density function is denoted by $\varphi(\cdot)$, that is,

$$\varphi(x) = \frac{1}{\sqrt{2\pi}} \exp\left(-\frac{x^2}{2}\right) \quad \text{for} \quad x \in \mathbb{R}, \tag{A.29}$$

and its shape is shown in Figure A.11. The cumulative distribution function of the standard normal distribution is denoted by $\Phi(\cdot)$, and it needs to be calculated as an integral of $\varphi(\cdot)$. A related function is the *error function* defined as

$$\text{erf}(x) = \frac{2}{\sqrt{\pi}} \int_0^x \exp\left(-t^2\right) dt = 2\Phi\left(x\sqrt{2}\right) - 1. \tag{A.30}$$

We can also write

$$\Phi(x) = \frac{1}{2} + \frac{1}{2}\text{erf}\left(\frac{x}{\sqrt{2}}\right). \tag{A.31}$$

Some values of $\Phi(x)$ are tabulated for $x \geq 0$ in Table A.3 in Section A.18. For negative arguments, one can use the formula $\Phi(-x) = 1 - \Phi(x)$. The upper (100α)th percentile of the standard normal distribution is denoted by $z(\alpha)$, and its value can be read from Table A.3.

A.14   POISSON DISTRIBUTION

Consider independent events happening randomly over time at a constant average rate. In medical imaging, *quanta*, such as X-rays, electrons, or light photons, arrive randomly at a given rate (everything else being constant). In such processes, the number X of events (or arriving quanta) happening in a fixed time interval follows the *Poisson distribution* with the parameter $\lambda > 0$ (under some mild assumptions). Hence, the number of X-rays recorded in a given image pixel follows the Poisson distribution. This is a discrete distribution with the probability mass function defined by

$$P(X = k) = \frac{\lambda^k}{k!} e^{-\lambda} \quad \text{for any integer} \quad k \geq 0. \tag{A.32}$$

The mean $E(X)$ is equal to λ, which is the average rate of the events. We also have the variance $\mathrm{Var}(X) = \lambda$, which tells us how the variability changes with the change in magnitude. For example, the relative noise in an X-ray pixel can be measured by the *coefficient of variation* (COV) calculated as

$$\mathrm{COV} = \frac{\mathrm{StDev}(X)}{E(X)} = \frac{\sqrt{\lambda}}{\lambda} = \frac{1}{\sqrt{\lambda}}, \tag{A.33}$$

which means that the relative noise gets smaller for a stronger signal with larger λ. This can also be expressed by the *signal-to-noise ratio* (SNR) calculated as[4]

$$\mathrm{SNR} = \frac{E(X)}{\mathrm{StDev}(X)} = \frac{\lambda}{\sqrt{\lambda}} = \sqrt{\lambda}, \tag{A.34}$$

which gets larger for a stronger signal. For large λ, the Poisson distribution can be approximated by the normal distribution.

A.15 t (STUDENT'S) DISTRIBUTION

Consider two independent random variables Y and W following the standard normal distribution and the chi-squared distribution with v degrees of freedom, respectively. The random variable

$$X = \frac{Y}{\sqrt{W/v}} \tag{A.35}$$

follows the *t-distribution* with v degrees of freedom. This is a distribution of many statistics, where a normally distributed random variable is standardized by an estimator of the standard deviation, which in turn has a distribution proportional to the square root of the chi-squared distribution. The density function of the t-distribution is given by

$$f(x) = \frac{\Gamma((v+1)/2)}{\sqrt{\pi v}\,\Gamma(v/2)} \left(1 + \frac{x^2}{v}\right)^{-(v+1)/2} \quad \text{for} \quad x \in \mathbb{R}, \tag{A.36}$$

where $\Gamma(\cdot)$ is the gamma function defined in equation (A.10). The shapes of the density functions with various degrees of freedom are shown in Figure A.12. The t

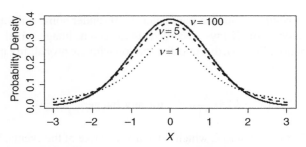

Figure A.12   The t-distribution density functions with various v degrees of freedom.

density with $v = 100$ is fairly close to the standard normal density function, and the two would overlap if plotted in Figure A.12. The difference between the two distributions can be assessed based on Figure 3.6.

The t-distribution is symmetric with respect to zero, which is also the distribution mean $E(X) = 0$ for $v > 1$. For k even, the kth moment μ_k of X about zero is infinite for $k \geq v$, and for $k < v$, it is given by

$$\mu_k = E\left[X^k\right] = \frac{(k - 1)(k - 3) \cdots 3 \cdot 1}{(v - 2)(v - 4) \cdots (v - k)} v^{k/2}, \tag{A.37}$$

which gives a recursive formula $\mu_k = v(k - 1)\mu_{k-2}/(v - k)$ for even $k \geq 4$ such that $k < v$. We also have $\mu_2 = \text{Var}(X) = v/(v - 2)$ for $v > 2$. For k odd, we have $E\left[X^k\right] = 0$ when $k < v$. The upper (100α)th percentile of the t-distribution with v degrees of freedom is denoted by $t_v(\alpha)$ and is tabulated in Section A.18.

The square X^2 of a t distributed random variable follows an F-distribution with 1 and v degrees of freedom as can be seen by comparing equations (A.35) and (A.18).

A.16   UNIFORM (RECTANGULAR) CONTINUOUS DISTRIBUTION

The distribution of X is called the *uniform* (or *rectangular*) distribution on the interval $[a, b]$ if its probability density function is constant on that interval, that is,

$$f(x) = \frac{1}{b - a} \quad \text{for} \quad a \leq x \leq b \tag{A.38}$$

and zero otherwise. The cumulative distribution function is given by $F(x) = (x - a)/(b - a)$ for $a \leq x \leq b$. According to the uniform distribution, each value in the interval $[a, b]$ is equally likely to occur. The distribution is symmetric with respect to the mean $E(X) = (a + b)/2$. The kth central moment of X is equal to $E\left[(X - E(X))^k\right] = [(b - a)/2]^k/(k + 1)$ for even k. In particular, we have

$\mathrm{Var}(X) = (b - a)^2/12$. For k odd, we have $E\left[(X - E(X))^k\right] = 0$ from the symmetry about $E(X)$.

A.17 WEIBULL DISTRIBUTION

The distribution of X is called the *two-parameter Weibull distribution* with the shape parameter $\alpha > 0$ and the scale parameter $\beta > 0$ if its probability density function is given by

$$f(x) = \frac{\alpha x^{\alpha-1}}{\beta^\alpha} \exp\left[-\left(\frac{x}{\beta}\right)^\alpha\right] \quad \text{for} \quad x \geq 0 \qquad (A.39)$$

and zero otherwise. Its cumulative distribution function is

$$F(x) = 1 - \exp\left[-\left(\frac{x}{\beta}\right)^\alpha\right] \quad \text{for} \quad x \geq 0. \qquad (A.40)$$

The kth moment of X about zero is equal to $E[X^k] = \beta^k \Gamma[1 + (k/\alpha)]$ for all integer k. In particular, we have $E(X) = \beta \cdot \Gamma[1 + (1/\alpha)]$, and the variance can be calculated as

$$\mathrm{Var}(X) = \beta^2 \left\{ \Gamma\left(1 + \frac{2}{\alpha}\right) - \left[\Gamma\left(1 + \frac{1}{\alpha}\right)\right]^2 \right\}. \qquad (A.41)$$

A power transformation $Y = (X/\beta)^\alpha$ generates Y following the standard exponential distribution. Consequently, the Weibull distribution with the shape parameter $\alpha = 1$ is the exponential distribution.

Note that for any $\alpha > 1$, the density at the point zero is equal to zero, that is, $f(0) = 0$. For $\alpha = 1$, that is, the exponential distribution, we have $f(0) = 1/\beta$, and for

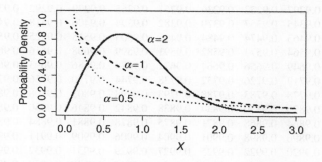

Figure A.13 Three Weibull density functions for various values of α and the scale parameter $\beta = 1$.

$\alpha < 1$, we have $\lim_{x \to 0^+} f(x) = \infty$. Figure A.13 shows three examples of Weibull density functions for various values of α and the scale parameter $\beta = 1$.

A.18   TABLES OF DISTRIBUTIONS

Here we provide tables for the following distributions:

- Normal distribution (Table A.3).
- Chi-squared distribution (Table A.4).
- t-distribution (Table A.5).
- F-distribution (Table A.6).

Table A.3   Areas Under the Standard Normal Curve, That is, Values of the Standard Normal Cumulative Distribution Function Given by $\Phi(z) = \int_{-\infty}^{z} (1/\sqrt{2\pi}) \exp(-t^2/2)\, dt$

z	0.00	0.01	0.02	0.03	0.04	0.05	0.06	0.07	0.08	0.09
0.0	0.5000	0.5040	0.5080	0.5120	0.5160	0.5199	0.5239	0.5279	0.5319	0.5359
0.1	0.5398	0.5438	0.5478	0.5517	0.5557	0.5596	0.5636	0.5675	0.5714	0.5753
0.2	0.5793	0.5832	0.5871	0.5910	0.5948	0.5987	0.6026	0.6064	0.6103	0.6141
0.3	0.6179	0.6217	0.6255	0.6293	0.6331	0.6368	0.6406	0.6443	0.6480	0.6517
0.4	0.6554	0.6591	0.6628	0.6664	0.6700	0.6736	0.6772	0.6808	0.6844	0.6879
0.5	0.6915	0.6950	0.6985	0.7019	0.7054	0.7088	0.7123	0.7157	0.7190	0.7224
0.6	0.7257	0.7291	0.7324	0.7357	0.7389	0.7422	0.7454	0.7486	0.7517	0.7549
0.7	0.7580	0.7611	0.7642	0.7673	0.7704	0.7734	0.7764	0.7794	0.7823	0.7852
0.8	0.7881	0.7910	0.7939	0.7967	0.7995	0.8023	0.8051	0.8078	0.8106	0.8133
0.9	0.8159	0.8186	0.8212	0.8238	0.8264	0.8289	0.8315	0.8340	0.8365	0.8389
1.0	0.8413	0.8438	0.8461	0.8485	0.8508	0.8531	0.8554	0.8577	0.8599	0.8621
1.1	0.8643	0.8665	0.8686	0.8708	0.8729	0.8749	0.8770	0.8790	0.8810	0.8830
1.2	0.8849	0.8869	0.8888	0.8907	0.8925	0.8944	0.8962	0.8980	0.8997	0.9015
1.3	0.9032	0.9049	0.9066	0.9082	0.9099	0.9115	0.9131	0.9147	0.9162	0.9177
1.4	0.9192	0.9207	0.9222	0.9236	0.9251	0.9265	0.9279	0.9292	0.9306	0.9319
1.5	0.9332	0.9345	0.9357	0.9370	0.9382	0.9394	0.9406	0.9418	0.9429	0.9441
1.6	0.9452	0.9463	0.9474	0.9484	0.9495	0.9505	0.9515	0.9525	0.9535	0.9545
1.7	0.9554	0.9564	0.9573	0.9582	0.9591	0.9599	0.9608	0.9616	0.9625	0.9633
1.8	0.9641	0.9649	0.9656	0.9664	0.9671	0.9678	0.9686	0.9693	0.9699	0.9706
1.9	0.9713	0.9719	0.9726	0.9732	0.9738	0.9744	0.9750	0.9756	0.9761	0.9767
2.0	0.9772	0.9778	0.9783	0.9788	0.9793	0.9798	0.9803	0.9808	0.9812	0.9817
2.1	0.9821	0.9826	0.9830	0.9834	0.9838	0.9842	0.9846	0.9850	0.9854	0.9857
2.2	0.9861	0.9864	0.9868	0.9871	0.9875	0.9878	0.9881	0.9884	0.9887	0.9890
2.3	0.9893	0.9896	0.9898	0.9901	0.9904	0.9906	0.9909	0.9911	0.9913	0.9916
2.4	0.9918	0.9920	0.9922	0.9925	0.9927	0.9929	0.9931	0.9932	0.9934	0.9936
2.5	0.9938	0.9940	0.9941	0.9943	0.9945	0.9946	0.9948	0.9949	0.9951	0.9952

Table A.4   The Upper (100α)th Percentiles $\chi^2_v(\alpha)$ from the Chi-Squared Distribution with v Degrees of Freedom

d.f. v	α Values									
	0.995	0.990	0.975	0.950	0.900	0.100	0.050	0.025	0.010	0.005
1	0.0000	0.0002	0.0010	0.0039	0.016	2.706	3.841	5.024	6.635	7.879
2	0.0100	0.0201	0.0506	0.1026	0.211	4.605	5.991	7.378	9.210	10.597
3	0.072	0.115	0.216	0.352	0.584	6.251	7.815	9.348	11.345	12.838
4	0.207	0.297	0.484	0.711	1.064	7.779	9.488	11.143	13.277	14.860
5	0.412	0.554	0.831	1.145	1.610	9.236	11.070	12.833	15.086	16.750
6	0.676	0.872	1.237	1.635	2.204	10.645	12.592	14.449	16.812	18.548
7	0.989	1.239	1.690	2.167	2.833	12.017	14.067	16.013	18.475	20.278
8	1.34	1.65	2.18	2.73	3.49	13.36	15.51	17.53	20.09	21.95
9	1.73	2.09	2.70	3.33	4.17	14.68	16.92	19.02	21.67	23.59
10	2.16	2.56	3.25	3.94	4.87	15.99	18.31	20.48	23.21	25.19
11	2.60	3.05	3.82	4.57	5.58	17.28	19.68	21.92	24.72	26.76
12	3.07	3.57	4.40	5.23	6.30	18.55	21.03	23.34	26.22	28.30
13	3.57	4.11	5.01	5.89	7.04	19.81	22.36	24.74	27.69	29.82
14	4.07	4.66	5.63	6.57	7.79	21.06	23.68	26.12	29.14	31.32
15	4.60	5.23	6.26	7.26	8.55	22.31	25.00	27.49	30.58	32.80
16	5.14	5.81	6.91	7.96	9.31	23.54	26.30	28.85	32.00	34.27
17	5.70	6.41	7.56	8.67	10.09	24.77	27.59	30.19	33.41	35.72
18	6.26	7.01	8.23	9.39	10.86	25.99	28.87	31.53	34.81	37.16
19	6.84	7.63	8.91	10.12	11.65	27.20	30.14	32.85	36.19	38.58
20	7.43	8.26	9.59	10.85	12.44	28.41	31.41	34.17	37.57	40.00
21	8.03	8.90	10.28	11.59	13.24	29.62	32.67	35.48	38.93	41.40
22	8.64	9.54	10.98	12.34	14.04	30.81	33.92	36.78	40.29	42.80
23	9.26	10.20	11.69	13.09	14.85	32.01	35.17	38.08	41.64	44.18
24	9.89	10.86	12.40	13.85	15.66	33.20	36.42	39.36	42.98	45.56
25	10.52	11.52	13.12	14.61	16.47	34.38	37.65	40.65	44.31	46.93
30	13.79	14.95	16.79	18.49	20.60	40.26	43.77	46.98	50.89	53.67
40	20.71	22.16	24.43	26.51	29.05	51.81	55.76	59.34	63.69	66.77
50	27.99	29.71	32.36	34.76	37.69	63.17	67.50	71.42	76.15	79.49
60	35.53	37.48	40.48	43.19	46.46	74.40	79.08	83.30	88.38	91.95
70	43.28	45.44	48.76	51.74	55.33	85.53	90.53	95.02	100.43	104.21
80	51.17	53.54	57.15	60.39	64.28	96.58	101.88	106.63	112.33	116.32
90	59.20	61.75	65.65	69.13	73.29	107.57	113.15	118.14	124.12	128.30
100	67.33	70.06	74.22	77.93	82.36	118.50	124.34	129.56	135.81	140.17

Table A.5   The Upper (100α)th Percentiles $t_v(\alpha)$ from the t-Distribution with v Degrees of Freedom

d.f. v	α Values								
	0.250	0.100	0.050	0.025	0.010	0.00833	0.00625	0.005	0.0025
1	1.000	3.078	6.314	12.706	31.821	38.190	50.923	63.657	127.321
2	0.816	1.886	2.920	4.303	6.965	7.649	8.860	9.925	14.089
3	0.765	1.638	2.353	3.182	4.541	4.857	5.392	5.841	7.453
4	0.741	1.533	2.132	2.776	3.747	3.961	4.315	4.604	5.598
5	0.727	1.476	2.015	2.571	3.365	3.534	3.810	4.032	4.773
6	0.718	1.440	1.943	2.447	3.143	3.287	3.521	3.707	4.317
7	0.711	1.415	1.895	2.365	2.998	3.128	3.335	3.499	4.029
8	0.706	1.397	1.860	2.306	2.896	3.016	3.206	3.355	3.833
9	0.703	1.383	1.833	2.262	2.821	2.933	3.111	3.250	3.690
10	0.700	1.372	1.812	2.228	2.764	2.870	3.038	3.169	3.581
11	0.697	1.363	1.796	2.201	2.718	2.820	2.981	3.106	3.497
12	0.695	1.356	1.782	2.179	2.681	2.779	2.934	3.055	3.428
13	0.694	1.350	1.771	2.160	2.650	2.746	2.896	3.012	3.372
14	0.692	1.345	1.761	2.145	2.624	2.718	2.864	2.977	3.326
15	0.691	1.341	1.753	2.131	2.602	2.694	2.837	2.947	3.286
16	0.690	1.337	1.746	2.120	2.583	2.673	2.813	2.921	3.252
17	0.689	1.333	1.740	2.110	2.567	2.655	2.793	2.898	3.222
18	0.688	1.330	1.734	2.101	2.552	2.639	2.775	2.878	3.197
19	0.688	1.328	1.729	2.093	2.539	2.625	2.759	2.861	3.174
20	0.687	1.325	1.725	2.086	2.528	2.613	2.744	2.845	3.153
21	0.686	1.323	1.721	2.080	2.518	2.601	2.732	2.831	3.135
22	0.686	1.321	1.717	2.074	2.508	2.591	2.720	2.819	3.119
23	0.685	1.319	1.714	2.069	2.500	2.582	2.710	2.807	3.104
24	0.685	1.318	1.711	2.064	2.492	2.574	2.700	2.797	3.091
25	0.684	1.316	1.708	2.060	2.485	2.566	2.692	2.787	3.078
26	0.684	1.315	1.706	2.056	2.479	2.559	2.684	2.779	3.067
27	0.684	1.314	1.703	2.052	2.473	2.552	2.676	2.771	3.057
28	0.683	1.313	1.701	2.048	2.467	2.546	2.669	2.763	3.047
29	0.683	1.311	1.699	2.045	2.462	2.541	2.663	2.756	3.038
30	0.683	1.310	1.697	2.042	2.457	2.536	2.657	2.750	3.030
40	0.681	1.303	1.684	2.021	2.423	2.499	2.616	2.704	2.971
60	0.679	1.296	1.671	2.000	2.390	2.463	2.575	2.660	2.915
120	0.677	1.289	1.658	1.980	2.358	2.428	2.536	2.617	2.860
∞	0.674	1.282	1.645	1.960	2.326	2.394	2.498	2.576	2.813

Table A.6 The Upper 5th Percentiles $F_{n,m}(0.05)$ from the F-Distribution with n and m Degrees of Freedom

m	\multicolumn{17}{c}{n}
	1	2	3	4	5	6	7	8	9	10	12	15	20	25	30	40	60
1	161.4	199.5	215.7	224.6	230.2	234.0	236.8	238.9	240.5	241.9	243.9	245.9	248.0	249.3	250.1	251.1	252.2
2	18.51	19.00	19.16	19.25	19.30	19.33	19.35	19.37	19.38	19.40	19.41	19.43	19.45	19.46	19.46	19.47	19.48
3	10.13	9.55	9.28	9.12	9.01	8.94	8.89	8.85	8.81	8.79	8.74	8.70	8.66	8.63	8.62	8.59	8.57
4	7.71	6.94	6.59	6.39	6.26	6.16	6.09	6.04	6.00	5.96	5.91	5.86	5.80	5.77	5.75	5.72	5.69
5	6.61	5.79	5.41	5.19	5.05	4.95	4.88	4.82	4.77	4.74	4.68	4.62	4.56	4.52	4.50	4.46	4.43
6	5.99	5.14	4.76	4.53	4.39	4.28	4.21	4.15	4.10	4.06	4.00	3.94	3.87	3.83	3.81	3.77	3.74
7	5.59	4.74	4.35	4.12	3.97	3.87	3.79	3.73	3.68	3.64	3.57	3.51	3.44	3.40	3.38	3.34	3.30
8	5.32	4.46	4.07	3.84	3.69	3.58	3.50	3.44	3.39	3.35	3.28	3.22	3.15	3.11	3.08	3.04	3.01
9	5.12	4.26	3.86	3.63	3.48	3.37	3.29	3.23	3.18	3.14	3.07	3.01	2.94	2.89	2.86	2.83	2.79
10	4.96	4.10	3.71	3.48	3.33	3.22	3.14	3.07	3.02	2.98	2.91	2.85	2.77	2.73	2.70	2.66	2.62
11	4.84	3.98	3.59	3.36	3.20	3.09	3.01	2.95	2.90	2.85	2.79	2.72	2.65	2.60	2.57	2.53	2.49
12	4.75	3.89	3.49	3.26	3.11	3.00	2.91	2.85	2.80	2.75	2.69	2.62	2.54	2.50	2.47	2.43	2.38
13	4.67	3.81	3.41	3.18	3.03	2.92	2.83	2.77	2.71	2.67	2.60	2.53	2.46	2.41	2.38	2.34	2.30
14	4.60	3.74	3.34	3.11	2.96	2.85	2.76	2.70	2.65	2.60	2.53	2.46	2.39	2.34	2.31	2.27	2.22
15	4.54	3.68	3.29	3.06	2.90	2.79	2.71	2.64	2.59	2.54	2.48	2.40	2.33	2.28	2.25	2.20	2.16
16	4.49	3.63	3.24	3.01	2.85	2.74	2.66	2.59	2.54	2.49	2.42	2.35	2.28	2.23	2.19	2.15	2.11
17	4.45	3.59	3.20	2.96	2.81	2.70	2.61	2.55	2.49	2.45	2.38	2.31	2.23	2.18	2.15	2.10	2.06
18	4.41	3.55	3.16	2.93	2.77	2.66	2.58	2.51	2.46	2.41	2.34	2.27	2.19	2.14	2.11	2.06	2.02
19	4.38	3.52	3.13	2.90	2.74	2.63	2.54	2.48	2.42	2.38	2.31	2.23	2.16	2.11	2.07	2.03	1.98
20	4.35	3.49	3.10	2.87	2.71	2.60	2.51	2.45	2.39	2.35	2.28	2.20	2.12	2.07	2.04	1.99	1.95
21	4.32	3.47	3.07	2.84	2.68	2.57	2.49	2.42	2.37	2.32	2.25	2.18	2.10	2.05	2.01	1.96	1.92
22	4.30	3.44	3.05	2.82	2.66	2.55	2.46	2.40	2.34	2.30	2.23	2.15	2.07	2.02	1.98	1.94	1.89
23	4.28	3.42	3.03	2.80	2.64	2.53	2.44	2.37	2.32	2.27	2.20	2.13	2.05	2.00	1.96	1.91	1.86
24	4.26	3.40	3.01	2.78	2.62	2.51	2.42	2.36	2.30	2.25	2.18	2.11	2.03	1.97	1.94	1.89	1.84
25	4.24	3.39	2.99	2.76	2.60	2.49	2.40	2.34	2.28	2.24	2.16	2.09	2.01	1.96	1.92	1.87	1.82

(continued)

347

Table A.6 (*Continued*)

m									n								
	1	2	3	4	5	6	7	8	9	10	12	15	20	25	30	40	60
26	4.23	3.37	2.98	2.74	2.59	2.47	2.39	2.32	2.27	2.22	2.15	2.07	1.99	1.94	1.90	1.85	1.80
27	4.21	3.35	2.96	2.73	2.57	2.46	2.37	2.31	2.25	2.20	2.13	2.06	1.97	1.92	1.88	1.84	1.79
28	4.20	3.34	2.95	2.71	2.56	2.45	2.36	2.29	2.24	2.19	2.12	2.04	1.96	1.91	1.87	1.82	1.77
29	4.18	3.33	2.93	2.70	2.55	2.43	2.35	2.28	2.22	2.18	2.10	2.03	1.94	1.89	1.85	1.81	1.75
30	4.17	3.32	2.92	2.69	2.53	2.42	2.33	2.27	2.21	2.16	2.09	2.01	1.93	1.88	1.84	1.79	1.74
40	4.08	3.23	2.84	2.61	2.45	2.34	2.25	2.18	2.12	2.08	2.00	1.92	1.84	1.78	1.74	1.69	1.64
60	4.00	3.15	2.76	2.53	2.37	2.25	2.17	2.10	2.04	1.99	1.92	1.84	1.75	1.69	1.65	1.59	1.53
120	3.92	3.07	2.68	2.45	2.29	2.18	2.09	2.02	1.96	1.91	1.83	1.75	1.66	1.60	1.55	1.50	1.43
∞	3.84	3.00	2.60	2.37	2.21	2.10	2.01	1.94	1.88	1.83	1.75	1.67	1.57	1.51	1.46	1.39	1.32

m = denominator degrees of freedom; n = numerator degrees of freedom.

APPENDIX B

Data Sets

B.1 INTRODUCTION

This appendix provides some background information and descriptions of data sets used throughout the book. The data set text files are available at

http://people.rit.edu/~pxbeqa/ImagingStat

where more details about data format are available.

B.2 PRINTING DATA

Printer manufacturers want to ensure high consistency of printing by their devices. There are various types of calibrations and tests that can be done on a printer. One of them is to print a page of random color patches such as those shown in Figure 1.3. The pattern of patches is chosen randomly, but only once, that is, the same pattern is typically used by a given manufacturer. The patches are in four basic colors of the CMYK color model used in printing: cyan, magenta, yellow, and black. In a given color, there are several gradations, from the maximum amount of ink to less ink, where the patch has a lighter color if printed on a white background. For a given gradation of color, there are several patches across the page printed in that color gradation (exactly eight patches for the test prints used here).

Printing data used here are a subset of a larger data set collected in an experiment, where a printer was calibrated several times and pages were printed between calibrations. The printer was also kept idle at various times. For the subset used here, three pages were printed immediately after the printer calibration. The printer was then idle for 14 h, and a set of 30 pages was printed, of which only 18 pages were utilized. This gives us a total of 21 pages, which were then measured by a scanning spectrophotometer. We use only the measurements of the eight cyan patches per page

---

at maximum color gradation. For each patch, the reflectances were recorded in 31 spectral bands in the visible spectrum. The bands are in the spectral range from 400 to 700 nm at 10 nm increments. The reflectance is the power of the light reflected by a surface divided by the power of the light incident upon the surface. The 31 reflectances define a reflectance spectrum, or a spectral reflectance curve. Visual density is a measure of print quality that is calculated from the reflectance spectrum.

Printing Spectra data set consists of 31 spectral reflectances for the eight patches for each of the 21 pages. For each patch, visual density was calculated and the values are stored as the Printing Density data set consisting of eight values for the eight cyan patches for each of the 21 pages.

B.3 FISH IMAGE DATA

Wold et al. (2006) describe a multispectral imaging near-infrared tranflectance system developed for online determination of crude chemical composition of highly hetero-geneous food and other biomaterials. The transflection measures the light penetrating the sample as opposed to reflectance that measures only the light reflected from the sample surface. The transflection is then well suited for nonhomogeneous materials that are not well characterized by simply observing their surface. The near-infrared tranflectance system was used for moisture determination of dried salted coalfish (bacalao). One of the multispectral images used in Wold et al. (2006) is used here. This is an image of fish on a conveyer belt. There are 45 pixels along the width of the conveyer belt and 1194 pixels along its length, for a total of 53,730 pixels. For each pixel, we have the transflected light intensity values for 15 near-infrared spectral bands. The values are stored in Fish Image data set. An average of those 15 spectral values was calculated for each image pixel and plotted in Figure 2.6.

B.4 EYE TRACKING DATA

Eye tracking devices are used to examine people's eye movements as people perform certain tasks (see Pelz et al. (2000)). This information is used in research on the human visual system, in psychology, in product design, and in many other applications. In eye tracking experiments, a lot of data are collected. In order to reduce the amount of data, fixation periods are identified when a shopper fixes her gaze at one spot. In a data collection effort described in Kinsman et al. (2010), 760 fixation images were identified. Here we use only one such image shown in Figure 1.1. The cross in the image shows the spot the shopper is looking at. This 128 by 128 pixel image was recorded with a camcorder in the RGB (red, green, and blue) channels. For each pixel, the intensity values (ranging from 0 to 1) for the three colors are given. This means that each pixel is represented by a mixture of the three colors. The Eye Tracking data set gives the RGB values for all 16,384 pixels from the image shown in Figure 1.1.

In Chapter 10, a subset of the Eye Tracking image is used. The subset is given in the file "Eye_Tracking_Subset.txt" with 196 rows representing the pixels chosen at random. The first column gives the pixel number within the whole image of 16,384 pixels. The next three columns are the intensity values in the RGB (red, green, and blue) channels, respectively.

B.5   LANDSAT DATA

The Landsat Program is a series of Earth-observing satellite missions jointly managed by NASA and the U.S. Geological Survey since 1972. Due to the long-term nature of the program, there is a significant interest in the long-term calibration of the results, so that measurements taken at different times can be meaningfully compared. One approach to this calibration problem is discussed by Anderson (2010). As part of the approach, Landsat measurements of a fixed desert site were collected. The desert site was confirmed to be sufficiently stable over time, so that the changes in measurements can be attributed to a drift of the measuring instrument, except for some factors such as the Sun position in the sky. The Landsat data set consists of 76 rows for observations taken at different times. There are eight variables given in columns. The first three columns are the day of the year, the solar elevation angle, and the solar azimuth angle. The next five columns are spectral reflectances in Bands 1–4 and then Band 7.

B.6   OPTICAL FIBER EXPERIMENTS

Two experiments were performed in order to find out how much power is lost when sending laser light signals through optical fiber. In both experiments, a laser light signal was sent from one end of optical fiber, and the output power was measured at the other end. The input power and the output power were recorded in mW.

In the first experiment, five pieces of 100 m length of optical fiber were tested, as described in Example 2.1. The resulting data are shown in Table 2.1 and are available as the Fibers Experiment 1 data. In the second experiment, one piece of 100 m length of optical fiber was tested at several levels of input power. The resulting data are available as the Optical Fiber Experiment 2 data given in the order of test runs.

B.7   SPECTROMETER DATA

An experiment was designed in order to investigate a potential drift or trend in spectrometer readings over time. Three tiles were chosen for the experiment—a white, a gray, and a black tile coded as 1, 2, and 3, respectively. Two spectrometers of the same type were chosen and were coded as 1 and 2, respectively. Two operators performed the measurements and were also coded. The first three columns in Spectrometer Data show the operator, spectrometer, and tile numbers for the 24 experimental runs (given in rows in time order) performed in the experiment. The subsequent 31 columns give the reflectance values in the 31 spectral bands from 400 to 700 nm at 10 nm increments.

B.8   TILES DATA

Spectral reflectance of 12 tiles in the BCRA II Series Calibration tiles was measured using an X-Rite Series 500 Spectrodensitometer. Each row in Tiles Data consists of 31 values of reflectance measured in 31 spectral bands over the spectral range from 400 to 700 nm at 10 nm increments. Each of the 12 tiles was measured four times for a total of 48 multivariate observations. Table 5.1 shows a subset of the whole data set. Each row represents an observation, with the first four rows representing the four repeated measurements of the first tile, followed by four measurements of the second tile, and so on. Each column represents one spectral band. The colors of the 12 tiles are shown in Table 5.2 and are given in the file Tile_Color_Names.txt.

B.9   PRINT-ON-DEMAND DATA

Phillips et al. (2010) describe an experiment to evaluate quality of print-on-demand books provided by various online vendors. Sixteen observers rated overall image quality of six print-on-demand books on a scale from 1 to 5 (with ratings being $1 =$ very low satisfaction, $5 =$ very high satisfaction). Those ratings are given as the first six columns in the Print-on-Demand Data. The observers were also asked how much they would be willing to pay for this quality of book as a memento of the observer's vacation. For each observer, the six prices in dollars for the six books are given in Columns 7–12. The final 13th column is the age of the observer.

B.10   MARKER DATA

In biomedical applications, radiopaque markers are used to observe motion of internal organ such as a heart. The marker is implanted into the body and then observed using X-rays. Here, we consider a simplified scenario of an implanted radiopaque marker, which is monitored by an orthogonal projection on two X-ray screens. Figure 7.1a shows a simplified two-dimensional scenario, where the X-ray screens are shown along the normalized vectors v_1 and v_2, and the 150 dots represent measurements collected over a 5-minute interval (once every 2 seconds). Marker Data contain the physical standard coordinates f_1 and f_2 of all 150 points. The unit length vectors v_1 and v_2 are given as $v_1 = [10, 1]/\sqrt{101}$ and $v_2 = [1, 5]/\sqrt{26}$.

B.11   AVIRIS DATA

A general introduction to remote sensing data can be found in Example 1.3. Here we introduce data from the Airborne Visible/Infrared Imaging Spectrometer (AVIRIS), which is a sensor collecting spectral radiance in the range of wavelengths from 400 to 2500 nm. It has been flown on various aircraft platforms, and many images of the Earth's surface are available. Figure 7.10 shows a 100 by 100 pixel AVIRIS image of an urban area in Rochester, NY, near the Lake Ontario shoreline. The scene has a wide range of natural and man-made material including a mixture of commercial/warehouse and residential neighborhoods, which adds a wide range of spectral diversity. Prior to processing, invalid bands (due to atmospheric water absorption) were removed, reducing the overall dimensionality to 152 bands. This image has been used in Bajorski et al. (2004) and Bajorski (2011a, 2011b). The first 152 values in the AVIRIS Data represent the spectral radiance values (a spectral curve) for the top left pixel in the image shown in Figure 7.10. This is followed by spectral curves of the pixels in the first row, followed by the next row, and so on.

B.12 HyMap Cooke City Data

This is also remote sensing data (see the previous section). The HyMap Cooke City image shown in Figure 7.15 has 280 by 800 pixels, where each pixel is described by a 126-band spectrum. More information about the image can be found in Snyder et al. (2008). The data set is available as the self-test image on the web site http://dirsapps. cis.rit.edu/blindtest/.

B.13 GRASS DATA

This is a spectral image of grass texture. Each pixel is represented by a spectral reflectance curve in 42 spectral bands with reflectance given in percent. Grass 64 by 64 data set describes a 64 by 64 pixel image of grass texture used in Chapter 9. The image in Figure 9.2 shows the areas of diseased and healthy grass. Denote by i the columns in that image. The rows are denoted by j, but they are counted from the bottom of the image rather than from the top. With this notation, the area of diseased grass considered in Example 9.2 is defined as all pixels with (i, j) coordinates such that $i = 49, 50, \ldots, 64$ and j goes from $[64 - 2(i - 49)]$ to 64. There are 256 pixels with distressed grass in total.

In Example 9.6, three groups of grass pixels are introduced. We define here the exact location of Groups 2 and 3. Using indexes i and j, Group 3 is identified as all pixels with (i, j) coordinates such that $i = 55, 56, \ldots, 64$ and j goes from $[64 - 2(i - 55)]$ to 64. There are 100 pixels in Group 3. Group 2 is identified as all pixels with (i, j) coordinates such that $i = 45, 46, \ldots, 64$ and j goes from $[64 - 2(i - 45)]$ to 64, but those that are not in Group 3. There are 300 pixels in Group 2.

A small 15 by 15 pixel subimage of grass texture is used in Chapter 8 and is provided as the Grass 15 by 15 data set.

B.14 ASTRONOMY DATA

Here, we describe a subset of infrared astronomy data used in Kastner et al. (2008), where one can find further references. There are 179 stars or star-like objects in our

Astronomy Data. The first column provides object number as used in the Large Magellanic Cloud (LMC) survey conducted by the Midcourse Space Experiment (MSX). The next three columns give infrared magnitudes obtained for those objects in the J (1.25 μm), H (1.65 μm), and K (2.17 μm) bands from the Two-Micron All-Sky Survey. The fifth column is the A (8.3 μm) band magnitude obtained from the MSX survey. The last column is the object's classification used in Kastner et al. (2008). The red supergiants (RSG) are coded as 1. Code 2 denotes the carbon-rich asymptotic giant branch (C AGB) stars, which are dying, sun-like stars (red giants). Code 4 denotes the so-called "H II regions," which are plasmas ionized by hot, massive young stars that are still deeply embedded in the molecular clouds out of which they were formed. The oxygen-rich asymptotic giant branch (O AGB) stars are coded as 5.

B.15   CIELAB DATA

This data set is based on the Tile Data discussed in Section B.8. For each tile, four spectral curve measurements were given in the Tile Data. An average of those four spectral curves was calculated as spectrum representing a given tile. Based on that spectrum, three-dimensional CIELAB color space coordinates were calculated. This scale describes a given color with three numbers. The L^* coordinate describes color lightness with the maximum value of 100 and the minimum of zero representing black. The remaining two coordinates a^* and b^* have no specific numerical limits. The negative a^* values indicate green, while positive values indicate red. The negative b^* values indicate blue, while positive values indicate yellow. The three-dimensional color space of L^*, a^*, and b^* values is approximately uniform in the sense that the perceptual difference between two colors is well approximated by the Euclidean distance between the two colors.

The measured reflectance spectrum of a given surface, such as a tile here, tells us the fraction of light that is reflected in various wavelengths. However, if little light at a given wavelength is illuminated at the surface, then not much can be reflected, even if the reflectance in high. This is why the color perception also depends on the light illuminated at the surface. Hence, the calculation of CIELAB color space coordinates based on the reflectance spectrum also depends on the illuminant used. For example, the colors may look different in daylight than under artificial light indoors. Here we used two illuminants, one representing the noon daylight with overcast sky (Illuminant D_{65}) and the other representing the incandescent or tungsten light source found in homes (Illuminant A).

The CIELAB data set consists of 12 rows representing 12 tiles with numbers listed in the first column. The second column gives the names of the tiles' colors. This is followed by three columns of L^*, a^*, and b^* values calculated based on Illuminant D_{65}. The next three columns give L^*, a^*, and b^* values calculated based on Illuminant A.

APPENDIX C

Miscellanea

C.1 SINGULAR VALUE DECOMPOSITION

In Supplement 4A, we define a spectral decomposition of a symmetric square matrix. A related decomposition for a more general matrix is defined by the following theorem.

Theorem C.1 Let $\mathcal{X}$ be an n by p matrix of real numbers. Then there exist an n by p matrix $\mathbf{U}$ with orthogonal columns (i.e., $\mathbf{U}^{\mathrm{T}}\mathbf{U} = \mathbf{I}_p$), a p by p diagonal matrix $\mathbf{\Delta}$ with nonnegative elements, and a p by p orthogonal matrix $\mathbf{V}$ such that

$$\mathcal{X} = \mathbf{U}\mathbf{\Delta}\mathbf{V}^T. \tag{C.1}$$

The above decomposition of $\mathcal{X}$ is called the *singular value decomposition*. The diagonal elements δ_j of $\mathbf{\Delta}$ are called singular values. We have the following properties.

1. The squares δ_j^2 of the singular values are the eigenvalues of $\mathcal{X}^{\mathrm{T}}\mathcal{X}$ and the corresponding eigenvectors are the columns of the matrix $\mathbf{V}$.
2. The squares δ_j^2 of the singular values are the eigenvalues of $\mathcal{X}\mathcal{X}^{\mathrm{T}}$ and the corresponding eigenvectors are the columns of the matrix $\mathbf{U}$.

When the singular value decomposition is applied to a symmetric (square) matrix, we obtain its spectral decomposition discussed in Supplement 4A. A more typical application of the singular value decomposition is on the matrix $\mathcal{X}$ of a p-dimensional data set consisting of n observations. In that case, the diagonal elements δ_j describe the variability of the data around zero. Since in statistics, we are usually interested in the variability around the mean, the singular value decomposition is often performed on the centered data, that is, on $\mathcal{X}_{\mathrm{c}} = \mathcal{X} - \mathbf{1}_n \cdot \bar{\mathbf{x}}^T$, where $\bar{\mathbf{x}} = (1/n)\mathcal{X}^{\mathrm{T}}\mathbf{1}_n$ is the vector of the column means and $\mathbf{1}_n$ is an n-dimensional

---

Statistics for Imaging, Optics, and Photonics, Peter Bajorski.
© 2012 John Wiley & Sons, Inc. Published 2012 by John Wiley & Sons, Inc.

vector with all coordinates equal to 1. Assume now the singular value decomposition of the centered data, that is,

$$\mathscr{X}_c = \mathbf{U}\boldsymbol{\Lambda}\mathbf{V}^\mathsf{T}. \tag{C.2}$$

Many calculations are easier to perform and are more computationally stable when using the above decomposition. For example, the sample variance–covariance matrix can be calculated as

$$\mathbf{S} = \frac{1}{n-1}\mathbf{V}\boldsymbol{\Lambda}^2\mathbf{V}^\mathsf{T}, \tag{C.3}$$

and its inverse and square root matrices are

$$\mathbf{S}^{-1} = (n-1)\mathbf{V}\boldsymbol{\Lambda}^{-2}\mathbf{V}^\mathsf{T}, \quad \mathbf{S}^{1/2} = \frac{1}{\sqrt{n-1}}\mathbf{V}\boldsymbol{\Lambda}\mathbf{V}^\mathsf{T}, \quad \mathbf{S}^{-1/2} = \sqrt{n-1}\cdot\mathbf{V}\boldsymbol{\Lambda}^{-1}\mathbf{V}^\mathsf{T}. \tag{C.4}$$

As an example of some other useful formulas, consider a task of calculating the Mahalanobis distances of all p-dimensional observations given as rows in an n by p matrix $\mathscr{X}$ from the mean vector $\bar{\mathbf{x}}$. This is equivalent to calculating a diagonal of the n by n matrix $\mathscr{X}_c\mathbf{S}^{-1}\mathscr{X}_c^\mathsf{T}$, which can be written as

$$\mathscr{X}_c\mathbf{S}^{-1}\mathscr{X}_c^\mathsf{T} = (n-1)\mathbf{U}\mathbf{U}^\mathsf{T}. \tag{C.5}$$

Since n is often large, we would like to avoid calculating the whole matrix, if only the diagonal of that matrix is needed. This can be done by using the following formula:

$$\mathrm{diag}\left(\mathscr{X}_c\mathbf{S}^{-1}\mathscr{X}_c^\mathsf{T}\right) = (n-1)\mathrm{diag}\left(\mathbf{U}\mathbf{U}^\mathsf{T}\right) = (n-1)(\mathbf{U}*\mathbf{U})\mathbf{1}_p, \tag{C.6}$$

where * stands for the element-by-element multiplication of matrices and the multiplication by $\mathbf{1}_p$ results in the summation of the row elements (so, it would typically be achieved by a summation in a computer procedure). In a more general setting, assume that we have two n by p matrices $\mathbf{A}$ and $\mathbf{B}$, and the task is to calculate the diagonal elements of $\mathbf{A}\mathbf{B}^\mathsf{T}$. We can then use the formula

$$\mathrm{diag}\left(\mathbf{A}\mathbf{B}^\mathsf{T}\right) = (\mathbf{A}*\mathbf{B})\mathbf{1}_p. \tag{C.7}$$

The singular value decomposition (of the centered data) shown in (C.2) can also be used in principal component analysis. The columns of the matrix $\mathbf{V}$ are the eigenvectors of the sample variance–covariance matrix $\mathbf{S}$. The matrix $\mathbf{V}$ is denoted as $\mathbf{P}$ in Chapter 7. If the vector of principal components is denoted as $\mathbf{Y}$, the sample variance–covariance matrix of $\mathbf{Y}$ can be calculated as

$$\widehat{\mathrm{Var}}(\mathbf{Y}) = \frac{1}{n-1}\boldsymbol{\Lambda}^2, \tag{C.8}$$

and the estimated variances of principal components are

$$\widehat{\text{Var}}(Y_j) = \lambda_j = \frac{\delta_j^2}{n-1},$$ (C.9)

where λ_j are the eigenvalues of **S**.

C.2 IMAGING RELATED SAMPLING SCHEME

In Section 7.6.2, we introduce Sampling Scheme C that is based on the linear mixing model describing the materials seen in a spectral image and the image noise. Here we give a justification for the correction coefficients b_j used in the Sampling Scheme C. In order to simplify our considerations, let us assume that the system of coordinates was shifted by $\bar{\mathbf{x}}$ and then rotated (by the matrix of eigenvectors), so that the values of the centered PCs are the coordinates of the observation vectors. It means that the (rotated) image spectra are the realizations of the random vector $(Y_1, \ldots, Y_p)$, where Y_j is the jth PC. The Sampling Scheme C assumes a deterministic structure within the subspace generated by the first k PCs. However, we also want to make sure that the random vector ε_i has some nontrivial components within that subspace. This is why we assumed the variance of the noise to be λ_{k+1} in the first k PC directions. In the linear mixing model (7.32), the a_{ij}'s are considered nonrandom. However, when performing PCA on the global covariance matrix, the resulting PCs measure the variability, as if a_{ij}'s were realizations of some random variables. In that sense, model (7.32) is conditional on the values a_{ij}. In the simplified notation of the random vector $(Y_1, \ldots, Y_p)$, the realizations of PCs Y_j are the equivalents of the coefficients a_{ij}. Let us now construct the random variable $Z_j = b_j Y_j + E_j$, where $b_j = \sqrt{1 - \lambda_{k+1}/\lambda_j}$, and E_j is independent of Y_j and normally distributed $N(0, \lambda_{k+1})$. The variable $T_j = b_j Y_j$ will form the jth coordinate of the deterministic part of the model (in the conditional sense) and E_j will be the jth coordinate of the error. For this model to be consistent with the original data, we want to have the unconditional variance of Z_j to be equal to λ_j. Based on the conditional variance formula, we have

$$\text{Var}(Z_j) = E(\text{Var}(Z_j|T_j)) + \text{Var}(E(Z_j|T_j))$$

$$= E(\lambda_{k+1}) + \text{Var}(T_j) = \lambda_{k+1} + b_j^2 \lambda_j = \lambda_j.$$ (C.10)

This justifies the use of the coefficients $b_j = \sqrt{1 - \lambda_{k+1}/\lambda_j}$.

C.3 APPROACHES TO CANNONICAL CORRELATION ANALYSIS

In Theorem 8.1, canonical variables are defined with the help of the matrix $\mathbf{G} = \Sigma_{XX}^{-1/2} \Sigma_{XY} \Sigma_{YY}^{-1} \Sigma_{XY}^{\text{T}} \Sigma_{XX}^{-1/2}$. On the other hand, some other sources, for example,

Schott (2007), use the matrix $\mathbf{G}^* = \mathbf{\Sigma}_{XX}^{-1}\mathbf{\Sigma}_{XY}\mathbf{\Sigma}_{YY}^{-1}\mathbf{\Sigma}_{XY}^{T}$ instead. We want to show here that the two approaches are essentially equivalent. In Section 8.2, we define the ith canonical variable as

$$U_i = \mathbf{e}_i^{T}\mathbf{\Sigma}_{XX}^{-1/2}\mathbf{X},\tag{C.11}$$

where $\mathbf{e}_i$ is the normalized eigenvector of $\mathbf{G}$ with an associated eigenvalue ρ_i^2. An alternative approach is to define

$$U_i^* = \mathbf{w}_i^{T}\mathbf{X},\tag{C.12}$$

where $\mathbf{w}_i$ is the normalized eigenvector of $\mathbf{G}^*$. In order to compare the two approaches, we need the following lemma.

Lemma C.1 The vector $\mathbf{\Sigma}_{XX}^{-1/2}\mathbf{e}_i$ is an eigenvector of $\mathbf{G}^*$ with the associated eigenvalue ρ_i^2.

Proof. $\mathbf{G}^*\left(\mathbf{\Sigma}_{XX}^{-1/2}\mathbf{e}_i\right) = \mathbf{\Sigma}_{XX}^{-1}\mathbf{\Sigma}_{XY}\mathbf{\Sigma}_{YY}^{-1}\mathbf{\Sigma}_{XY}^{T}\mathbf{\Sigma}_{XX}^{-1/2}\mathbf{e}_i = \mathbf{\Sigma}_{XX}^{-1/2}\mathbf{G}\mathbf{e}_i = \mathbf{\Sigma}_{XX}^{-1/2}\rho_i^2\mathbf{e}_i = \rho_i^2\left(\mathbf{\Sigma}_{XX}^{-1/2}\mathbf{e}_i\right).$ $\square$

Hence, if $\mathbf{w}_i$ is taken as $\mathbf{\Sigma}_{XX}^{-1/2}\mathbf{e}_i$, we get exactly the same solution in both cases. However, $\mathbf{w}_i$ is often taken as the normalized eigenvector. In that case, U_i^* is a scaled version of U_i, and its variance is not equal to 1. The variance of this "not-quite-canonical" variable U_i^* is

$$\left\|\mathbf{\Sigma}_{XX}^{-1/2}\mathbf{e}_i\right\|^{-2} = \left(\mathbf{e}_i^{T}\mathbf{\Sigma}_{XX}^{-1}\mathbf{e}_i\right)^{-1}.$$

We can also calculate it as

$$\mathrm{Var}\left(U_i^*\right) = \mathbf{w}_i^{T}\mathbf{\Sigma}_{XX}\mathbf{w}_i.$$

In a similar fashion, we can define

$$V_i^* = \mathbf{z}_i^{T}\mathbf{X}\tag{C.13}$$

as a scaled version of V_i, where $\mathbf{z}_i$ is the normalized eigenvector of the matrix $\mathbf{\Sigma}_{YY}^{-1}\mathbf{\Sigma}_{YX}\mathbf{\Sigma}_{XX}^{-1}\mathbf{\Sigma}_{YX}^{T}$. The resulting pairs $\left(U_i^*, V_i^*\right)$ have the same canonical correlations (see formula (8.9)) and the same properties (8.12) of canonical variables, except that their variances are not equal to 1.

C.4 CRITICAL VALUES FOR THE RYAN–JOINER TEST OF NORMALITY

The following table gives the c_α critical values for the Ryan–Joiner test of normality defined in Section 3.6.4.

Sample Size	α		
	0.10	0.05	0.01
5	0.9026	0.8793	0.8260
6	0.9106	0.8886	0.8379
7	0.9177	0.8974	0.8497
8	0.9240	0.9052	0.8605
9	0.9294	0.9120	0.8701
10	0.9340	0.9179	0.8786
11	0.9381	0.9230	0.8861
12	0.9417	0.9276	0.8928
13	0.9449	0.9316	0.8987
14	0.9477	0.9352	0.9040
15	0.9503	0.9384	0.9088
16	0.9526	0.9413	0.9132
17	0.9547	0.9439	0.9171
18	0.9566	0.9463	0.9207
19	0.9583	0.9484	0.9240
20	0.9599	0.9504	0.9270
21	0.9614	0.9523	0.9297
22	0.9627	0.9540	0.9323
23	0.9640	0.9556	0.9347
24	0.9652	0.9571	0.9369
25	0.9663	0.9584	0.9390
26	0.9673	0.9597	0.9409
27	0.9683	0.9609	0.9427
28	0.9692	0.9620	0.9444
29	0.9700	0.9631	0.9460
30	0.9708	0.9641	0.9475
31	0.9716	0.9651	0.9489
32	0.9723	0.9660	0.9503
33	0.9730	0.9668	0.9516
34	0.9736	0.9676	0.9528
35	0.9742	0.9684	0.9539
36	0.9748	0.9691	0.9550
37	0.9754	0.9698	0.9560
38	0.9759	0.9705	0.9570
39	0.9764	0.9711	0.9580
40	0.9769	0.9717	0.9589
41	0.9774	0.9723	0.9598
42	0.9778	0.9728	0.9606
43	0.9782	0.9734	0.9614
44	0.9786	0.9739	0.9621
45	0.9790	0.9744	0.9629

Sample Size	α		
	0.10	0.05	0.01
46	0.9794	0.9748	0.9636
47	0.9798	0.9753	0.9642
48	0.9801	0.9757	0.9649
49	0.9805	0.9762	0.9655
50	0.9808	0.9766	0.9661
51	0.9811	0.9770	0.9667
52	0.9814	0.9773	0.9673
53	0.9817	0.9777	0.9678
54	0.9820	0.9781	0.9683
55	0.9823	0.9784	0.9688
56	0.9825	0.9787	0.9693
57	0.9828	0.9791	0.9698
58	0.9831	0.9794	0.9703
59	0.9833	0.9797	0.9707
60	0.9835	0.9800	0.9711
61	0.9838	0.9802	0.9716
62	0.9840	0.9805	0.9720
63	0.9842	0.9808	0.9724
64	0.9844	0.9810	0.9728
65	0.9846	0.9813	0.9731
66	0.9848	0.9815	0.9735
67	0.9850	0.9818	0.9738
68	0.9852	0.9820	0.9742
69	0.9854	0.9822	0.9745
70	0.9856	0.9825	0.9748
71	0.9857	0.9827	0.9752
72	0.9859	0.9829	0.9755
73	0.9861	0.9831	0.9758
74	0.9862	0.9833	0.9761
75	0.9864	0.9835	0.9764
80	0.9871	0.9844	0.9777
90	0.9884	0.9859	0.9799
100	0.9894	0.9872	0.9818
200	0.9943	0.9931	0.9904
300	0.9960	0.9952	0.9934
400	0.9969	0.9964	0.9950
600	0.9979	0.9975	0.9966
800	0.9984	0.9981	0.9974
1000	0.9987	0.9985	0.9979
2000	0.9993	0.9992	0.9989

C.5   LIST OF ABBREVIATIONS AND MATHEMATICAL SYMBOLS

$n!$	n factorial
$\lvert x \rvert$	absolute value
$\lVert \mathbf{x} \rVert$	Euclidean norm
$\lVert \mathbf{x} \rVert_m$	L_m Minkowski norm
$\mathbf{X}^\mathrm{T}$	transpose operation of a matrix $\mathbf{X}$
$\lvert \mathbf{A} \rvert$	determinant of a square matrix $\mathbf{A}$
$\overline{X}$	sample mean
$\overline{\overline{\mathbf{x}}}$	overall sample mean
$\displaystyle\bigcup_{i=1}^{k} A_i$	union of k sets
$\displaystyle\bigcap_{i=1}^{k} A_i$	intersection of k sets
$A \cap B$	intersection of two sets
$B^c = \mathcal{S} \setminus B$	complement of the set B
$\widehat{\mu}_{1\cdot}$	the dot indicates an average over that index
$\mathbf{1}_n$	an n-dimensional vector with all coordinates equal to 1
$\chi_v^2(\alpha)$	the upper (100α)th percentile from the chi-squared distribution with v degrees of freedom
ε	error term in a model
$(\lambda_i, \mathbf{e}_i)$	pair of an eigenvalue and a normalized eigenvector
π_i	the ith population (in Chapter 9 on classification)
Φ	cumulative distribution function of the standard normal distribution
φ	density function of the standard normal distribution
θ	general notation for an arbitrary parameter
$\widehat{\theta}$	estimator of the parameter θ
Σ	population variance–covariance matrix
σ	population standard deviation (as a parameter)
η_p	$(100p)$th percentile of a distribution
$(\tau\beta)_{jk}$	a term for an interaction between two factors with main effects denoted by τ and β
μ	population mean (as a parameter)
Γ	gamma function
γ_1	coefficient of skewness
γ_2	excess kurtosis
APER	apparent error rate
CCA	canonical correlation analysis
CCR	canonical correlation regression
CDF	cumulative distribution function

$\text{Cov}(X, Y)$	covariance of two random variables
$\text{Cov}(\mathbf{X}, \mathbf{Y})$	covariance matrix of two random vectors
$\mathbf{D}$	diagonal matrix of variances
$\mathbf{d}$	deviation vector (in Chapter 5)
$d(\mathbf{x}, \mathbf{y})$	distance between vectors $\mathbf{x}$ and $\mathbf{y}$
$E(X)$	expected value of X
$\mathbf{E}_i$	the ith residual vector (in Chapter 7)
e_i	the ith residual (in Chapter 4)
ECM	expected cost of misclassification
EER	estimated error rate
ESS	error sum of squares (for clusters)
ETE	estimated test error
$\exp(x) = e^x$	exponential function with the base $e \approx 2.71$
F	cumulative distribution function
F_n	empirical cumulative distribution function
$F_{n,m}(\alpha)$	the (100α)th upper percentile from the F-distribution with n and m degrees of freedom
f	probability density function
GSV	generalized sample variance
$\mathbf{H}$	hat matrix (in Chapter 4)
h_{ii}	the ith diagonal element of the hat matrix $\mathbf{H}$
$\mathbf{I}$	identity matrix
kGSV	k-dimensional generalized sample variance
$\text{Kurt}(X)$	kurtosis
L_m	Minkowski metric
L_∞	Chebyshev distance
$\ln(x)$	natural log of x
MLE	maximum likelihood estimator
MSE	mean squared error
MVU	minimum variance unbiased (estimator)
$N(\mu, \sigma^2)$	normal distribution with the mean μ and variance σ^2
$N_p(\boldsymbol{\mu}, \boldsymbol{\Sigma})$	p-dimensional normal distribution with the mean vector $\boldsymbol{\mu}$ and the variance–covariance matrix $\boldsymbol{\Sigma}$
n	usually the sample size
$P_\mathbf{v}(\mathbf{x})$	projection of $\mathbf{x}$ on $\mathbf{v}$
$P(A|B)$	probability of A given B
p	number of random components or variables
PCA	principal component analysis
$\mathbb{R}$	set of real numbers
$\mathbf{R}$	sample correlation matrix
R_i	the ith classification region (in Chapter 9)
R^2	coefficient of determination
r	sample correlation coefficient
RSS	residual sum of squares
$\mathcal{S}$	sample space

S	sample variance–covariance matrix
S$_{\text{pooled}}$	pooled estimated variance–covariance matrix
s, s^2	sample standard deviation and variance
s_{jk}	sample covariance between the jth and kth variables
SE$\left(\widehat{\theta}\right)$	standard error of the estimator $\widehat{\theta}$
SS$_{\text{Regr}}$	regression sum of squares
SS$_{\text{Res}}$	residual sum of squares
SS$_{\text{Total}}$	total sum of squares
StDev(X)	standard deviation of X
$t_v(\alpha)$	the upper (100α)th percentile of the t-distribution with v degrees of freedom
TPM	total probability of misclassification
TVR	total variability in residuals
Var(X)	variance of X
Var$(\mathbf{X})$	variance–covariance matrix of a random vector
$\mathscr{X}$	matrix X
$\mathbf{X} = \left[X_1, X_2, \ldots, X_p\right]^{\text{T}}$	random vector of p components
Z	standardized variable
$z(\alpha)$	the upper (100α)th percentile of the standard normal distribution

References

Abe, S. (2005). *Support Vector Machines for Pattern Classification*. Springer.

Agresti, A. (2002). *Categorical Data Analysis*, 2nd edition. Wiley.

Anderson, C.R. (2010). "Refinement of the Method for Using Pseudo-Invariant Sites for Long Term Calibration Trending of Landsat Reflective Bands." Ph.D. Thesis, Rochester Institute of Technology.

Anderson, T.W. (2003). *An Introduction to Multivariate Statistical Analysis*, 3rd edition. Wiley.

Antony, J. (2003). *Design of Experiments for Engineers and Scientists*. Butterworth–Heinemann.

Arthur, D. and Vassilvitskii, S. (2007). "*K*-means ++ : The Advantages of Careful Seeding." *Proceedings of the 18th Annual ACM–SIAM Symposium on Discrete Algorithms*, pp. 1027–1035.

Bajorski, P. (2011a). "Second Moment Linear Dimensionality as an Alternative to Virtual Dimensionality." *IEEE Transactions on Geoscience and Remote Sensing*, Vol. 49, No.2, pp. 672–678.

Bajorski, P. (2011b). "Statistical Inference in PCA for Hyperspectral Images." *IEEE Journal of selected topics in Signal Processing*, Vol. 5, Issue 3, pp. 438–445.

Bajorski, P., Ientilucci E., and Schott, J. (2004). "Comparison of Basis-Vector Selection Methods for Target and Background Subspaces as Applied to Subpixel Target Detection." *Algorithms and Technologies for Multispectral, Hyperspectral, and Ultraspectral Imagery X*, Orlando, FL. *Proceedings of SPIE*, Vol. 5425, pp. 97–108.

Balakrishnan, N. and Nevzorov, V.B. (2003). *A Primer on Statistical Distributions*. Wiley.

Barrentine, L.B. (1999). *An Introduction to Design of Experiments: A Simplified Approach*. ASQ Quality Press.

Basener, W.F. and Messinger, D.W. (2009). "Enhanced Detection and Visualization of Anomalies in Spectral Imagery," *Proceedings of SPIE*, Vol. 7334, pp. 73341Q-1–73341Q-10.

Berns, R.S. (2000). *Billmeyer and Saltzman's Principles of Color Technology*, 3rd edition. Wiley.

Statistics for Imaging, Optics, and Photonics, Peter Bajorski.
© 2012 John Wiley & Sons, Inc. Published 2012 by John Wiley & Sons, Inc.

Bickel, P.J. and Doksum, K.A. (2001). *Mathematical Statistics: Basic Ideas and Selected Topics*, Vol. 1, 2nd edition. Prentice Hall.

Box, G.E.P. (1949). "A General Distribution Theory for a Class of Likelihood Criteria." *Biometrika*, Vol. 36, pp. 317–346.

Box, G.E.P. (1950). "Problems in the Analysis of Growth and Wear Curves." *Biometrics*, Vol. 6, pp. 362–389.

Box, G.E.P. and Draper, N.R. (1969). *Evolutionary Operation: A Statistical Method for Process Improvement*. Wiley.

Box, G.E.P., Hunter, J.S., and Hunter, W.G. (2005). *Statistics for Experimenters: Design, Innovation, and Discovery*, 2nd edition. Wiley–Interscience.

Caefer, C.E., Silverman, J., Orthal, O., Antonelli, D., Sharoni, Y., and Rotman, S.R. (2008). "Improved Covariance Matrices for Point Target Detection in Hyperspectral Data." *Optical Engineering*, Vol. 47, p. 076402.

Canty, M.J. (2010). *Image Analysis, Classification, and Change Detection in Remote Sensing*. CRC Press.

Clarke, B., Fokoué, E., and Zhang, H.H. (2009). *Principles and Theory for Data Mining and Machine Learning*. Springer.

Davison, A.C. and Hinkley, D.V. (1997). *Bootstrap Methods and Their Application*. Cambridge University Press.

Devore, J.L. (2004). *Probability and Statistics for Engineering and the Sciences*, 6th edition. Thomson.

Draper, N.R. and Smith, H. (1998). *Applied Regression Analysis*, 3rd edition. Wiley.

Efron, B. and Tibshirani, R.J. (1994). *An Introduction to the Bootstrap*. Chapman & Hall/CRC.

Feng, X., Schott, J.R., and Gallagher, T.W. (1994). "Modeling and Testing of a Modular Imaging Spectrometer Instrument." *Proceedings of SPIE*, Vol. 2224, p. 215.

Fisher, R.A. (1938). "The Statistical Utilization of Multiple Measurements." *Annals of Eugenics*, Vol. 8, pp. 376–386.

Forbes, C., Evans, M., Hastings, N., and Peacock, B. (2010). *Statistical Distributions*, 4th edition. Wiley.

Good, P.I. (2005). *Resampling Methods: A Practical Guide to Data Analysis*. Birkhäuser Boston.

Grubbs, F. (1969). "Procedures for Detecting Outlying Observations in Samples." *Technometrics*, Vol. 11, No.1, pp. 1–21.

Hadi, A.S. (1996). *Matrix Algebra as a Tool*. Duxbury Press.

Hamilton, L.C. (1992). *Regression with Graphics: A Second Course in Applied Statistics*. Brooks/Cole Publishing Company.

Hardle, W. and Simar, L. (2007). *Applied Multivariate Statistical Analysis*, 2nd edition. Springer.

Hartigan, J.A. and Wong, M.A. (1979). "A *K*-Means Clustering Algorithm." *Applied Statistics*, Vol. 28, pp. 100–108.

Harville, D.A. (1997). *Matrix Algebra from a Statistician's Perspective*. Springer.

Hastie, T., Tibshirani, R., and Friedman, J. (2001). *The Elements of Statistical Learning, Data Mining, Inference, and Prediction*. Springer.

Healey, G. and Slater, D. (1999). "Models and Methods for Automated Material Identification in Hyperspectral Imagery Acquired Under Unknown Illumination and Atmospheric Conditions." *IEEE Transactions on Geoscience and Remote Sensing*, Vol. 37, No.6, pp. 2706–2717.

Hernandez-Baquero, E. (2001). "Comparison of Statistical Inversion Techniques for Atmospheric Sounding." *Proceedings of the 2001 IEEE International Geoscience and Remote Sensing Symposium (IGARSS 2001)* Vol. 4, pp. 1705–1707.

Hernandez-Baquero, E.D. and Schott, J.R. (2000). "Atmospheric and Surface Parameter Retrievals from Multispectral Thermal Imagery via Reduced-Rank Multivariate Regression." *Proceedings of the 2000 IEEE International Geoscience and Remote Sensing Symposium (IGARSS 2000)* Vol. 4, pp. 1525–1527.

Hosmer, D.W. (2000). *Applied Logistic Regression*, 2nd edition. Wiley.

Ientilucci, E.J. (2003). "Comparison and Usage of Principal Component Analysis (PCA) and Noise Adjusted Principal Component (NAPC) Analysis or Maximum Noise Fraction (MNF)." Technical Report, Rochester Institute of Technology, see http://www.cis.rit.edu/user/32.

Ientilucci, E.J. and Bajorski, P. (2008). "Stochastic Modeling of Physically Derived Signature Spaces." *Journal of Applied Remote Sensing*, Vol. 2, p. 023532.

Jackson, J.E. (1991). *A User's Guide to Principal Components*. Wiley.

Johnson, N.L., Kotz, S., and Balakrishnan, N. (1994). *Continuous Univariate Distributions*, Vol. 1, 2nd edition. Wiley.

Johnson, N.L., Kotz, S., and Balakrishnan, N. (1995). *Continuous Univariate Distributions*, Vol. 2, 2nd edition. Wiley.

Johnson, R.A. and Wichern, D.W. (2007). *Applied Multivariate Statistical Analysis*, 6th edition. Prentice Hall.

Kastner, J., Thorndike, S., Romanczyk, P., Buchanan, C., Hrivnak, B., Sahai, R., and Egan, M. (2008). "The Large Magellanic Cloud's Top 250: Classification of the Most Luminous Compact 8 μm Sources in the Large Magellanic Cloud." *Astronomical Journal*, Vol. 136, pp. 1221–1241.

Kinsman, T., Bajorski P., and Pelz, J.B. (2010). "Hierarchical Image Clustering for Analyzing Eye Tracking Videos." *Image Processing Workshop (WNYIPW)*, pp. 58–61.

Krishnamoorthy, K. and Yu, J. (2004). "Modified Nel and Van der Merwe Test for the Multivariate Behrens–Fisher Problem." *Statistics and Probability Letters*, Vol. 66, pp. 161–169.

Kutner, M.H., Nachtsheim, C.J., Nether, J., and Li, W. (2005). *Applied Linear Statistical Models*, 5th edition. McGraw-Hill.

Lawley, D.N. (1956). "Tests of Significance for the Latent Roots of Covariance and Correlation Matrices." *Biometrika*, Vol. 43, pp. 128–136.

Levy, P.S. and Lemeshow, S. (2009). *Sampling of Populations: Methods and Applications*, 4th edition. Wiley.

Lohr, S.L. (2009). *Sampling: Design and Analysis*, 2nd edition. Duxbury Press.

Manolakis, D. and Shaw, G. (2002). "Detection Algorithms for Hyperspectral Imaging Applications." *IEEE Signal Processing Magazine*, Vol. 19, No.1, pp. 29–43.

Marchetti, C.E., Mudholkar, G.S., and Wilding, G.E. (2002). "Testing Goodness-of-Fit of the Gamma Models." *Recent Advances in Statistical Methods: Proceedings of Statistics 2001 Canada, The 4th Conference in Applied Statistics*, pp. 207–217.

Mendenhall, W., Beaver, R.J., and Beaver, B.M. (2006). *Introduction to Probability and Statistics*, 12th edition. Thomson.

Montgomery, D.C. (2008). *Design and Analysis of Experiments*, 7th edition. Wiley.

Montgomery, D.C., Peck, E.A., and Vining, G.G. (2006). *Introduction to Linear Regression Analysis*, 4th edition. Wiley.

Nel, D.G. and Van der Merwe, C.A. (1986). "A Solution to the Multivariate Behrens–Fisher Problem." *Communications in Statistics—Theory and Methods*, Vol. 15, pp. 3719–3735.

Pelz, J.B., Canosa, R., Babcock, J., Kucharczyk, D., Silver, A., and Konno, D. (2000). "Portable Eyetracking: A Study of Natural Eye Movements." *Proceedings of the SPIE, Human Vision and Electronic Imaging*, San Jose, CA.

Phillips, J., Bajorski, P., Burns, P., Fredericks, E., and Rosen, M. (2010). "Comparing Image Quality of Print-on-Demand Books and Photobooks from Web-Based Vendors." *Journal of Electronic Imaging*, Vol. 19, p. 011013.

R Development Core Team (2010). R: A language and environment for statistical computing. R Foundation for Statistical Computing, Vienna, Austria. ISBN 3-900051-07-0, URL http://www.R-project.org.

Rencher, A.C. (2002). *Methods of Multivariate Analysis*, 2nd edition. Wiley.

Ross, S. (2002). *The First Course in Probability*. Prentice Hall.

Scharf, L.L. (1991). *Statistical Signal Processing*. Addison-Wesley.

Schaum, A. (2007). "Hyperspectral Anomaly Detection Beyond RX." *Proceedings of SPIE*, Vol. 6565, p. 656502.

Scheaffer, R.L., Mendenhall, W., Ott, R.L., and Gerow, K.G. (2011). *Elementary Survey Sampling*, 7th edition. Duxbury Press.

Schlamm, A., Messinger, D., and Basener, W. (2008). "Geometric Estimation of the Inherent Dimensionality of a Single Material Cluster in Multi- and Hyperspectral Imagery." *Algorithms and Technologies for Multispectral, Hyperspectral, and Ultraspectral Imagery XII*, Shen, S.S. and Lewis, P.E., eds., *Proceedings of SPIE*, Vol. 6966.

Schott, J.R. (2007). *Remote Sensing: The Image Chain Approach*, 2nd edition. Oxford University Press.

Small, N.J.H. (1978). "Plotting Squared Radii." *Biometrika*, Vol. 65, pp. 657–658.

Snyder, D., Kerekes, J., Fairweather, I., Crabtree, R., Shive, J., and Hager, S. (2008). "Development of a Web-Based Application to Evaluate Target Finding Algorithms." *Proceedings of the 2008 IEEE International Geoscience and Remote Sensing Symposium (IGARSS 2008)* Boston, MA, Vol. 2, pp. 915–918.

Srivastava, M.S. (1984). "A Measure of Skewness and Kurtosis and a Graphical Method for Assessing Multivariate Normality." *Statistics and Probability Letters*, Vol. 2, pp. 263–267.

Srivastava, M.S. (2002). *Methods of Multivariate Statistics*. Wiley–Interscience.

Thompson, S.K. (2002). *Sampling*, 2nd edition. Wiley.

Tiku, M.L. and Singh, M. (1982). "Robust Statistics for Testing Mean Vectors of Multivariate Distributions." *Communications in Statistics—Theory and Methods*, Vol. 11, No. 9, pp. 985–1001.

Tso, B. and Mather, P.M. (2009). *Classification Methods for Remotely Sensed Data*. CRC Press.

Tufte, E.R. (2001). *The Visual Display of Quantitative Information*, 2nd edition. Graphics Press.

Ward, J.H., Jr., (1963). "Hierarchical Grouping to Optimize an Objective Function." *Journal of the American Statistical Association*, Vol. 58, pp. 236–244.

Wold, J.P., Johansen, I.-R., Haugholt, K.H., Tschudi, J., Thielemann, J., Segtnan, V.H., Narum, B., and Wold, E. (2006). "Non-Contact Transflectance Near Infrared Imaging for Representative On-line Sampling of Dried Salted Coalfish (Bacalao)." *Journal of Near Infrared Spectroscopy*, Vol. 14, pp. 59–66.

Index

Absolute deviation, 56
Abstract population, 52
Additive model, 118
Adjusted norm, 221
Agglomerative hierarchical Clustering, 304
Alpha risk, 64
Alternative hypothesis, 64
    one-sided/ two-sided, 65
Analysis of variance (ANOVA), 92
Analysis of variance (ANOVA) table, 101, 117, 123
Angle between vectors, 126
Anomaly detection, 231
ANOVA (analysis of variance), 92
ANOVA (analysis of variance) table, 101, 117, 123
Apparent error rate (APER), 272
Astronomy data, 71, 261, 353
Average linkage Clustering, 315
AVIRIS data, 352
AVIRIS image, 204, 230

Balanced design, 116
Bayes' theorem, 29
Best linear unbiased estimator (BLUE), 98
Beta distribution, 329
Beta risk, 65
Between-group variability, 286
Bias, 55, 80, 237
Bias-adjusted estimate, 56, 81
Biased estimator, 55

Bidirectional reflectance distribution function (BRDF), 112
Binomial distribution, 330
Bivariate distribution, 35
Bonferroni Confidence interval, 178
Bonferroni inequality, 177
Bootstrap, 79, 236
    hybrid method, 81–82
    nonparametric, 80, 229, 236
    parametric, 80, 236
    percentile method, 81
    percentile-reversal method, 81–82, 236
Box plot 23
Box's M statistic, 188
Broken-stick stopping rule for PCA, 211

Canonical correlation, 243
    analysis (CCA), 241–251
    regression (CCR), 251–256
    variables, 243
Cauchy distribution, 331
CCA (Canonical correlation analysis), 241–251
CCR (Canonical correlation regression), 251–256
Central composite design, 115
Central limit theorem, 43
Central moment, 33
Centroid, 320
Chebyshev's distance, 299
Chebyshev's inequality, 41

Statistics for Imaging, Optics, and Photonics, Peter Bajorski.
© 2012 John Wiley & Sons, Inc. Published 2012 by John Wiley & Sons, Inc.

WILEY SERIES IN PROBABILITY AND STATISTICS

ESTABLISHED BY WALTER A. SHEWHART AND SAMUEL S. WILKS

Editors: *David J. Balding, Noel A. C. Cressie, Garrett M. Fitzmaurice,*
Harvey Goldstein, Iain M. Johnstone, Geert Molenberghs, David W. Scott,
Adrian F. M. Smith, Ruey S. Tsay, Sanford Weisberg
Editors Emeriti: *Vic Barnett, J. Stuart Hunter, Joseph B. Kadane, Jozef L. Teugels*

The *Wiley Series in Probability and Statistics* is well established and authoritative. It covers many topics of current research interest in both pure and applied statistics and probability theory. Written by leading statisticians and institutions, the titles span both state-of-the-art developments in the field and classical methods.

Reflecting the wide range of current research in statistics, the series encompasses applied, methodological and theoretical statistics, ranging from applications and new techniques made possible by advances in computerized practice to rigorous treatment of theoretical approaches.

This series provides essential and invaluable reading for all statisticians, whether in academia, industry, government, or research.

† ABRAHAM and LEDOLTER · Statistical Methods for Forecasting
AGRESTI · Analysis of Ordinal Categorical Data, *Second Edition*
AGRESTI · An Introduction to Categorical Data Analysis, *Second Edition*
AGRESTI · Categorical Data Analysis, *Second Edition*
ALTMAN, GILL, and McDONALD · Numerical Issues in Statistical Computing for the
    Social Scientist
AMARATUNGA and CABRERA · Exploration and Analysis of DNA Microarray and
    Protein Array Data
ANDĚL · Mathematics of Chance
ANDERSON · An Introduction to Multivariate Statistical Analysis, *Third Edition*
* ANDERSON · The Statistical Analysis of Time Series
ANDERSON, AUQUIER, HAUCK, OAKES, VANDAELE, and WEISBERG ·
    Statistical Methods for Comparative Studies
ANDERSON and LOYNES · The Teaching of Practical Statistics
ARMITAGE and DAVID (editors) · Advances in Biometry
ARNOLD, BALAKRISHNAN, and NAGARAJA · Records
* ARTHANARI and DODGE · Mathematical Programming in Statistics
* BAILEY · The Elements of Stochastic Processes with Applications to the Natural
    Sciences
BAJORSKI · Statistics for Imaging, Optics, and Photonics
BALAKRISHNAN and KOUTRAS · Runs and Scans with Applications
BALAKRISHNAN and NG · Precedence-Type Tests and Applications
BARNETT · Comparative Statistical Inference, *Third Edition*
BARNETT · Environmental Statistics
BARNETT and LEWIS · Outliers in Statistical Data, *Third Edition*
BARTOSZYNSKI and NIEWIADOMSKA-BUGAJ · Probability and Statistical Inference
BASILEVSKY · Statistical Factor Analysis and Related Methods: Theory and
    Applications
BASU and RIGDON · Statistical Methods for the Reliability of Repairable Systems
BATES and WATTS · Nonlinear Regression Analysis and Its Applications
BECHHOFER, SANTNER, and GOLDSMAN · Design and Analysis of Experiments for
    Statistical Selection, Screening, and Multiple Comparisons
BEIRLANT, GOEGEBEUR, SEGERS, TEUGELS, and DE WAAL · Statistics of
    Extremes: Theory and Applications

*Now available in a lower priced paperback edition in the Wiley Classics Library.
†Now available in a lower priced paperback edition in the Wiley–Interscience Paperback Series.

BELSLEY · Conditioning Diagnostics: Collinearity and Weak Data in Regression
† BELSLEY, KUH, and WELSCH · Regression Diagnostics: Identifying Influential
  Data and Sources of Collinearity
BENDAT and PIERSOL · Random Data: Analysis and Measurement Procedures,
  Fourth Edition
BERNARDO and SMITH · Bayesian Theory
BERRY, CHALONER, and GEWEKE · Bayesian Analysis in Statistics and
  Econometrics: Essays in Honor of Arnold Zellner
BHAT and MILLER · Elements of Applied Stochastic Processes, *Third Edition*
BHATTACHARYA and WAYMIRE · Stochastic Processes with Applications
BIEMER, GROVES, LYBERG, MATHIOWETZ, and SUDMAN · Measurement Errors
  in Surveys
BILLINGSLEY · Convergence of Probability Measures, *Second Edition*
BILLINGSLEY · Probability and Measure, *Third Edition*
BIRKES and DODGE · Alternative Methods of Regression
BISGAARD and KULAHCI · Time Series Analysis and Forecasting by Example
BISWAS, DATTA, FINE, and SEGAL · Statistical Advances in the Biomedical Sciences:
  Clinical Trials, Epidemiology, Survival Analysis, and Bioinformatics
BLISCHKE AND MURTHY (editors) · Case Studies in Reliability and Maintenance
BLISCHKE AND MURTHY · Reliability: Modeling, Prediction, and Optimization
BLOOMFIELD · Fourier Analysis of Time Series: An Introduction, *Second Edition*
BOLLEN · Structural Equations with Latent Variables
BOLLEN and CURRAN · Latent Curve Models: A Structural Equation Perspective
BOROVKOV · Ergodicity and Stability of Stochastic Processes
BOSQ and BLANKE · Inference and Prediction in Large Dimensions
BOULEAU · Numerical Methods for Stochastic Processes
BOX · Bayesian Inference in Statistical Analysis
BOX · Improving Almost Anything, *Revised Edition*
BOX · R. A. Fisher, the Life of a Scientist
BOX and DRAPER · Empirical Model-Building and Response Surfaces
* BOX and DRAPER · Evolutionary Operation: A Statistical Method for Process
  Improvement
BOX and DRAPER · Response Surfaces, Mixtures, and Ridge Analyses, *Second Edition*
BOX, HUNTER, and HUNTER · Statistics for Experimenters: Design, Innovation,
  and Discovery, *Second Editon*
BOX, JENKINS, and REINSEL · Time Series Analysis: Forcasting and Control, *Fourth
  Edition*
BOX, LUCEÑO, and PANIAGUA-QUIÑONES · Statistical Control by Monitoring
  and Adjustment, *Second Edition*
BRANDIMARTE · Numerical Methods in Finance: A MATLAB-Based Introduction
† BROWN and HOLLANDER · Statistics: A Biomedical Introduction
BRUNNER, DOMHOF, and LANGER · Nonparametric Analysis of Longitudinal Data in
  Factorial Experiments
BUCKLEW · Large Deviation Techniques in Decision, Simulation, and Estimation
CAIROLI and DALANG · Sequential Stochastic Optimization
CASTILLO, HADI, BALAKRISHNAN, and SARABIA · Extreme Value and Related
  Models with Applications in Engineering and Science
CHAN · Time Series: Applications to Finance with R and S-Plus®, *Second Edition*
CHARALAMBIDES · Combinatorial Methods in Discrete Distributions
CHATTERJEE and HADI · Regression Analysis by Example, *Fourth Edition*
CHATTERJEE and HADI · Sensitivity Analysis in Linear Regression
CHERNICK · Bootstrap Methods: A Guide for Practitioners and Researchers,
  Second Edition
CHERNICK and FRIIS · Introductory Biostatistics for the Health Sciences

*Now available in a lower priced paperback edition in the Wiley Classics Library.
†Now available in a lower priced paperback edition in the Wiley–Interscience Paperback Series.

---

*Now available in a lower priced paperback edition in the Wiley Classics Library.

†Now available in a lower priced paperback edition in the Wiley–Interscience Paperback Series.

---

*Now available in a lower priced paperback edition in the Wiley Classics Library.
†Now available in a lower priced paperback edition in the Wiley–Interscience Paperback Series.

---

*Now available in a lower priced paperback edition in the Wiley Classics Library.

†Now available in a lower priced paperback edition in the Wiley–Interscience Paperback Series.

---

*Now available in a lower priced paperback edition in the Wiley Classics Library.
†Now available in a lower priced paperback edition in the Wiley–Interscience Paperback Series.

*Now available in a lower priced paperback edition in the Wiley Classics Library.
†Now available in a lower priced paperback edition in the Wiley–Interscience Paperback Series.

MYERS, MONTGOMERY, VINING, and ROBINSON · Generalized Linear Models. With Applications in Engineering and the Sciences, *Second Edition*

† NELSON · Accelerated Testing, Statistical Models, Test Plans, and Data Analyses

† NELSON · Applied Life Data Analysis

NEWMAN · Biostatistical Methods in Epidemiology

OCHI · Applied Probability and Stochastic Processes in Engineering and Physical Sciences

OKABE, BOOTS, SUGIHARA, and CHIU · Spatial Tesselations: Concepts and Applications of Voronoi Diagrams, *Second Edition*

OLIVER and SMITH · Influence Diagrams, Belief Nets and Decision Analysis

PALTA · Quantitative Methods in Population Health: Extensions of Ordinary Regressions

PANJER · Operational Risk: Modeling and Analytics

PANKRATZ · Forecasting with Dynamic Regression Models

PANKRATZ · Forecasting with Univariate Box-Jenkins Models: Concepts and Cases

PARDOUX · Markov Processes and Applications: Algorithms, Networks, Genome and Finance

* PARZEN · Modern Probability Theory and Its Applications

PEÑA, TIAO, and TSAY · A Course in Time Series Analysis

PIANTADOSI · Clinical Trials: A Methodologic Perspective

PORT · Theoretical Probability for Applications

POURAHMADI · Foundations of Time Series Analysis and Prediction Theory

POWELL · Approximate Dynamic Programming: Solving the Curses of Dimensionality, *Second Edition*

PRESS · Bayesian Statistics: Principles, Models, and Applications

PRESS · Subjective and Objective Bayesian Statistics, *Second Edition*

PRESS and TANUR · The Subjectivity of Scientists and the Bayesian Approach

PUKELSHEIM · Optimal Experimental Design

PURI, VILAPLANA, and WERTZ · New Perspectives in Theoretical and Applied Statistics

† PUTERMAN · Markov Decision Processes: Discrete Stochastic Dynamic Programming

QIU · Image Processing and Jump Regression Analysis

* RAO · Linear Statistical Inference and Its Applications, *Second Edition*

RAO · Statistical Inference for Fractional Diffusion Processes

RAUSAND and HØYLAND · System Reliability Theory: Models, Statistical Methods, and Applications, *Second Edition*

RAYNER · Smooth Tests of Goodnes of Fit: Using R, *Second Edition*

RENCHER · Linear Models in Statistics

RENCHER · Methods of Multivariate Analysis, *Second Edition*

RENCHER · Multivariate Statistical Inference with Applications

* RIPLEY · Spatial Statistics

* RIPLEY · Stochastic Simulation

ROBINSON · Practical Strategies for Experimenting

ROHATGI and SALEH · An Introduction to Probability and Statistics, *Second Edition*

ROLSKI, SCHMIDLI, SCHMIDT, and TEUGELS · Stochastic Processes for Insurance and Finance

ROSENBERGER and LACHIN · Randomization in Clinical Trials: Theory and Practice

ROSS · Introduction to Probability and Statistics for Engineers and Scientists

ROSSI, ALLENBY, and McCULLOCH · Bayesian Statistics and Marketing

† ROUSSEEUW and LEROY · Robust Regression and Outlier Detection

ROYSTON and SAUERBREI · Multivariate Model Building: A Pragmatic Approach to Regression Analysis Based on Fractional Polynomials for Modeling Continuous Variables

* RUBIN · Multiple Imputation for Nonresponse in Surveys

RUBINSTEIN and KROESE · Simulation and the Monte Carlo Method, *Second Edition*

*Now available in a lower priced paperback edition in the Wiley Classics Library.
†Now available in a lower priced paperback edition in the Wiley–Interscience Paperback Series.

RUBINSTEIN and MELAMED · Modern Simulation and Modeling

RYAN · Modern Engineering Statistics

RYAN · Modern Experimental Design

RYAN · Modern Regression Methods, *Second Edition*

RYAN · Statistical Methods for Quality Improvement, *Third Edition*

SALEH · Theory of Preliminary Test and Stein-Type Estimation with Applications

SALTELLI, CHAN, and SCOTT (editors) · Sensitivity Analysis

* SCHEFFE · The Analysis of Variance

SCHIMEK · Smoothing and Regression: Approaches, Computation, and Application

SCHOTT · Matrix Analysis for Statistics, *Second Edition*

SCHOUTENS · Levy Processes in Finance: Pricing Financial Derivatives

SCHUSS · Theory and Applications of Stochastic Differential Equations

SCOTT · Multivariate Density Estimation: Theory, Practice, and Visualization

* SEARLE · Linear Models

† SEARLE · Linear Models for Unbalanced Data

† SEARLE · Matrix Algebra Useful for Statistics

† SEARLE, CASELLA, and McCULLOCH · Variance Components

SEARLE and WILLETT · Matrix Algebra for Applied Economics

SEBER · A Matrix Handbook For Statisticians

† SEBER · Multivariate Observations

SEBER and LEE · Linear Regression Analysis, *Second Edition*

† SEBER and WILD · Nonlinear Regression

SENNOTT · Stochastic Dynamic Programming and the Control of Queueing Systems

* SERFLING · Approximation Theorems of Mathematical Statistics

SHAFER and VOVK · Probability and Finance: It's Only a Game!

SHERMAN · Spatial Statistics and Spatio-Temporal Data: Covariance Functions and Directional Properties

SILVAPULLE and SEN · Constrained Statistical Inference: Inequality, Order, and Shape Restrictions

SINGPURWALLA · Reliability and Risk: A Bayesian Perspective

SMALL and McLEISH · Hilbert Space Methods in Probability and Statistical Inference

SRIVASTAVA · Methods of Multivariate Statistics

STAPLETON · Linear Statistical Models, *Second Edition*

STAPLETON · Models for Probability and Statistical Inference: Theory and Applications

STAUDTE and SHEATHER · Robust Estimation and Testing

STOYAN, KENDALL, and MECKE · Stochastic Geometry and Its Applications, *Second Edition*

STOYAN and STOYAN · Fractals, Random Shapes and Point Fields: Methods of Geometrical Statistics

STREET and BURGESS · The Construction of Optimal Stated Choice Experiments: Theory and Methods

STYAN · The Collected Papers of T. W. Anderson: 1943–1985

SUTTON, ABRAMS, JONES, SHELDON, and SONG · Methods for Meta-Analysis in Medical Research

TAKEZAWA · Introduction to Nonparametric Regression

TAMHANE · Statistical Analysis of Designed Experiments: Theory and Applications

TANAKA · Time Series Analysis: Nonstationary and Noninvertible Distribution Theory

THOMPSON · Empirical Model Building, *Second Edition*

THOMPSON · Sampling, *Second Edition*

THOMPSON · Simulation: A Modeler's Approach

THOMPSON and SEBER · Adaptive Sampling

THOMPSON, WILLIAMS, and FINDLAY · Models for Investors in Real World Markets

TIAO, BISGAARD, HILL, PEÑA, and STIGLER (editors) · Box on Quality and Discovery: with Design, Control, and Robustness

---

*Now available in a lower priced paperback edition in the Wiley Classics Library.

†Now available in a lower priced paperback edition in the Wiley–Interscience Paperback Series.

Printed and bound by CPI Group (UK) Ltd, Croydon, CR0 4YY

16/04/2025

14658371-0001